Springer Tracts in Electrical and Electronics Engineering

Series Editors

Brajesh Kumar Kaushik, Department of Electronics and Communication Engineering, Indian Institute of Technology Roorkee, Roorkee, Uttarakhand, India

Mohan Lal Kolhe, Faculty of Engineering and Sciences, University of Agder, Kristiansand, Norway

Springer Tracts in Electrical and Electronics Engineering (STEEE) publishes the latest developments in Electrical and Electronics Engineering - quickly, informally and with high quality. The intent is to cover all the main branches of electrical and electronics engineering, both theoretical and applied, including:

- Signal, Speech and Image Processing
- Speech and Audio Processing
- Image Processing
- Human-Machine Interfaces
- Digital and Analog Signal Processing
- Microwaves, RF Engineering and Optical Communications
- Electronics and Microelectronics, Instrumentation
- Electronic Circuits and Systems
- Embedded Systems
- Electronics Design and Verification
- Cyber-Physical Systems
- Electrical Power Engineering
- Power Electronics
- Photovoltaics
- Energy Grids and Networks
- Electrical Machines
- Control, Robotics, Automation
- Robotic Engineering
- Mechatronics
- Control and Systems Theory
- Automation
- Communications Engineering, Networks
- Wireless and Mobile Communication
- Internet of Things
- Computer Networks

Within the scope of the series are monographs, professional books or graduate textbooks, edited volumes as well as outstanding PhD theses and books purposely devoted to support education in electrical and electronics engineering at graduate and post-graduate levels.

Review Process

The proposal for each volume is reviewed by the main editor and/or the advisory board. The books of this series are reviewed in a single blind peer review process.

Ethics Statement for this series can be found in the Springer standard guidelines here https://www.springer.com/us/authors-editors/journal-author/journal-author-hel pdesk/before-you-start/before-you-start/1330#c14214

Bharat Bhushan · Sudhir Kumar Sharma ·
Raghvendra Kumar · Ishaani Priyadarshini
Editors

5G and Beyond

 Springer

Editors
Bharat Bhushan
Department of Computer Science
and Engineering, School of Engineering
and Technology
Sharda University
Greater Noida, India

Raghvendra Kumar
Department of Computer Science
and Engineering
GIET University
Gunupur, Odisha, India

Sudhir Kumar Sharma
Department of Computer Science
Institute of Information Technology
and Management
New Delhi, India

Ishaani Priyadarshini
School of Information
University of California, Berkeley
Berkeley, CA, USA

This title is freely available in an open access edition with generous support from the Library of the University of California, Berkeley.

ISSN 2731-4200   ISSN 2731-4219 (electronic)
Springer Tracts in Electrical and Electronics Engineering
ISBN 978-981-99-3667-0   ISBN 978-981-99-3668-7 (eBook)
https://doi.org/10.1007/978-981-99-3668-7

This Springer imprint is published by the registered company Springer Nature Singapore Pte Ltd.
The registered company address is: 152 Beach Road, #21-01/04 Gateway East, Singapore 189721, Singapore

Preface

The fifth-generation (5G) networks continue to emerge through evolved technologies and new models of cellular operations. The key advances over preceding generations include multi-interface access, spectrum extensions, network softwarization powered by Network Function Virtualization (NFV), Software-Defined Networking (SDN), etc. Recent years have witnessed an unprecedented rise in the use of portable computing devices such as laptops, tablets, smartphones, and machine-type devices interconnected via Internet of Things (IoT). The emergence of such innovative online services and mobile applications poses unprecedented challenges to the 5G networks. Therefore, there is a need for novel solutions (such as New Radio (NR) interface in millimeter-wave (mmWave) bands, beamforming, and massive Multiple-Input Multiple-Output (massive MIMO)) to optimize the use of finite radio spectrum resources. Further, the future networks are expected to meet several requirements such as increased network capacity, latency, energy efficiency, reliability, heterogeneity, and Quality-of-Service (QoS) demands.

Owing to the exponential rise in the number of devices being connected to the Internet, it is anticipated that the wireless data traffic might increase to 10000-fold by the year 2030. Predictions evidently indicate the skyrocketing demand on data traffic and applications for machine-type communication such as smart cities, industries, healthcare monitoring and self-driving vehicles along with traditional human-centric communications. Such coexistence of such machine-type and human-centric services will tend to make the next-generation wireless networks more complex and diverse. In order to provide better support to the IoT applications, numerous challenges such as advanced signal processing, network resource allocation and network architectures need to be overcome in 5G and beyond. As 5G research is maturing toward a global standard, there is a need for the development of beyond-5G solutions, that is, 6G and beyond. 6G might entail technologies that are beyond the scope of 5G, especially the way in which data is collected, processed, and transmitted within the wireless network. These next-generation wireless networks must support low latency, ultra-reliable communication, and intelligently managed IoTs or even future IoT devices in a highly dynamic real-time environment. Furthermore, the paradigm shift to machine-oriented communications from people-centric communication makes the future wireless networks even more complex. Therefore, enabling the vision requires addressing a myriad of practical and theoretical challenges.

This book aims to highlight the coming surge of 5G network-based applications and predicts that the centralized networks and its current capacity will be incapable to meet the demands. The main aim of this book is to outline the major benefits as well as challenges associated with integration of 5G networks with varied applications. Further, the book aims to gather and investigate the most recent 5G-based research solutions that handle the security and privacy threats while considering the resource-constrained wireless devices. The information, applications and recent advances discussed in this book will serve to be of immense help for practitioners, database professional, and researchers.

Greater Noida, India   Bharat Bhushan
New Delhi, India   Sudhir Kumar Sharma
Gunupur, India   Raghvendra Kumar
Berkeley, USA   Ishaani Priyadarshini

Contents

Editors and Contributors

About the Editors

Bharat Bhushan is an Assistant Professor of the Department of Computer Science and Engineering (CSE) at the School of Engineering and Technology, Sharda University, Greater Noida, India. He is an alumnus of Birla Institute of Technology, Mesra, Ranchi, India. He received his B.Tech. in Computer Science and Engineering in 2012, an M.Tech. in Information Security in 2015, and a Ph.D. in Computer Science and Engineering in 2021 from Birla Institute of Technology, Mesra, India. He has published over 80 research papers in various renowned International conferences and Journals. He has also contributed over 25 book chapters and has edited 11 books. He is a member of numerous renowned bodies such as IEEE, IAENG, CSTA, SCIEI, IAE, and UACEE.

Sudhir Kumar Sharma is currently a Professor and Head of the Department of Computer Science, Institute of Information Technology and Management affiliated with GGSIPU, New Delhi, India. He has extensive experience of over 21 years in the field of Computer Science and Engineering. He obtained his Ph.D. degree in Information Technology in 2013 from USICT, Guru Gobind Singh Indraprastha University, New Delhi, India. Dr. Sharma received his M.Tech. degree in Computer Science & Engineering in 1999 from the Guru Jambheshwar University, Hisar, India, and an M.Sc. degree in Physics from the University of Roorkee (now IIT Roorkee), Roorkee, in 1997. His research interests are machine learning, data mining, and security. Dr. Sharma has published over 60 research papers in various prestigious International Journals and Conferences. He authored and edited 7 books in the fields of IoT, WSN, blockchain, and cyber-physical systems.

Raghvendra Kumar is an Associate Professor in Computer Science and Engineering Department at GIET University, India. He received B.Tech., M.Tech., and Ph.D. in Computer Science and Engineering, India, and Postdoc Fellow from the Institute of Information Technology, Virtual Reality, and Multimedia, Vietnam. He has published a number of research papers in international journals and conferences. He also published 13 chapters in edited books published by renowned publishers. His research areas are computer networks, data mining, cloud computing and secure multiparty computations, theory of computer science, and design of algorithms. He authored and edited 23 computer science books on IoT, data mining, biomedical engineering, big data, and robotics.

Ishaani Priyadarshini is a lecturer at the School of Information, UC Berkeley, USA, Ph.D., and Master's Degree in Cybersecurity from the University of Delaware, USA. Prior to that, she completed her Bachelor's degree in Computer Science Engineering and a Master's degree in Information Security from Kalinga Institute of Industrial Technology, India. She has authored several book chapters for reputed publishers and is also an author of several publications for SCIE-indexed journals. As a certified reviewer, she conducts peer reviews of research papers for prestigious publishers and is a part of the Editorial Board for the International Journal of Information Security and Privacy (IJISP). Her areas of research include cybersecurity, artificial intelligence, and HCI.

Contributors

Abdullah Al Mahfuj Shaan Department of ECE, North South University, Dhaka, Bangladesh

N. Ambika Department of Computer Science and Applications, St. Francis College, Bangalore, India

G. R. Anantha Raman Department of Computer Science and Engineering, Malla Reddy Institute of Engineering and Technology Maisammaguda, Hyderabad, Telangana, India

V. S. Anoop Smith School of Business, Queen's University, Kingston, ON, Canada

Kande Archana Research Scholar, Department of Computer Science and Engineering, Jawaharlal Nehru Technological University, Hyderabad, Telangana, India; Assistant Professor, Department of Computer Science and Engineering, Malla Reddy Institute of Engineering and Technology Maisammaguda, Hyderabad, Telangana, India

M. Ashok Department of Computer Science and Engineering, Malla Reddy Institute of Engineering and Technology Maisammaguda, Hyderabad, Telangana, India

Parma Nand Astya Department of Computer Science and Engineering, School of Engineering and Technology (SET), Sharda University, Greater Noida, India

A. K. M. Bahalul Haque LUT School of Engineering Science, LUT University, Lappeenranta, Finland;
Department of ECE, North South University, Dhaka, Bangladesh

Saurabh Bhatt Department of Computer Science and Engineering, School of Engineering and Technology, Sharda University, Greater Noida, India

Surbhi Bhatia Khan Department of Data Science, School of Science, Engineering and Environment, University of Salford, Manchester, United Kingdom

Surbhi Bhatia College of Computer Science and Information Technology, King Faisal University, Hofuf, Saudi Arabia

Diptendu Bhattacharya Department of Computer Science and Engineering, National Institute of Technology, Agartala, Agartala, Tripura, India

Bharat Bhushan Department of Computer Science and Engineering, School of Engineering and Technology, Sharda University, Greater Noida, India

Rajasekhar Chaganti University of Texas, San Antonio, USA

Ritwesh Chatterjee Adamas Tech Consulting, Bangalore, India

Aninda Chowdhury St. Placid's School and College, Chittagong, Bangladesh

Tibor Cinkler Budapest University of Technology and Economics, Budapest, Hungary

Anwesha Das Adamas University, Kolkata, India

V. Kamakshi Prasad Department of Computer Science and Engineering, Jawaharlal Nehru Technological University, Hyderabad, Telangana, India

Rekha Kashyap Noida Institute of Engineering and Technology (NIET), Greater Noida, India

Sumit Kumar Indian Institute of Management, Kozhikode, India

Parijata Majumdar Department of Computer Science and Engineering, National Institute of Technology, Agartala, Agartala, Tripura, India

Ayasha Malik Delhi Technical Campus (DTC), GGSIPU, Greater Noida, India

Sanjoy Mitra Department of Computer Science and Engineering, Tripura Institute of Technology, Agartala, Agartala, Tripura, India

Saydul Akbar Murad Faculty of Computing, Universiti Malaysia Pahang, Pekan, Malaysia

Abhaya Nand IIMT College of Management, Greater Noida, India

Tasfia Nausheen Department of ECE, North South University, Dhaka, Bangladesh

Md. Oahiduzzaman Mondol Zihad Department of Electrical and Computer Engineering, North South University, Dhaka, Bangladesh

Veena Parihar KIET Group of Institutions Delhi-NCR, Ghaziabad, India

Husam Rajab Budapest University of Technology and Economics, Budapest, Hungary

Nitin Rakesh Department of Computer Science and Engineering, School of Engineering and Technology, Sharda University, Greater Noida, India

Md. Rifat Hasan Department of Computer Science & Engineering, Fareast International University, Dhaka, Bangladesh

Smriti Sachan Department of Electronics and Communication, SRM Institute of Science and Technology, NCR Campus, Ghaziabad, India

Amit Sehgal School of Engineering and Technology, Sharda University, Greater Noida, India

Astha Sharma GL Bajaj Institute of Technology and Management, Greater Noida, India

Rohit Sharma Department of Electronics and Communication, SRM Institute of Science and Technology, NCR Campus, Ghaziabad, India

Riya Sil Adamas University, Kolkata, India

Suryabhan Pratap Singh Institute of Engineering and Technology, Deen Dayal Upadhyaya Gorakhpur University, Gorakhpur, India

Mukesh Soni Department of CSE, University Centre for Research & Development Chandigarh University, Mohali, Punjab, India

Tanya Srivastava Department of Computer Science and Engineering, School of Engineering and Technology, Sharda University, Greater Noida, India

Chapter 1
Evolution of Next-Generation Communication Technology

Riya Sil and Ritwesh Chatterjee

Abstract In this technological era of wireless communication, various Internet devices and Wi-Fi zone play a significant role in the fast growth of data usage. The communication of people has been revolutionized by mobile communication systems. The rapid increase in the number of users, higher data rate, and requirement for higher bandwidth and voluminous data has become a challenge for Internet service providers. All these requirements are expected to be met by the next-generation communication network. The evolution of next-generation communication technology aims to emphasize the user terminal development that will provide access to various technologies and combine several flows from numerous technologies. In this paper, the authors have provided a detailed overview of the various wireless communication technologies and how it can be enhanced in future. A comparative study of the different generations of communication technologies has been done which includes 1G which has contented the elementary mobile voice, and 2G which has familiarized us with capacity and coverage. The quest continued with 3G for high-speed data that in turn provided a true experience to mobile bandwidth with 4G and finally the next-generation communication technology—5G. The varied variety of telecommunication services such as advanced mobile services, with the help of mobile and fixed networks, can be accessed with the help of the Fourth Generation (4G). The technology in the Fifth Generation (5G) is more advanced, and intelligent and it interconnects with the entire world. Moreover, in this paper, the authors have discussed the various issues and challenges faced in the existing cellular network, the limitation of conventional cellular systems, and the reason for the need for next-generation communication technology. Also, a detailed study on 5G network communication has been deliberated.

Keywords Wireless communications · Fifth-generation communication · Mobile broadband · Technology · Networks

R. Sil (✉)
Adamas University, Kolkata 700040, India
e-mail: riyasil1802@gmail.com

R. Chatterjee
Adamas Tech Consulting, Bangalore 560070, India

© The Author(s) 2023
B. Bhushan et al. (eds.), *5G and Beyond*, Springer Tracts in Electrical and Electronics Engineering, https://doi.org/10.1007/978-981-99-3668-7_1

1

Introduction

The rapid evolution of architecture, communication services, and technologies has stimulated due to the demand for new applications, research innovations, and other foremost possibilities for enhancements at various levels (Rappaport et al. 2014). These changes are driving a fundamental shift in the process of designing and delivering infrastructure and services. Communication systems have generally been envisioned and run as centralized secure utility services, with fewer options for specialization and customization, especially at the network edge (Holma et al. 2015). Supportive infrastructure is built and delivered consistently, according to a set of specific designs, resulting in hardened structures that should last for many years (Stallings 2007). In an environment where service requirements are constantly changing, this method is somewhat limited. Traditional rigid designs make it difficult to create, enhance, deploy, and customize services.

The constant increase in wireless user data usage, devices, and the desire for an experience of improved quality have overall affected the advancement of cellular network generations. At the end of 2020, there were over 50 billion linked devices that were using cellular network services, which has resulted in a massive increase in data traffic in comparison to 2014. On the other hand, the solutions are insufficient to address the aforementioned challenges (Bangerter et al. 2014). In brief, the growth of 5G networks is aided by a rise in 3D (i.e., "D"ata, "D"evice, and "D"ata transfer rate). The fifth-generation cellular networks precisely address and highlight three main perspectives: (i) user-centric (i.e., deliver device connectivity 24×7, continuous communication assistance, and pleasing user experience); (ii) provider-centric service (provide services that include sensors, intelligent transportation systems that are connected); (iii) network-operator-centric; and (iv) network-operator (provide energy-efficient, programmable, low-cost, scalable). As a result, the three key features described below are expected to appear in 5G networks: (i) ubiquitous connectivity: A wide range of devices will be able to communicate and deliver a uniform user experience. In fact, ubiquitous connectivity will allow for a user-centric viewpoint. (ii) Zero latency: 5G networks should achieve zero latency or very low latency on the order of one millisecond. Thus, zero latency will make the service-provider-centric strategy a reality. (iii) Gigabit connection: To achieve zero latency, a fast gigabit connection for rapid data transmission and reception, on the gigabits/second to users and machines order, might be employed (Andrews et al. 2014; Khan et al. 2012; Adhikari 2008; Intelligence 2014).

The cellular wireless generation includes a shift in the service's core foundation, new frequency bands, backward-compatible, and non-backward transmission technologies. Since the initial move from the analog (1G) to the analog (2G) network in 1981, new generations (G) have appeared every 10 years, with 3G multimedia capability following in 2011. Recently, the wireless industry has experienced spectacular growth. A distinct shift from fixed phones to mobile cellular phones has been noticed from the beginning of the century. There were more than four times as many mobile cellular subscribers as landline telephone lines at the end of 2010.

Manufacturers along with mobile network operators acknowledge the significance of effective networks and efficient design. As a result, network design and optimization services are increasingly becoming popular (Chen and Zhao 2014). 4G networks, or next-generation mobile networks, are envisioned as a collection of heterogeneous systems connected by a horizontal IP-centric architecture (3GPP 2015). These new technologies, together with the aforementioned requirements, provide a number of roadblocks to 5G development.

Table 1.1 compares the major characteristics and limitations of each generation of communication technology (Ericsson 2015; Qualcomm Technologies Inc. 2014; Huawei 2013; NTT Docomo 2015; Nokia Networks 2014).

The scope of this paper includes discussions related to the existing cellular network and its limitations, 5G networks vision, proposed architectures, advantages, issues related to implementation, applications, and a detailed discussion on next-generation network. In Sect. 1.2, authors have discussed the existing cellular network and its challenges. Section 1.3 explains the conventional cellular systems limitation. In the next section, i.e., Sect. 1.4, the authors have provided a clear idea about the vision and mission of 5G network. The architecture of the 5G network has been demonstrated in Sect. 1.5. Next, Sect. 1.6 explains the applications of next-generation network. Finally, in Sect. 1.7, the overall conclusion and future scope are discussed.

Table 1.1 Generations of communication technology

Sl. no	Year	Generation	Key features	Limitations
1	1980s	1G	• Voice signals • Analog cellular phones • NMT, AMPS	Low security
2	1990s	2G	• Text messages • Voice signal • Digital signal • GSM, TDMA, CDMA, GPRS, EDGE	Low internet support
3	1998s	3G	• Voice signals • Data signals • Video signals • Wireless and fixed internet access • W-CDMA, UMTS	Low high-speed internet support
4	2008s	4G	• Higher data rate • Interoperability protocol • Mobile IP	Does not help with the connected devices are more than 50 billion
5	2019s	5G	• Increased connectivity • IoT • Low latency • Greater transmission speed	

Existing Cellular Network and Its Challenges

As per reports related to the statistics of wireless network, review reveals that global mobile traffic had increased roughly by 70% in 2014. 26% of smartphones (out of all mobile devices worldwide) account for 88% of total mobile data traffic (Samsung Electronics Co, 2015). As the number of people using smartphones grow, so does the amount of mobile video traffic. Video traffic has accounted for over half of all mobile traffic since 2012. As per reports, the typical mobile user had downloaded 1 terabyte of data per year in 2020. In today's 4G LTE cellular systems, supporting this massive and quick rise in data demand and connection is a huge challenge. To increase capacity and data rates, the LTE cellular network is pursuing several research and development options such as MIMO, tiny cells, HetNets, multiple antennas, and coordinated multipoint transmission. However, this current traffic surge is unlikely to persist in the long run. As a result, the key problem in mobile broadband communications is to meet the exponential growth in user and traffic capacity (5G-Infrastructure Public–Private Partnership 2013; Osseiran et al. 2014; European Commission 2011).

First Generation

The first generation of mobile systems depends on analog transmission for voice services. NTT or Nippon Telephone and Telegraph in Tokyo, Japan, launched the world's foremost cellular system in the year 1979. After 2 years, the cellular period came in Europe. The Advanced Mobile Phone System (AMPS) was established in 1982 in the United States. Nordic Mobile Telephones (NMT) and Total Access Communication Systems (TACS) are the two most widespread analog systems (Kallnichev 2001). For AMPS, the Federal Communications Commission (FCC) allocated a 40-MHz bandwidth in the 800 to 900 MHz frequency range. As a result, for AMPS, a seven-cell reuse pattern has been established. The Frequency Modulation (FM) technology is used by AMPS and TACS for radio transmission. Frequency division multiple access technology is used to multiplex traffic (Lai et al. 2015; Agyapong et al. 2014; Lara et al. 2014).

Second Generation

Second generation (2G) of mobile system was launched around the end of the 1980s. In contrast to the first-generation (1G) systems, the 2G systems use digital multiple access technologies which includes TDMA (time division multiple access) and CDMA (code division multiple access). Thus, 2G systems outperformed first-generation systems in terms of data services, spectrum effectiveness, and roaming capabilities (Cho et al. 2014). In the United States, there were three distinct streams of development for second-generation digital cellular networks. The first digital system, the IS-54 (North America TDMA Digital Cellular), launched in 1991, and a second version IS-136 with expanded services was produced in 1996. Meanwhile, IS-95 (CDMA One) was adopted in 1993 (Arslan et al. 2015). 2G connection is often used to link GSM (global system for mobile) services. The most general packet radio service (GPRS) and GSM are commonly used to power 2.5G networks (Checko et al. 2015; Cvijetic 2014; Chen and Duan 2011).

Third Generation

W-CDMA, CDMA2000, and TD-SCDMA: 3G employs wide brand wireless network with which clarity is boosted. High-volume data movement was achievable in EDGE, but the packet transfer over the air interface still operates like a circuit switching call. The information transfer rate of 3G telecommunication networks is at least 2 Mbps (Liu et al. 2014). As a result, in the circuit switch scenario, some of the packet connection efficiency is lost. Furthermore, different portions of the world have varied criteria for establishing networks. As a result, 3G was born. 3G is not a single standard; it is a collection of standards that can all communicate with one another. The development has been continued by the Third-Generation Partnership Project (3GPP), which has defined a mobile system that meets the IMT-2000 standard. It was known in Europe as UMTS (Universal Terrestrial Mobile System), which is governed by the ETSI. The ITU-T nomenclature for the third-generation system is IMT-2000, but the American 3G variation is cdma2000. The air-interface technology for UMTS is W-CDMA. 3G networks allow network operators to provide a broader choice of more advanced services to users while increasing network capacity through increased spectral efficiency. On October 1, 2001, NTT DoCoMo in Japan branded FOMA launched the first commercial for 3G network that was based on W-CDMA technology (Banikazemi et al. 2013; Rost et al. 2014; Zhou and Yu 2014).

Fourth Generation

On June 23, 2005, the first successful 4G field testing was performed in Tokyo, Japan on June 23, 2005. In the downlink, NTT DoCoMo was able to achieve 1 Gbps real-time packet transmission at a pace of roughly 20 km/h (Zhang et al. 2015). Base stations emit signaling messages for service subscription to mobile stations on a regular basis in modern GSM systems. Because of the variations in wireless technology and access protocols, this procedure becomes more challenging in 4G heterogeneous systems. Terminal mobility is required in 4G infrastructure to deliver wireless services at any time and from any location (5G Training and Certification 2014; 5G Forum 2015; Wunder et al. 2014). Mobile clients can travel across wireless network, geographic borders due to terminal mobility. The two most important challenges in terminal mobility are handoff management and location management. The system tracks and locates a mobile terminal for prospective integration with location management. Location management entails managing all information regarding roaming terminals, including their initial and current locations, authentication information, and so on. When the terminal roams, handoff management, on the other hand, keeps the lines of communication open (Wunder et al. 2014). For IPv6 wireless systems, Mobile IPv6 (MIPv6) is a standardized IP-based mobility protocol (Pirinen 2014; Boccardi et al. 2014). Each terminal has an IPv6 home address in this arrangement. After the local network is left by the terminal, the home address turns into invalid, and thus a new IPv6 address (known as a care-of address) is assigned to the terminal on the visiting network (Rappaport et al. 2013a, b; Olsson et al. 2013).

The Third-Generation Partnership Project (3GPP) created the fundamentals for future Long-Term Evolution (LTE) advanced standards. The 3GPP candidate is for 4G designing and optimizing forthcoming radio access methods and further evolution of the present system. In downlink and uplink transmission, peak spectrum efficiency targets for LTE advanced systems were, respectively, established at 30 bps/Hz and 15 bps/Hz (Taori and Sridharan 2014).

Fifth Generation

WiMAX, WWWW, RAT: 5G can provide limitless wireless connectivity, bringing an ideal real-world wireless web—the World-Wide Wireless Web (WWWW). Beyond the 4G/IMT-advanced standards, 5G refers to the next significant phase of mobile telecommunication standards. At this time, 5G is not an official word for any explicit specification or document yet made public by telecommunication corporations or standardization groups like 3GPP, WiMax Forum, or ITU-R. Each new update will improve system performance while also introducing new features and application areas. Home automation, smart transportation, security, and e-books are some of the other applications that benefit from mobile connection (Pi and Khan 2011a, b; Korakis et al. 2003; Bae et al. 2014). The Institute of Electrical and Electronics Engineers (IEEE) has approved a set of wireless broadband standards known as IEEE 802.16 (IEEE).

The WiMAX Forum industry organization has marketed it under the name "WiMAX" (from "Worldwide Interoperability for Microwave Access"). The air interface and related services connected with wireless local loop are standardized by IEEE 802.16. The utilization of cell phones inside extremely high bandwidth has altered due to 5G mobile technology. Bluetooth technology and Pico nets have just been accessible on the market for children's rocking pleasure. Users may also connect their 5G technology cell phones to their laptops to have access to high-speed Internet. Camera, MP3 recording, video player, huge phone capacity, dialing speed, audio player, and much more are all part of 5G technology. Network design in the fifth generation consists of a user terminal (which plays an important part in the new architecture) and a variety of independent, autonomous radio access technologies (RAT) (Rajagopal et al. 2011; Feng and Zhang 1998; Roh et al. 2014). The 5G mobile system is an all-IP interoperability concept for wireless and mobile networks.

Conventional Cellular Systems—Limitations

4G networks are insufficient to serve a large number of low latencies linked devices and high spectral effectiveness that will be critical in the future. In this part, authors have discussed over a few key areas where traditional cellular networks fall short, prompting the development of 5G networks. Heavy data transmission isn't supported. Various mobile applications send messages to their servers and occasionally request a high data transmission speed for a brief period of time (Cardieri and Rappaport 2001). With increased heavy data in the network, such sorts of data transfer may drain the battery life of (mobile) User Equipment's (UEs), potentially crashing the core network. However, in today's networks, only one sort of signaling/control mechanism is built for all forms of traffic, resulting in substantial overhead for heavy traffic (Abd El-atty and Gharsseldien 2013). The processing power of a Base Station (BS) can only be used by its associated UEs in contemporary cellular networks, and they are intended to accommodate peak time traffic. When a BS is lightly loaded, however, its processing power may be dispersed across a vast geographical region. On weekends or holidays, BSs in residential areas are overloaded while BSs in business areas are almost empty. However, because practically idle BSs require the same amount of power as over-subscribed BSs, the network's overall cost rises (Huq et al. 2013).

A typical cellular network employs two distinct channels: one for transmission from a UE to a BS, known as uplink (UL), and the other for transmission from a UE to a BS, known as downlink (DL). A UE being assigned to two separate channels is not an efficient use of the frequency spectrum. However, if both channels run at the same frequency, as in a full duplex wireless radio, co-channel interference (interference between signals utilizing the same frequency) in the UL and DL channels becomes a big concern in 4G networks (Wang et al. 2013; Hossain et al. 2014; Sanguinetti et al. 2015).

It also hinders network densification or the deployment of a large number of BSs in a given region. Heterogeneous wireless networks are not supported. Heterogeneous wireless networks (HetNets) are wireless networks that use a variety of access technologies, such as third generation (3G), fourth generation (4G), wireless local area networks (WLAN), Bluetooth, and Wi-Fi (Goyal et al. 2021). In 4G, HetNets are already standardized, but the underlying architecture was not designed to accommodate them. Furthermore, existing cellular networks only enable a UE to have a DL channel, a UL channel must be coupled with a single BS, preventing HetNets from being fully utilized. For performance improvement in HetNets, a UE might choose a UL channel and a DL channel from two separate BSs that belong to two different wireless networks.

Vision and Mission—5G

The convergence of growing mm-wave spectrum availability and new application-specific needs will usher in the next major advancement in wireless communications, i.e., 5G. Substantial increase in wireless data speeds, bandwidth, coverage, and connection is expected with 5G wireless communications as well as massive reductions in round trip latency and energy usage. The following is a high-level summary of the 5G standardization process. The first standard is projected to be completed by 2020, according to the report. The Group Special Mobile Association (GSMA) is working with its partners to shape 5G communication to its full potential (Nam et al. 2014; Talwar et al. 2014; Arora et al. 2020; Galinina et al. 2014).

The eight major needs of future generation 5G networks have been determined by combining the many research activities by industry and academia:

(i) Real-world data rates of 110 Gbps: This is about ten times faster than the theoretical peak transmission rate of 150 Mbps for standard LTE networks.

(ii) 1 ms round trip latency: This is about a tenfold improvement over 4G's 10 ms round trip delay.

(iii) High bandwidth per unit area: A large number of linked devices with greater bandwidths for longer periods of time are required in a particular space.

(iv) Massive number of connected devices: To achieve the IoT goal, new 5G networks must be able to link thousands of devices.

(v) 99.999% perceived availability: 5G envisions a network that is both practical and efficient.

(vi) Near-total coverage for "anytime, anyplace" connectivity: 5G wireless networks must provide comprehensive coverage regardless of the location of users.

(vii) Near-90% reduction in energy consumption: Green technology development is already being examined by standard authorities. With 5G wireless's high data speeds and huge connection, this will be even more critical.

(viii) Long battery life: In future 5G networks, device power consumption reduction is critical. Wireless industry, universities, and research groups have begun working on various parts of 5G wireless networks in response to the eight needs listed above (Lee et al. 2014).

According to Ericsson, 5G development will begin with existing 4G LTE networks and backward compatibility. This will aid in the continuation of traditional device services utilizing the same carrier frequency. Ericsson has also worked with SK Telecom, the market leader in South Korea, to demonstrate 5G networks at the 2018 Winter Olympics. Qualcomm is working on 4G and 5G at the same time in order to maximize their potential. The single platform should help save costs and save energy while enabling a wide range of new services. Huawei is working with international trade groups, a number of universities, government agencies, and ecosystem partners to develop critical 5G advancements.

The Docomo network has detected two key trends: (i) global cellular connectivity and (ii) extended real-time rich content delivery. It believes that the key to

5G implementation is the integration of both the higher and lower frequency bands. Basic coverage will be provided by the lower frequencies, while high data rates will be provided by the higher frequencies. Nokia's 5G wireless realization focuses on optimizing spectrum utilization, breakthrough advancements in 5G, dense tiny cells, and better performance. Samsung's vision for 5G is billions of autonomously linked heterogeneous gadgets, ushering in the Internet of Things. The European Union has launched and sponsored two major 5G research projects (Xu et al. 2014).

Architecture—5G

Cellular networks are on the edge of breaching the BS-centric network paradigm. This is due to the excessive demand of capacity limits and sub-millisecond latency in conventional wireless spectrum. Due to rise in demand in the wireless sector, the initial macro-hexagonal coverage was replaced by considerably smaller cell installations. Researchers are focusing their efforts these days on how to construct user-centric networking (Shen and Yu 2014). The user is expected to participate in network storage, content distribution, and processing, rather than being the wireless network's final resolution. Future networks are expected to connect a wide range of nodes that are near to one another. Thus, there would be a lot of co-channel interference in dense 5G networks. The use of directional (energy focused) and sectorized antennas rather of the more traditional omni-directional antennas. As a result, the use of Space Division Multiple Access (SDMA) and effective antenna design is critical. The basis for 5G systems is planned to be strengthened by decoupling the user and control planes, as well as seamless interoperability between diverse networks (Bhushan and Sahoo 2017). The needs for 5G network architecture, modifications in the air interface, and smart antenna design are all discussed in this section. SDN, Cloud-RAN, and HetNets are among the newer technologies addressed (Hu and Qian 2014).

There are two parts in a mobile communication network that includes (i) core network and (ii) radio access network (RAN). Services are provided to the users in core network whereas an RAN links individual devices to their core networks via radio connections. In comparison to LTE's EPS (evolved packet system) design, the key advancement of 5G architecture is the widespread use of virtualization technologies and cloud to offer a wide range of diverse and adaptable services. Existing mobile network designs were primarily built to fulfil the needs of voice and Internet services, which has proven to be inadequately adaptable in 5G, which includes a diverse set of nodes, interfaces, and services. This becomes one of the driving forces for 5G's software architecture (Wu et al. 2015). Because SDN (software-defined networking) and NFV (network function virtualization) technologies can support and administer the underlying physical infrastructure, network services may be virtualized and moved to the cloud, where central control, processing, and management can be performed. Compared to previous cellular networks, which use a wide range of proprietary nodes and specific hardware appliances, the software architecture can lower equipment and deployment costs while increasing administration and evolution flexibility and availability (Pozar 2005). Furthermore, network slicing allows for the creation of separate virtual networks dedicated to certain services as needed, such as a vehicle network service, over a single physical architecture, therefore meeting the diverse needs of varied services.

These network tasks are virtualized and software based in 5G, thus making the services that can be easily incorporated into cloud infrastructure. The access network and the core network are described in more depth below.

C-Ran

Usage of centralized radio access network (C-RAN) in 5G can be done for the radio access network, by leveraging cloud and virtualization technologies to centralize and virtualize some base station functions in the cloud, lowering the cost of deployment and management of the greatly increased and densified base stations. A cloud center and dispersed locations make up the RAN (Violette et al. 1988). Some RAN non-real-time functions in the upper layers with low latency requirements, including cell selection/reselection, intercell handover, and user-plane encryption, might be moved to the cloud, where the information can be interchanged and resources can be shared.

This RAN cloudification will have an impact on other components of the network. Many RAN services that were previously implemented in hardware with specific hardware support, such as IP cores, will be able to implement in a software environment in 5G, according to C-RAN. In this instance, it's critical to ensure their effectiveness. One example is the development of secrecy and integrity algorithms, which is one of the reasons why new software-efficient algorithms for 5G use should be considered.

SBA-Based Core Network

The main network architecture in 5G is Service-Based Architecture (SBA), with system functionality described as a collection of network functions, such as the Session Management Function (SMF) and the Access and Mobility Management Function (AMF). These NFs use standard Service-Based Interfaces (SBI) to deliver services to other approved NFs. The core network introduces a specific network function called the NF Repository Function (NRF) to deal with service registration and discovery, as well as maintain the NF pro-file and accessible NF instances, so that NFs may discover and access each other. The use of network slicing technology to construct optimal networks for individual services with varied performance needs is possible with such a service-based architecture. Subscriber keying materials, such as long-term keys and home network private keys, are stored in Unified Data Management (UDM). It also houses data management capabilities such as the Authentication

Credential Repository and Processing Function (ARPF), as well as the Subscription Identifier De-Concealing Function (SIDF). During an authentication, ARPF is in charge of determining the authentication technique based on the subscriber identification and specified policy, as well as computing the 5G Home Environment Authentication Vector (HEAV). The SIDF offers decryption services for a user's Subscription Hidden Identifier (SUCI) in order to get the user's long-term identity Subscription Permanent Identifier (SUPI) (Kyro et al. 2012).

Next-Generation Network—Applications

The next-generation applications will be emerging in a multiplatform environment. 4G applications are offered on a variety of wireless technologies, including printers, LTE, e-readers, Wi-Fi, cell phones, digital cameras, laptops, and other devices. Although 4G apps are anticipated to be expanded and better versions of present 3G services, the capacity of 4G is yet unknown in the mobile industry.

Example of Next-Generation Network Applications

In this section, authors have provided some of the examples of next-generation network application:

(i) Virtual Presence: This refers to 4G and 5G's ability to deliver services 24 × 7 to the users, even while they are off-site.

(ii) Virtual navigation: 4G offers virtual navigation, which allows users to access a database of various places including buildings, streets, and various other landmarks in big cities (Jaitly et al. 2017).

(iii) Telemedicine: 4G and 5G will allow for patient monitoring remotely. A user may obtain video conference without going to hospital for help from a doctor at any point of time and place.

(iv) Tele-geoprocessing applications: This is a hybrid of GPS (global positioning system) and GIS (geographical information system) that allows a user to query the position.

(v) Crisis Management: Natural catastrophes can disrupt communication networks, which can lead to a crisis.

(vi) Education: 4G provides numerous prospect for people who want to continue learning throughout their lives (Ranvier et al. 2009). People from various parts of the world can save money by continuing with their education online.

(vii) Artificial Intelligence: As human life becomes increasingly surrounded by artificial sensors capable of communicating with mobile phones, more applications combining artificial intelligence (AI) will emerge.

(viii) Traveling: Familiarization to new mobile phone applications, and usage of smartphones with NFC technology and Bluetooth in the passenger travel process. Over the next decade, technology is expected to play a leading role such as experience a location virtually before traveling or to seek inspiration and exchange information live (Xu et al. 2000; Dillard et al. 2004).

 (ix) Security: This layer crosses all layers of the 4G and 5G network architecture, performing functions such as authentication, encryption, authorization, and implementation of service policy agreements between vendors.

  (x) Economic growth: It is aided by technological advancements that permit consumers and organizations to take use of content services and high-value wireless data. 5G networks are projected to support a wide range of applications and services because of its low latency and fast data transfer speeds.

 (xi) Smart grids: It decentralizes energy distribution and also improves the analysis of energy consumption (Rappaport et al. 2012). Smart grids would be able to enhance their efficiency and economic advantages as a result of this. The 5G networks would enable for regular statistical data observation, analyzing them, and retrieval from far sensors, as well as change the energy distribution as needed (Vook et al. 2014).

(xii) Automation: In the near future there will be availability of self-driving vehicles that will be required to connect and communicate in real time. Furthermore, they would communicate with other devices on the roadways, residences, and businesses with a need of virtually zero latency. As a result, a linked vehicular environment would allow for a secure and effective incorporation with other data systems.

(xiii) Healthcare systems: Medical services can benefit from dependable, secure, and quick mobile communication, such as regular data transfers from patients' bodies to the cloud or healthcare facilities (Rajagopal 2012; Anderson and Rappaport 2004; Collonge et al. 2004). As a result, medical treatments that are relevant and urgent may be forecasted and supplied to patients extremely quickly.

(xiv) Industrial applications: 5G networks' zero-latency capability would allow robots, mobile devices, sensors, drones, and data collector devices to get a real-time data without delay, allowing industrial functions to be managed and operated swiftly while conserving energy (Wang et al. 2014; Sinha et al. 2017; Jungnickel et al. 2014; Rappaport 1996; Ahmad et al. 2020).

Conclusion and Future Scope

The advancement of next-generation applications has emerged in a multiplatform environment. Many wireless technologies support 4G application which include LTE and Wi-Fi. Traditionally, networks and communication services have been built and supplied as secure resources with limited potential for customization, improvement, and specialization. Thus, this approach is not adaptable enough to fulfil a varied range of new application needs or to get benefit from a slew of new research-based advances. As a consequence, in some of the selected areas, a new communication design model is prototyped, developed, and put into production. Other than providing a permanent infrastructure, this approach views communication resources as a flexible, programmable environment that can be continually updated to meet new requirements.

In the mobile sector, Mobile Wireless Communication Technology will be leading the new phase. Nowadays, offices are at the fingertips or on phones due to the emergence of Personal Data Assistants (PDAs) and mobile phones. There is a huge scope in the future for 5G technology as it is able to handle most of the modern technologies and supply clients with excellent handsets. 5G will be providing assistance to the idea of Super Core, where all network operators are linked through a single core and are a part of a common infrastructure, independent of their access methods. 4G and 5G techniques provide lower battery consumption, low probability (more coverage), and cheaper or no infrastructural implementation costs along with effective user services. In 5G systems, every mobile phone consists of a permanent "Home IP address" and "care-of address" that refers to its current location. A packet is sent to home address's server when a computer on the Internet wants to communicate with a mobile phone. It then sends a packet to the real location via the tunnel. Cloud computing is a system that uses the Internet and a central distant server to keep data and apps up to date. This central distant service is the content provider in 5G network. Thus, it is due to cloud computing that consumers and companies are able to use programs and therefore access their personal files from any computer with an Internet connection without installing.

References

3GPP (2015) The mobile broadband standard [Online]. Available: http://www.3gpp.org/news-eve nts/3gpp-news/1674-timeline_5g

5G Forum (2015) Make it happen: creating new values together [Online]. Available: http://www. 5gforum.org/

5G-Infrastructure Public-Private Partnership (2013). [Online]. Available: http://5g-ppp.eu/

5G Training and Certification (2014) An initiative project in preparing 5G competence [Online]. Available: http://www.ieee-5g.org/about/

Abd El-atty SM, Gharsselldien ZM (2013) On performance of HetNet with coexisting small cell technology. In: Proceedings of the IEEE conference on wireless mobile network, pp 1–8

Adhikari P (2008) Understanding millimeter wave wireless communication. Loea Corp., White paper

Agyapong P, Iwamura M, Staehle D, Kiess W, Benjebbour A (2014) Design considerations for a 5G network architecture. IEEE Commun Mag 52(11):65–75

Ahmad A, Bhushan B, Sharma N, Kaushik I, Arora S (2020) Importunity & evolution of IoT for 5G. In: 2020 IEEE 9th international conference on communication systems and network technologies (CSNT). https://doi.org/10.1109/csnt48778.2020.9115768

Anderson CR, Rappaport TS (2004) In-building wideband partition loss measurements at 2.5 and 60 GHz. IEEE Trans Wireless Commun 3(3):922–928

Andrews JG et al (2014) What will 5G be? IEEE J Sel Areas Commun 32(6):1065–1082

Arora S, Sharma N, Bhushan B, Kaushik I, Ahmad A (2020) Evolution of 5G wireless network in IoT. In: 2020 IEEE 9th international conference on communication systems and network technologies (CSNT). https://doi.org/10.1109/csnt48778.2020.9115773

Arslan M, Sundaresan K, Rangarajan S (2015) Software-defined networking in cellular radio access networks: potential and challenges. IEEE Commun Mag 53(1):150–156

Bae J, Choi YS, Kim JS, Chung MY (2014) Architecture and performance evaluation of mmWave based 5G mobile communication system. In: Proceedings of the international conference on information and communication technology convergence (ICTC), pp 847–851

Bangerter B, Talwar S, Arefi R, Stewart K (2014) Intel networks and devices for the 5G era. IEEE Commun Mag 52(2):90–96

Banikazemi M, Olshefski D, Shaikh A, Tracey J, Wang G (2013) Meridian: an SDN platform for cloud network services. IEEE Commun Mag 51(2):120–127

Bhushan B, Sahoo G (2017) A comprehensive survey of secure and energy efficient routing protocols and data collection approaches in wireless sensor networks. In: 2017 international conference on signal processing and communication (ICSPC). https://doi.org/10.1109/cspc.2017.8305856

Boccardi F, Heath RW, Lozano A, Marzetta TL, Popovski P (2014) Five disruptive technology directions for 5G. IEEE Commun Mag 52(2):74–80

Cardieri P, Rappaport TS (2001) Application of narrow-beam antennas and fractional loading factor in cellular communication systems. IEEE Trans Veh Technol 50(2):430–440

Checko A et al (2015) Cloud RAN for mobile networks-a technology overview. IEEE Commun Surv Tuts 17(1):405–426

Chen K, Duan R (2011) C-RAN: the road towards green RAN. China Mobile Research Institute, Beijing, White paper

Chen S, Zhao J (2014) The requirements, challenges, and technologies for 5G of terrestrial mobile telecommunication. IEEE Commun Mag 52(5):36–43

Cho HH, Lai CF, Shih TK, Chao HC (2014) Integration of SDR and SDN for 5G. IEEE Access 2:1196–1204

Collonge S, Zaharia G, Zein GE (2004) Influence of the human activity on wide-band characteristics of the 60 GHz indoor radio channel. IEEE Trans Wireless Commun 3(6):2396–2406

Cvijetic N (2014) Optical network evolution for 5G mobile applications and SDN-based control. In: Proceedings of the international telecommunication network strategy planning symposium, pp 1–5

Dillard CL, Gallagher TM, Bostian CW, Sweeney DG (2004) Rough surface scattering from exterior walls at 28 GHz. IEEE Trans Antennas Propag 52(12):3173–3179

Ericsson (2015) 5G radio access. White paper

European Commission (2011) HORIZON 2020, the EU framework programme for research and innovation [Online]. Available: https://ec.europa.eu/programmes/horizon2020/.

Feng Z, Zhang Z (1998) Dynamic spatial channel assignment for smart antenna. Wireless Pers Commun 11(1):79–87. Agiwal et al.: Next Generation 5G Wireless Networks 1651

Galinina O et al (2014) Capturing spatial randomness of heterogeneous celular/WLAN deployments with dynamic traffic. IEEE J Sel Areas Commun 32(6): 1083–1099

Goyal S, Sharma N, Kaushik I, Bhushan B, Kumar N (2021) A green 6G network era: architecture and propitious technologies. Data Analytics Manage. https://doi.org/10.1007/978-981-15-833 5-3_7

GSMA Intelligence (2014) Understanding 5G: perspectives on future technological advancements in mobile. White paper

Holma H, Toskala A, Reunanen J (2015) LTE small cell optimization: 3GPP evolution to release 13. Wiley, Hoboken, NJ, USA

Hossain E, Rasti M, Tabassum H, Abdelnasser A (2014) Evolution toward 5G multi-tier cellular wireless networks: an interference management perspective. IEEE Wireless Commun 21(3):118–127

Hu RQ, Qian Y (2014) An energy efficient and spectrum efficient wireless heterogeneous network framework for 5G systems. IEEE Commun Mag 52(5):94–101

Huawei (2013) 5G a technology vision. White paper

Huq KMS, Mumtaz S, Alam M, Rodriguez J, Aguiar RL (2013) Frequency allocation for HetNet CoMP: energy efficiency analysis. In: Proceedings of the international symposium wireless communications and systems, pp 1–5

Jaitly S, Malhotra H, Bhushan B (2017) Security vulnerabilities and countermeasures against jamming attacks in Wireless Sensor Networks: a survey. In: 2017 international conference on computer, communications and electronics (Comptelix). https://doi.org/10.1109/comptelix.2017.8004033

Jungnickel V et al (2014) The role of small cells, coordinated multipoint, and massive MIMO in 5G. IEEE Commun Mag 52(5):44–51

Kallnichev V (2001) Analysis of beam-steering and directive characteristics of adaptive antenna arrays for mobile communications. IEEE Antennas Propag Mag 43(3):145–152

Khan F, Pi Z, Rajagopal S (2012) Millimeter-wave mobile broadband with large scale spatial processing for 5G mobile communication. In: Proceedings of the 50th annual allerton conference on communication and control computing (Allerton), pp 1517–1523

Korakis T, Jakllari G, Tassiulas L (2003) A MAC protocol for full exploitation of directional antennas in ad-hoc wireless networks. In: Proceedings of the ACM international symposium mobile ad hoc networking and computing, pp 97–108

Kyro M, Kolmonen V, Vainikainen P (2012) Experimental propagation channel characterization of mm-wave radio links in urban scenarios. IEEE Antennas Wireless Propag. Lett. 11:865–868

Lai CF, Hwang RH, Chao HC, Hassan M, Alamri A (2015) A buffer-aware HTTP live streaming approach for SDN-enabled 5G wireless networks. IEEE Netw 29(1):49–55

Lara A, Kolasani A, Ramamurthy B (2014) Network innovation using openflow: a survey. IEEE Commun Surv Tuts 16(1):493–512

Lee YL, Chuah TC, Loo J, Vinel A (2014) Recent advances in radio resource management for heterogeneous LTE/LTE-A networks. IEEE Commun Surv Tuts 16(4):2142–2180

Liu C, Wang J, Cheng L, Zhu M, Chang GK (2014) Key microwavephotonics technologies for next-generation cloud-based radio access networks. J Lightw Technol 32(20):3452–3460

Nam W, Bai D, Lee J, Kang I (2014) Advanced interference management for 5G cellular networks. IEEE Commun Mag 52(5):52–60

Nokia Networks (2014) Looking ahead to 5G: building a virtual zero latency gigabit experience. White paper

NTT Docomo (2015) 5G radio access: Requirements, concepts technologies. White paper

Olsson M, Cavdar C, Frenger P, Tombaz S, Sabella D, Jantti R (2013) 5GrEEn: towards green 5G mobile networks. In: Proceedings of the IEEE international conference on wireless mobile computing, networking and communications, pp 212–216

Osseiran A et al (2014) Scenarios for 5G mobile and wireless communications: the vision of the METIS project. IEEE Commun Mag 52(5):26–35

Pi Z, Khan F (2011a) System design and network architecture for a millimeter-wave mobile broadband (MMB) system. In: Proceedings of the IEEE Sarnoff Symposium, pp 1–6

Pi Z, Khan F (2011b) An introduction to millimeter-wave mobile broadband systems. IEEE Commun Mag 49(6):101–107

Pirinen P (2014) A brief overview of 5G research activities. In: Proceedings of the 1st international conference on 5G ubiquitous connectivity (5GU), pp 17–22

Pozar DM (2005) Microwave engineering. Wiley, Hoboken, NJ, USA

Qualcomm Technologies, Inc (2014) Qualcomm's 5G vision. White paper

Rajagopal S (2012) Beam broadening for phased antenna arrays using multibeam subarrays. In: Proceedings of the IEEE international conference on communications, pp 3637–3642

Rajagopal S, Abu-Surra S, Pi Z, Khan F (2011) Antenna array design for multi-gbpsmmwave mobile broadband communication. In: Proceedings of the global telecommunications conference (Globecom), pp 1–6

Ranvier S,.Kyro M, Haneda K, Mustonen T, Icheln C, Vainikainen P (2009) VNA-based wideband 60 GHz MIMO channel sounder with 3-D arrays. In: Proceedings of the IEEE Radio Wireless Symposium, pp 308–311

Rappaport TS (1996) Wireless communications: principles and practice. Prentice-Hall, Englewood Cliffs, NJ, USA

Rappaport TS et al (2013a) Millimeter wave mobile communications for 5G cellular: it will work! IEEE Access 1:335–345

Rappaport TS, Gutierrez F, Ben-Dor E, Murdock JN, Qiao Y, Tamir JI (2013b) Broadband millimeter wave propagation measurements and models using adaptive beam antennas for outdoor urban cellular communications. IEEE Trans Antennas Propag 61(4):1850–1859

Rappaport TS, Roh W, Cheun K (2014) Wireless engineers long considered high frequencies worthless for cellular systems. They couldn't be more wrong. IEEE Spectr 51(9):34–58

Rappaport TS, Ben-Dor E, Murdock JN, Qiao Y (2012) 38 GHz and 60 GHz angle-dependent propagation for cellular & peer-to-peer wireless communications. In: Proceedings of the IEEE international conference on communications, pp 4568–4573

Roh W et al (2014) Millimeter-wave beamforming as an enabling technology for 5G cellular communications: theoretical feasibility and prototype results. IEEE Commun Mag 52(2):106–113

Rost P et al (2014) Cloud technologies for flexible 5G radio access networks. IEEE Commun Mag 52(5):68–76

Samsung Electronics Co. (2015) 5G vision, white paper

Sanguinetti L, Moustakas AL, Debbah M (2015) Interference management in 5G reverse TDD HetNets with wireless backhaul: a large system analysis. IEEE J Sel Areas Commun 33(6):1187–1200

Shen K, Yu W (2014) Distributed pricing-based user association for downlink heterogeneous cellular networks. IEEE J Sel Areas Commun 32(6):1100–1113

Sinha P, Jha VK, Rai AK, Bhushan B (2017) Security vulnerabilities, attacks and countermeasures in wireless sensor networks at various layers of OSI reference model: a survey. In: 2017 international conference on signal processing and communication (ICSPC). https://doi.org/10.1109/cspc.2017.8305855

Stallings W (2007) Data and computer communications. Pearson/Prentice-Hall, Upper Saddle River, NJ, USA

Talwar S, Choudhury D, Dimou K, Aryafar E, Bangerter B, Stewart K (2014) Enabling technologies and architectures for 5G wireless. In: Proceedings of the MTT-S international microwave symposium (IMS), pp 1–4

Taori R, Sridharan A (2014) In-band, point to multi-point, mm-wave backhaul for 5G networks. In: Proceedings of the IEEE international conference on communications workshops, pp 96–101

Violette EJ, Espeland RH, DeBolt RO, Schwering FK (1988) Millimeter-wave propagation at street level in an urban environment. IEEE Trans Geosci Remote Sens 26(3):368–380

Vook FW, Ghosh A, Thomas TA (2014) MIMO and beamforming solutions for 5G technology. In: Proceedings of the IEEE microwave symposium (IMS), pp 1–4

Wang CX et al (2014) Cellular architecture and key technologies for 5G wireless communication networks. IEEE Commun Mag 52(2):122–130

Wang Z, Li H, Wang H, Ci S (2013) Probability weighted based spectral resources allocation algorithm in Hetnet under Cloud-RAN architecture. In: Proceedings of the international conference on communications in China workshops, pp 88–92

Wu T, Rappaport TS, Collins CM (2015) Safe for generations to come: considerations of safety for millimeter waves in wireless communications. IEEE Microw Mag 16(2):65–84

Wunder G et al (2014) 5GNOW: non-orthogonal, asynchronous waveforms for future mobile applications. IEEE Commun Mag 52(2):97–105

Wunder G et al (2014) 5GNOW: intermediate frame structure and transceiver concepts. In: Proceedings of the globecom workshops, pp 565–570

Xu H, Rappaport TS, Boyle RJ, Schaffner JH (2000) Measurements and models for 38-GHz point-to-multipoint radiowave propagation. IEEE J Sel Areas Commun 18(3):310–321

Xu J et al (2014) Cooperative distributed optimization for the hyper-dense small cell deployment. IEEE Commun Mag 52(5):61–67

Zhang N, Cheng N, Gamage AT, Zhang K, Mark JW, Shen X (2015) Cloud assisted HetNets toward 5G wireless networks. IEEE Commun Mag 53(6):59–65

Zhou Y, Yu W (2014) Optimized backhaul compression for uplink cloud radio access network. IEEE J Sel Areas Commun 32(6):1295–1307

Chapter 2
Third Industrial Revolution: 5G Wireless Systems, Internet of Things, and Beyond

Anwesha Das, Aninda Chowdhury, and Riya Sil

Abstract Commercial 5G mobile communication installations are currently ongoing. A variety of reasons, notably rising business and consumer needs as well as the advent of much more cheap equipment, are driving 5G and IoT growth. Substantial carrier investments in 5G networks, frequency, and infrastructure, as well as the adoption of international standards, are indeed assisting in driving development and increasing investor interest in IoT. Today's modern 5G mobile cellular systems are emerging beyond current 4G technology, which will remain to fulfill diverse applications. 5G, which is expected to last a long time, may meet present needs like intelligent power applications while also forecasting future use cases like self-driving automobiles. Mobile operators would need to guarantee to ensure its added versatility simultaneously present as well as future use cases need as companies oversee the growth of technology. Cautious providers would control their expenditures to assure customer service as infrastructures migrate to 5G. The majority of 5G use case scenarios fall into three broad segments: improved mobile broadband (eMBB), enormous IoT, as well as critical communications, within each set of performance, and bandwidth, including delay needs. While 4G would remain to be utilized for so many consumers and commercial IoT scenarios, 5G offers IoT features that 4G as well as other networks do not. This would include 5G's capacity to accommodate a massive amount of fixed and portable IoT systems with variable speeds, capacity, and service level needs. As the Internet of Things develops, the adaptability of 5G would become increasingly more important for organizations wanting to satisfy the stringent needs of vital connectivity. Because of 5G's ultra-reliability as well as reduced latency, self-driving vehicles, intelligent power infrastructures, better industrial automation, and some other demanding technologies are becoming a possibility. While 5G increases Internet bandwidth, cloud services, machine intelligence, as well as cloud technologies would all assist to manage huge data quantities created by IoT. Additional 5G advancements, like low latency, and non-public networking, including the core of

A. Das · R. Sil (✉)
Adamas University, Kolkata 700126, India
e-mail: riyasil1802@gmail.com

A. Chowdhury
St. Placid's School and College, Chittagong, Bangladesh

© The Author(s) 2023
B. Bhushan et al. (eds.), *5G and Beyond*, Springer Tracts in Electrical and Electronics Engineering, https://doi.org/10.1007/978-981-99-3668-7_2

5G, would eventually help realize the goals of an IoT network that is worldwide and capable of sustaining connectivity that is larger in size.

Keywords Wireless communications · IoT · Fifth-generation communication · Mobile broadband · Technology · Networks

Introduction

Fifth-generation (5G) connections become more widely available as an important driver of the expansion of IoT systems. Today's modern researchers and professionals are analogous to Christopher Columbus, who started to establish that the sphere no longer is analog (Alsamhi et al. 2021). Numerous emerging innovations have resulted in the emergence of the technological age, but nobody has had a greater influence than portable technological advances. Personally, such technology has transformed the everyday lives people connect, but collectively, they dramatically revolutionized the reality in which the people live by generating a "blue ocean" of a perspective that is new. A blue ocean, in basic words, seems to be the emergence of a completely fresh sector or developments in an established sector that modify the bounds of rivalry, resulting in a marketplace devoid of competition. Traditionally, blue seas have indeed been industry specific and also the consequence of a particular firm's breakthroughs, like Apple, Netflix, Starbucks, Uber, etc. The utilization of mobile technological advances, on the other hand, has transformed the blue ocean together into a sea of the IoT. Whenever opportunities emerge, the integration of various innovations has an impact across all sectors at the very same time.

IoT and Its Devices

The IoT refers to any things (i.e., items) which are linked to the Web and may be accessed via ubiquitous technologies. The IoT has given rise to plenty of innovative "intelligent" devices (i.e., Internet enabled). People are presently living amid a smart transformation wherein numerous items in their daily lives are connected via Internet (Ali et al. 2019; Goyal et al. 2021a; Chettri and Bera 2019). A few instances of advanced devices that have resulted in the emergence combining mobile computing as well as technologies of IoT have been shown in Table 2.1 (Peral-Rosado et al. 2018; Ahmad et al. 2020; French and Shim 2016; Ghendir et al. 2019; Arsh et al. 2021; Hussein et al. 2018).

Table 2.1 Examples of IoT devices in different sectors

Sl. no	Residential devices	Fitness devices	Attire devices	Gadget devices
1	Smart lock for door	Monitor for BP measurement	Smart watch	Smart stoves
2	System of hydroponic	Monitor for cholesterol measurement	Smart socks	Smart AC
3	Smart tank of propane	Monitor for blood glucose level measurement	Smart shirt	Smart washer for dishes
4	Smart control of sprinkler	Smart system for sleeping	Insoles that are enabled via Bluetooth	Smart machine for washing
5	Smart security for home	Smart cardio	Glasses of technology	Smart refrigerator

A 5G and IoT

As humans move closer to 5G possibilities, the simplicity, as well as the efficiency with which IoT links may be established, would improve, allowing for further technological improvements (Iannacci 2018; Arora et al. 2020; Islam et al. 2021; Jungnickel et al. 2019). Nevertheless, the implications of mobile computing, as well as IoT, go further than the implementation of novel technical prowess. Humans participate in the data that is collected by adding multimedia elements to each and all things humans contact within their ordinary activities. As a consequence, big information is no longer huge—massive, it is indeed and it'll only expand as businesses begin to move forward into the increasing connection between many things and people. As they continue to evolve opportunities for information analysis, the implications for data professionals are enormous (Sethi et al. 2021; Knieps 2019; Li et al. 2018). Businesses of all sizes must answer customer demands for even more connection among persons, computers, as well as things.

For achieving the objective, the authors have emphasized the implications of 5G mobile technology on the "Internet of Things (IoT)". A detailed discussion on the ubiquitous computing and 5G technology has been discussed taking into consideration the IoT and 5G technology individually. The neediness and the challenges of wireless 4G networks and requirements, vision (in the context of both research and industrial perspective), unification of technologies, and technology drivers for the 5G-enabled IoT technologies have been discussed descriptively. Finally, the challenges of the research and the respective trends for the future have been presented in the study related to the topic "Internet of Things (IoT) towards 5G Wireless Systems".

Section "5G and Ubiquitous Computing" provides a clear view of 5G and its ubiquitous computing. Section "History and Present Researches on IoT as Well as 5G" discusses the history and the current research on IoT and 5G. Section "Requirements of IoT and Shortcomings of Wireless 4G Network" gives a detailed description of the requirements of IoT and the various shortcomings of wireless 4G networks (Rathi et al. 2020; Mei et al. 2019; Migabo et al. 2020). Section "Needs for IoT that is 5G Enabled" focuses on the need for 5G-enabled IoT. Section "The Visionof 5G IoT: Industrial and Research Context" highlights the vision of 5G IoT in an industrial and research context. Section "Unification of Technologies" gives a detailed overview of the unification of technologies. Section "Conclusion and Future Scope" concludes the paper and discusses the future scope of the work.

5G and Ubiquitous Computing

People and corporations have commonly embraced the IoT and the analytics of big data in today's modern pervasive computing age, with the next evolution of cellular technologies, 5G networking, just at the vanguard (Malik and Bhushan 2022; Poncha et al. 2018). Deloitte Reviews, MIS Quarterly, Proceedings of the ACM, as well as "Information Systems Research", among others, have dedicated special issues to IoT, analytics of big data, including 5G. According to a new Bain & Company analysis, Europe, as well as the US, would add approximately usd8 trillion to world GDP through 2020. This section will discuss IoT as well as 5G.

The Internet of Things

Around 1999, British innovator and entrepreneur Kevin Ashton invented the phrase "Internet of things". IoT provides improved gadgets and network, including service connectedness which extends above machine-to-machine interactions (M2M) as well as encompasses a wide range of interfaces and areas, including activities (Santos et al. 2018; Wijethilaka and Liyanage 2021a, b). For describing IoT, two key concepts may be used: things and also the Web. An item has to be able to send data or orders to some other item over a connection to also be IoT capable. Human relationships or sensing could cause IoT-enabled items to undertake activities, resulting together in a linked network comprising things having pervasive management. The connection might be personalized, corporate, or governmental, while the Web has been the most commonly imagined foundation underlying IoT (Xing 2020). IoT is sometimes confused with advanced devices, which applies to just about any device having Internet access. Smart technology consists of devices that really can access the web, while IoT expands this paradigm to also include things that can be controlled from anywhere using online services (Yan 2019; Zikria et al. 2018; Agiwal et al. 2019). A smartphone, for instance, could access the Internet, however, the device should be used in person. An IoT-enabled device, on the other hand, may be accessed and controlled from every place and at any time (Fig. 2.1).

Machines could monitor and catalog all items and persons throughout regular living when they have unique identification (Albreem et al. 2021). Modern smartphones, smart watches, smart automobiles, cargo containers, as well as other devices

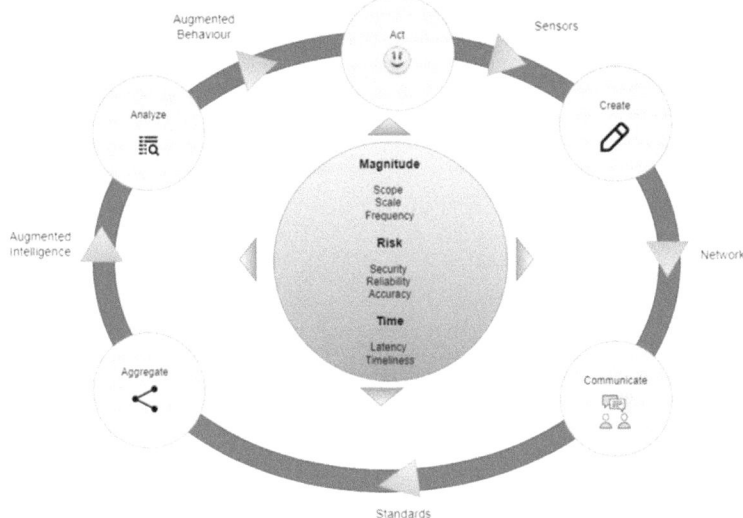

Fig. 2.1 Value loop for information

are becoming more linked than ever before. To now, the much more common uses include home automation, wearable technology, intelligent buildings, smart grids, linked cars, and even linked health care (Zheng et al. 2020). The Internet of Things can help you "monitor and tally, watch and recognize, analyze and respond in situations". The intelligent supply cycle depicts the enterprise value steps (i.e., generate, transmit, collect, analyze, as well as react) which must be completed to produce worth (Aman et al. 2020). Detectors, networking, norms, augmented cognition, and enhanced behavior are all present at each level. Aside from RFID, items can be tagged utilizing techniques such as base station communications, QR codes, barcodes, as well as electronic copyrighting.

A 5G Network

5G would be the next step in the evolution of the mobile world. The US reclaimed dominance within the mobile market through fourth-generation (4G) installations. European, Japan, and Korea lead the third-generation (3G) globe in the 2000s. Every area wants to be the global leader in the 5G network (Sicari et al. 2020). Although 5G is still very much in its initial stages of development, the "International Telecommunication Union (ITU)" has started work on the "International Mobile Telecommunications (IMT)" spectrum needs for 2020 and even beyond (Sodhro et al. 2020). The Fig. 2.2 depicts a probable timeline for the advancement of 5G networks (i.e., 5G study, development, and testing till 2016; 5G standards through mid-2018; 5G products till the 2020 preceding 5G implementation in 2021). This development of

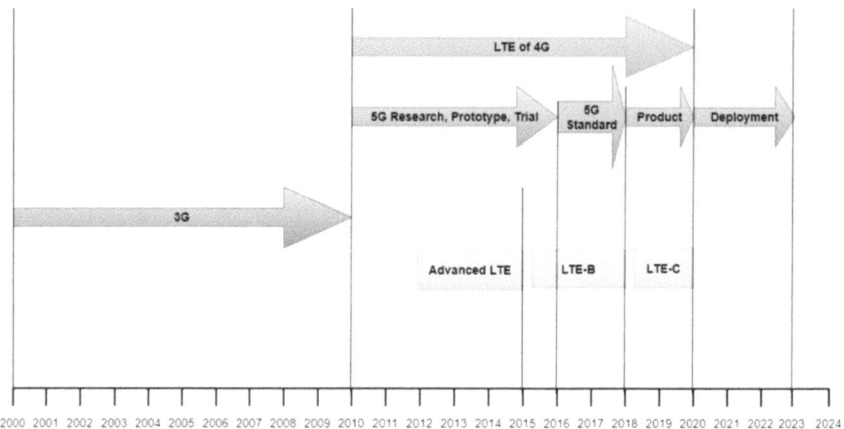

Fig. 2.2 Evolution of mobile technology

the 5G technique requires LTE-A, LTE-B, through LTE-C as elements of the "3rd Generation Partnership Project (3GPP)".

Even though there is no agreement on 5G, most business experts believe in the sorts of quality standards (for example, latency, network availability, energy efficiency, huge "multiple-in multiple-out (MIMO)", energy usage, linked devices, exposure, and increased security needs). Verizon intends to have been the first US operator to provide a 5G network pending implementation (Worlu et al. 2019; Xia et al. 2019; Qiu et al. 2020). Many other nations, including Japan and South Korea, are making arrangements to provide 5G field trials in the years ahead. As previously said, the expansion and sustainability of 5G would be dependent on the performance of the whole informational, communications, and technological (ICT) environment (Sadique et al. 2018; Ni et al. 2019; Oughton et al. 2018; Liu et al. 2020a). The whole ICT environment would play an important role in value generation as well as preservation.

History and Present Researches on IoT as Well as 5G

A variety of wireless systems, including 2G, 3G, or 4G; Bluetooth; Wi-Fi connectivity; and others, have indeed been employed in diverse systems of IoT, wherein billions of devices would be linked by the wireless system (Khanna and Kaur 2020). The 2G systems (which presently serve 90% of the global total population) are intended for speech, the 3G systems (which presently serve 65% of the global total population) for the phone as well as information, and the 4G networks for cable broadband services. Although 3G and 4G networks are commonly utilized for IoT, they are not entirely suited for IoT systems (Khatua et al. 2020). 4G has substantially improved the capability of mobile networks in terms of providing Web access to

IoT devices. Since 2012, the "lifelong evolution" (LTE) to 4G connection has been the quickest and also the most constant variation of 4G when contrasted to rival technologies like ZigBee, LoRa, WiMAX b, Sigfox, and many others (Husain et al. 2018). As its next networking, 5G networking and standards are anticipated to tackle issues that 4G networks faced, including more intricate communications, gadget computing capabilities, among intellects, etc., to meet the demands of intelligent devices, Industrial 4.0, and so forth.

The graph depicts the progression of mobile networks between 3G- and 5G-powered IoT. The 5G's growth would indeed be founded on the basis established with LTE of 4G, which also would give users the phone, information, and Internet connectivity (Kadhim et al. 2020). 5G will greatly enhance speed and reliability to enable dependable and fast connection to upcoming IoT devices. The present LTE technology of 4G could deliver a rate of transmission of 1 Gbps, although the connection of 4G can indeed be readily interrupted by Wi-Fi transmissions, structures, microwaves, and other conclusions (Amin and Hossain 2020; Athanasiadou et al. 2020; Ahad et al. 2020). 5G connections may give customers greater speeds over 4G technology, up to ten Gbps, even while providing dependable connectivity to thousands of devices simultaneously.

The image indicates something in IoT, massive machine-type communication (MTC) applications in intelligent urban, health systems, as well as other areas necessitate vast connectivity connections, resulting in a large heterogeneity of IoT and also many implementation issues (Ali et al. 2020). Several M2M communication techniques have been employed over the last two decades, which include short-range MTC like "Low Energy Bluetooth (BLE v4.0)", ZigBee, Wi-Fi, and others, as well as long-range MTC like "Low-Power wide-area (LPWA)", Ingenu "random phase multiple access (RPMA)", Sigfox, LoRa, and many other (Lu et al. 2018). The groups of three partnerships project (3GPP) recommended "Enhanced Machine-Type Communication (EMTC)", "Extended Coverage-Global System for Mobile Communications for the IoT (EC GSM-IoT)", and "Narrowband-IoT (NB-IoT)" as cellular-based LPWA technologies for such IoT to guarantee M2M capabilities (Mavromoustakis et al. 2016). Available communication technology remains varied, therefore meeting the needs of IoT applications will be a problem again for fifth-generation (5G) cellular operators.

Requirements of IoT and Shortcomings of Wireless 4G Network

Whereas the legacy connectivity is focused on H2H functionality over extended journeys, current interaction is trying to shift into a broader sense M2M console (Porambage et al. 2018; Salam 2020; Ye et al. 2019). The diversity of diversified requirements represents a problem to co-operative priority scheduling among multiple things, as well as information sharing and interaction among items extra broadly (Zhang et al.

2018). As a result, it's indeed necessary to explore legacy wireless communication from the standpoint of IoT.

Massive Connectivity—The basic concept of IoT transforms the volume and variety of linked devices (Zikria et al. 2019; Liu et al. 2020b; Painuly et al. 2020; Khan and Javaid 2021). Globally, 212 bn smart objects are estimated to also be installed by 2020. LTE wireless networks, on the other hand, were intended for limited "Radio Resource Control (RRC)" connected consumers. Previously, cables supported the company's aim of independent connection (Khanna and Kaur 2019). Even so, the immense parallelization anticipated inside the IoT scenery can be discussed by wireless systems. This technological divide, along with billions of networked diverse items, would eventually coax disruptive innovation through older systems (Bektas et al. 2018). Furthermore, the old Wi-Fi access method will struggle with congestion as well as overloading as a result of numerous demands from massive equipment. The network efficiency would've been degraded if there were a massive amount of MTC devices conducting concurrent remote access (Boursianis et al. 2022). Additionally, conventional network computation approaches will fall well short of extracting the needed information from the vast quantity, speed, as well as diversity of interconnections.

Long Life Time of Battery—The bulk of smart IoT technology is required to be battery powered to support wireless communication. Replacing or recharging the battery may not even be simple or cost-effective (Awoyemi et al. 2020). Furthermore, IoT-enabled devices are powered by small batteries. As a result, as previously stated, the requirement for improved battery life is an approaching problem in IoT implementation that cannot be disregarded. Common M2M travel patterns demonstrate that now the power needed for the transmission process is often low (Bana et al. 2019). Despite the addition of a power-saving option for MTC communications in 3GPP Release 12 assuring extra battery lifetime in IoT systems remains a daunting prospect for multiplexing OFDM division-based LTE.

Constraints of Infrequent Congestion as well as Orthogonality-Random-access synchronization processes are required in 4G LTE channels to confront orthogonality restrictions. Although synchronization maintains senders' temporal coherence, orthogonality reduces crosstalk. Regular time–frequency synchronization in short data packets results in signaling complexity (Hu et al. 2018). In reality, the volume of information info in sequential "Orthogonal Frequency-Division Multiple Access (OFDMA)-based" MTC approaches is similar and much less than elevated signaling bandwidth. Furthermore, in an IoT future, common smart things will have become huge traffic producers as well as recipients (Pedersen et al. 2018). As a result, future wireless communication is projected to just be intermittent, providing a significant problem for service-based IoT design.

Delayed Aware Solutions as well as Delayed Considerate Solutions—In IoT devices, limiting battery capacity as well as bandwidth constraints promote periodic communication (Sharma et al. 2020). To a certain degree, delay-tolerant connectivity is appropriate for certain workloads. Nevertheless, applications such as healthcare coverage, automated vehicles, as well as monitoring are major priorities and time sensitive. Additionally, haptic online, which is fast gaining popularity for apps at the fingers, is indeed a major driver for low bandwidth broadband Internet (Yang and Alouini 2019). 4G networks have such a round-trip latency of 10–15 ms (due to uplink scheduling requests), which is dubious for vital communications, autonomous cars, as well as other time-sensitive activities.

Narrowband Transmission—Demands for long battery life, low bandwidth M2M connectivity, especially stochastic flow are incompatible with traditional broadband wireless connections (Hossein Motlagh et al. 2020). Conventional LTE methods, which were built for Internet activities, are thus over-engineered for reduced, many-delay-tolerant applications that are envisioned in the IoT environment. 3GPP has recently added "narrowband IoT (NB-IoT) in Release 13" specifications (Hui et al. 2020). In contrast to short-range unregistered systems, such as ZigBee, Bluetooth, and others, NB-IoT technology enables minimal wattage and a vast range of commu-nication with an available band. It is conceivable to implement NB-IoT with just a limited bandwidth of around 200 kHz (Kaur 2020). Furthermore, it offers increased range, higher energy effectiveness enabling larger battery, and reduced complications with low-cost gadgets. Although conventional LTE networks employ a sub-carrier of 15 kHz, NB-IoT introduces subcarriers of 3.75 kHz again for uplink architecture (Aman et al. 2020). Nevertheless, tests have shown that a 3.75 kHz transmitted signal has certain detrimental consequences on cohabitation only with LTE's 15 kHz sub-carrier width. As a result, narrowband functioning is among the important criteria that have to be investigated further than just low-data workloads as well as adaptable IoT installation.

Transcend Human Interaction—IoT may be viewed as just a highly distributed communication network that interfaces well with the physical domain at the system level (Habibi et al. 2019). Gadgets observe physical processes, thus IoT connectivity systems include detectors, controllers, meters, utilities, communications, etc. As a result, a new issue has emerged that links not just persons but also technologies (Nguyen et al. 2021). Unlike H2H connections, the primary IoT need is the ability to link a wide range of devices remotely at such a low cost. Moreover, a physical device connection needs sufficient Internet bandwidth, extended battery life, as well as enhanced penetration so that the gadgets could access difficult areas. This pursuit of broad sensing application is projected to get to be a vital impediment to traditional wireless systems that is human oriented (Shi et al. 2020). The perspectives that are things oriented imply something other than personal interaction. Furthermore, as IoT gets more complex, objects and people would connect more frequently and seamlessly. As a result, the Internet of Things' need of connecting communications well with the physical domain cannot indeed be overlooked.

Needs for IoT that is 5G Enabled

The IoT is transforming daily life by enabling a lot of unique services that run on ecosystems of intelligent and extremely diverse gadgets (Teli et al. 2018). Numerous research works have been undertaken in recent years on several tough subjects for such 5G IoT, as well as the key criteria of IoT encompass:

(i)   With increased data speed, future IoT systems like HD streaming content "virtual reality (VR)" or rather "augmented reality (AR)" would demand greater data rates of roughly 25 Mbps to obtain satisfactory performances.

(ii) High-scalability and perfectly all right systems are required for 5G IoT to allow fine-grained front-haul networking breakdown through NFV.

(iii) Extremely reduced delay is required in 5G IoT services like haptic Web, AR, video gaming, and so on.

(iv) With reliability and robustness, 5G IoT necessitates enhanced availability and transition effectiveness for consumers of IoT devices and applications.

(v)   Safety, unlike typical security strategies that safeguard connection and privacy protection, the upcoming IoT payment service, as well as online wallet services, create a greater safety approach to increase information security.

(vi) Extended battery life: To handle billions of low-power as well as low-cost IoT systems in 5G IoT, 5G-powered IoT requires reduced energy technologies.

(vii) Connectivity density, a very large number of sensors would be linked together during 5G IoT, requiring 5G to facilitate the effective transmission of messages in a specific time and region.

(viii) Agility, the 5G IoT ought to be capable of handling a large number of device-to-device connections while being mobile.

The current state of IoT involves posting as well as saving all basic information generated via IoT systems to the cloud, where it would be analyzed via cloud storage to derive relevant knowledge via analysis techniques.

The Vision of 5G IOT: Industrial and Research Context

5G IoT's vision, as well as its goal, is to link a wide variety of devices inside the same system architecture (Atiqur et al. 2020). Numerous advanced 5G wireless technologies, such as smart cities, "Internet of vehicle (IoV)", advanced factories, and smart farming, including smart health care, contribute to the IoT boom (Farhan et al. 2018). A few of the major cellular, semiconductors, and network operators having outstanding research centers are performing laboratory and site experiments to make a 5G wireless network available by 2030. 5G studies and experimentation are being conducted at several research centers having world-class lab facilities (Hassan et al. 2020). The most recent advancements and upgrades in cellular technologies offer to address the requirement for fast broadband, improved spectrum utilization,

Table 2.2 Vision of 5G-enabled IoT for various telecommunication industries

Sl. no	Name of the industries	Vision
1	Intel	Intel is working on a new essential system that will allow 5G HetNets while also maximizing the effective usage of spectrum resources (Awin et al. 2019). Intel is developing new technologies, like licensed accessible access (LAA), to improve speed
2	Samsung	Samsung's ambition is really to link everything on the planet (Chen and Okada 2020). Samsung anticipates that almost all IoT platform gadgets will be linked to one another. Effective collaboration is required for the realization of 5G IoT fields like home automation, smart cities, smart manufacturing, smart health care, smart farming, transportation, and so on. Samsung has made significant contributions to the IoT accessible cloud service, which allows employees to access household equipment. The remote controller is programmed for Samsung equipment such as the air conditioner, washer, and refrigerators
3	Nokia	Nokia recently announced the creation of a cross-domain framework to enable 5G technologies. Nokia is focusing on system modernization, which serves to maintain overall power consumption stability by decreasing the usage of energy that is not connected directly to data transfer (He et al. 2021). They are focusing on many big chances to improve ground station fuel efficiency

long-range connectivity, higher reliability, and also the connectivity of intelligent devices (Narayanan et al. 2020). IoT together in the 5G context has the potential for being the most transformative technology in the world of information technology. As per the studies, 5G wireless communication would be available in several nations by 2030 (Table 2.2).

Unification of Technologies

Technologies are always progressing toward unity. Whenever the Web became commercialized within the 1990s, new tools developed, triggering a chain reaction of technological advancement (Goyal et al. 2021b). Items did not have Internet access in the 1990s. This era's technology worked individually with one another. As smartphones and televisions gained Web access in the 2000s, the latest craze of connected phones emerged. During this period, technology started to shift by incorporating functionalities that needed the Internet (Sun et al. 2020). As networking improved, data speeds rose, as did the capacities of connected phones, giving a boost to the Web of things when humanity moved from such a theoretical viewpoint to realize those skills in the 2010s.

Items today do indeed have Internet access, but they can communicate with one another and share data. Utilizing sensing as well as connected devices, things may communicate with one another, sharing information and giving new services to the

Fig. 2.3 Intelligent technologies evolution

public everywhere. As even more devices become IoT enabled, they get closer to an interlinked future in which all items may interact including all things accessible by people anyone at any time from any location. This transition to IoT opens up new opportunities, like social media of smart items which are closely attached. The Fig. 2.3 depicts the progression of advanced devices from standalone products to IoT social networking sites.

Enabling Technology Drivers in 5G IoT

5G's major properties for enabling various real-time multimedia include quick pace, reduced latency, as well as high throughput. As a result, 5G enablement technologies must be developed (Kim et al. 2019). Device-to-Device (D2D) communications, Machine-to-Machine (M2M) interaction, Millimeter Waves, "Quality of Service (QoS)", "Network Function Virtualization (NFV)", Vehicle-to-Everything (V2X), Full-Duplex, as well as Green Interaction are among the core technology solutions utilized in 5G technology. 5G system enables data transmission rates of 10–20 Gbps, which is 100 times faster than 4G technologies, enabling new IoT robotic surgical capabilities.

Challenges and Trends of Future of the Research

5G has characteristics that can fulfill the criteria for future IoT, but it also created a new series of exciting research problems on 5G IoT design, trustworthy interactions among gadgets, privacy difficulties, and so on (Pedersen et al. 2018). The 5G IoT incorporates several techniques and therefore is having a big influence on IoT systems. In this part, prospective research problems, as well as future developments in 5G IoT, are discussed.

Challenges

(i) Bands of frequencies—In contrast with 4G LTE, which runs on known bandwidths under 6 GHz, 5G needs frequencies up beyond 300 GHz. Certain frequencies, called mm waves, may transport significantly greater bandwidth and provide a 20-fold improvement in potential bandwidth above LTE.

(ii) Scope and deployments—Although 5G provides considerable increases in speed and capacity, it's a much more short scope which would necessitate infrastructures to a greater extent (Khujamatov et al. 2020). The wavelengths that are longer in length allow for direction in extreme manner radio signals that might be focused or directed—which is known as beamforming. The major difficulty is that, whereas 5G antenna can manage more user activity, they can only shoot out across short ranges.

(iii) Expenses of construction and acquisition—Establishing a network is costly; operators would fund it by providing consumers revenues (Qiu et al. 2020). Much the same as LTE packages had a high capital cost, 5G is ready to imitate a similar pattern. It's not simply adding a level to a current network; it's establishing the basis want something different.

(iv) Device compatibility—There's now a lot of information about 5G-enabled smartphones and various other instruments. However, its accessibility would be determined by how costly devices are for producing and how rapidly the system is employed. Various operators in the US, South Korea, including Japan have indeed commenced 5G projects for trial in various regions, and manufacturers have stated that suitable mobile phones would indeed be accessible in 2019.

(v) Data security—5G deployment, just like every data-driven innovation, would face either basic risks of cybersecurity that is of advanced level. Despite the information that 5G is somewhat covered by the "Authentication and Key Agreement (AKA)", which is a method meant to create confidence among providers, it is presently easy to follow individuals utilizing smartphones or perhaps even eavesdropping on ongoing phone calls. The obligation, as that is today, would be on operators and networking conglomerates to offer a virtual support system for consumers.

Trends of Future

Omdia believes that the confluence of AI as well as edge computing will boost the importance and influence propositions of IoT. Edge features on equipment within fields decrease delay, energy consumption, as well as expenses associated with data transport to the clouds (Liu et al. 2021). This provides a pathway for the analysis of more complicated types of data. As per "Fortune Business Insights", the worldwide IoT industry will be worth $1.1 trillion by 2026. It's being used for monitoring people, towns, agriculture, and whatever else one could conceive of for the next years.

Conclusion and Future Scope

5G offers several advantages not accessible using technological breakthroughs. This would include 5G's ability to accommodate an overwhelming amount of fixed and portable IoT systems with varying speed, capacity, and quality service needs. The adaptability of 5G would become increasingly more important for companies as the Internet of Things develops. Important connections would be supported by 5G, which will have much more stringent performance targets. The ultra-reliability, as well as reduced latency of 5G, might aid in the realization of self-driving vehicles, intelligent power networks, better industrial automation, as well as other sophisticated technologies.

5G networks could safely manage the large volume of data generated by IoT gadgets in addition to serving a very large number of devices including their different service needs. Cloud technology, machine intelligence, as well as cloud technologies would also aid in the management of IoT large datasets. 5G is indeed a worldwide technology that is being deployed following 3GPP international specifications. It was created to assure IoT compatibility, and it is still evolving through continual improvements to agreed-upon criteria. Expanding on the connectivity for IoT provided by 4G, Release 15 through 16 of something like the 3GPP standards will give additional assistance for IoT devices using 5G capabilities such as ultra-reliability as well as reduced latency. Further 5G advancements, like low latency, non-public networking, including 5G core, are believed to assist in attaining the goals of a worldwide IoT network able to support large connectivity with varying movement and accessible needs.

Appendix 1: Wireless Technology Features

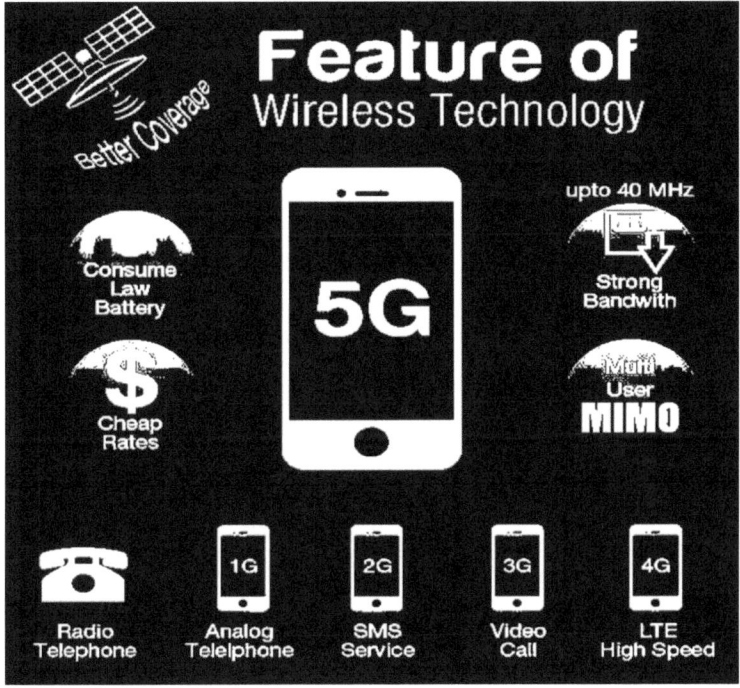

(*Source* https://i.pinimg.com/originals/9f/a1/dd/9fa1dd19560fe8b1d86bcb6db3c f31c1.jpg)

Appendix 2: Information Speed Increment

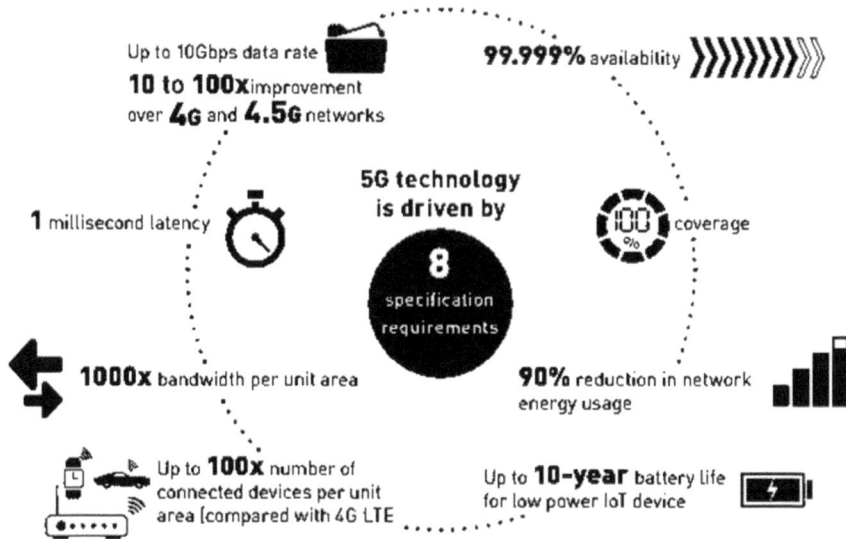

(*Source* https://www.thalesgroup.com/sites/default/files/gemalto/Image002.jpg)

Appendix 3: 5G Network Architecture

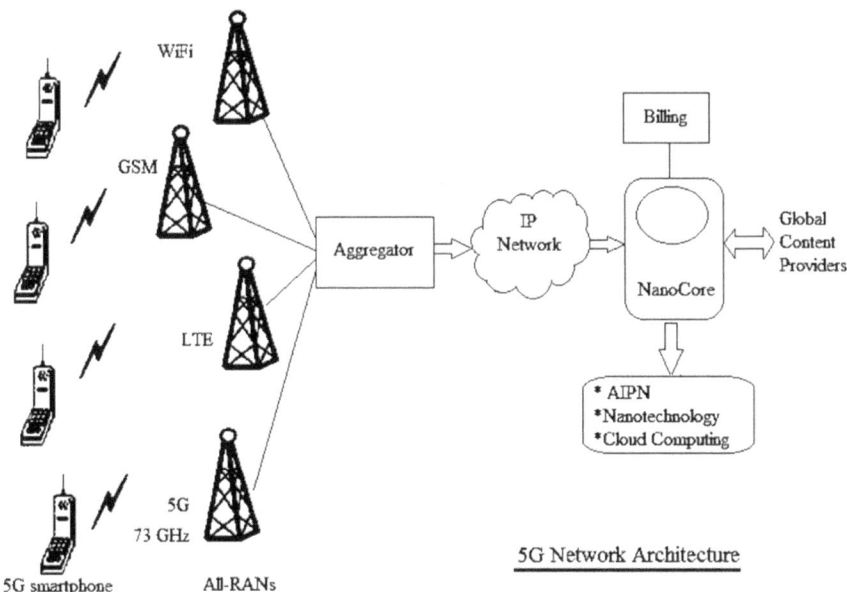

(*Source* https://www.rfwireless-world.com/images/5G-network-architecture.jpg)

Appendix 4: 5G Tower Signals

(*Source*   https://c8.alamy.com/zooms/9/92ef2e93fe594b4c906052e36f966231/
2bde4b7.jpg)

References

Agiwal, M., Saxena, N., & Roy, A. (2019). Towards connected living: 5G enabled internet of things (IoT). *IETE Technical Review, 36*(2), 190–202. Retrieved From: https://www.tandfonline.com/doi/abs/https://doi.org/10.1080/02564602.2018.1444516 [Retrieved On: 05.05.2022]
Ahad, A., Tahir, M., Aman Sheikh, M., Ahmed, K. I., Mughees, A., & Numani, A. (2020). Technologies trend towards 5G network for smart health-care using IoT: A review. *Sensors, 20*(14), 4047. Retrieved From: https://www.mdpi.com/1424-8220/20/14/4047 [Retrieved On: 05.05.2022]

A. Ahmad, B. Bhushan, N. Sharma, I. Kaushik, and S. Arora, "Importunity & Evolution of IoT for 5G," 2020 IEEE 9th International Conference on Communication Systems and Network Technologies (CSNT), 2020, pp. 102–107, DOI: https://doi.org/10.1109/CSNT48778.2020.911 5768.

Albreem, M. A., Sheikh, A. M., Alsharif, M. H., Jusoh, M., & Yasin, M. N. M. (2021). Green Internet of Things (IoT): applications, practices, awareness, and challenges. *IEEE Access, 9,* 38833–38858. Retrieved From: https://ieeexplore.ieee.org/abstract/document/9361680 [Retrieved On: 05.05.2022]

Ali, K., Nguyen, H. X., Shah, P., Vien, Q. T., & Ever, E. (2019). Internet of things (IoT) considerations, requirements, and architectures for the disaster management system. In *Performability in the internet of things* (pp. 111–125). Springer, Cham. Retrieved From: https://link.springer.com/chapter/https://doi.org/10.1007/978-3-319-93557-7_7 [Retrieved On: 05.05.2022]

Ali, F., He, Y., Shi, G., Sui, Y., & Yuang, H. (2020). Future Generation Spectrum Standardization for 5G and Internet of Things. *J. Commun., 15*(3), 276–282. Retrieved From: http://www.jocm.us/uploadfile/2020/0214/20200214025020987.pdf [Retrieved On: 05.05.2022]

Alsamhi, S. H., Afghah, F., Sahal, R., Hawbani, A., Al-qaness, M. A., Lee, B., & Guizani, M. (2021). Green internet of things using UAVs in B5G networks: a review of applications and strategies. *Ad Hoc Networks, 117,* 102505. Retrieved From: https://www.sciencedirect.com/science/article/pii/S1570870521000639 [Retrieved On: 05.05.2022]

Aman, A. H. M., Yadegaridehkordi, E., Attarbashi, Z. S., Hassan, R., & Park, Y. J. (2020). A survey on-trend and classification of internet of things reviews. *Ieee Access, 8,* 111763–111782. Retrieved From: https://ieeexplore.ieee.org/abstract/document/9119087 [Retrieved On: 05.05.2022]

Amin, S. U., & Hossain, M. S. (2020). Edge intelligence and Internet of Things in healthcare: A survey. *IEEE Access, 9,* 45–59. Retrieved From: https://ieeexplore.ieee.org/abstract/document/9294145 [Retrieved On: 05.05.2022]

S. Arora, N. Sharma, B. Bhushan, I. Kaushik, and A. Ahmad, "Evolution of 5G Wireless Network in IoT," 2020 IEEE 9th International Conference on Communication Systems and Network Technologies (CSNT), 2020, pp. 108–113, DOI: https://doi.org/10.1109/CSNT48778.2020.911 5773.

Arsh, M., Bhushan, B., Uppal, M. (2021). Internet of Things (IoT) Toward 5G Network: Design Requirements, Integration Trends, and Future Research Directions. In: Hassanien, A.E., Bhattacharyya, S., Chakrabati, S., Bhattacharya, A., Dutta, S. (eds) Emerging Technologies in Data Mining and Information Security. Advances in Intelligent Systems and Computing, vol 1286. Springer, Singapore. https://doi.org/10.1007/978-981-15-9927-9_85

Athanasiadou, G. E., Fytampanis, P., Zarbouti, D. A., Tsoulos, G. V., Gkonis, P. K., & Kaklamani, D. I. (2020). Radio network planning towards 5G mmWave standalone small-cell architectures. *Electronics, 9*(2), 339. Retrieved From: https://www.mdpi.com/2079-9292/9/2/339 [Retrieved On: 05.05.2022]

Atiqur, R., Wu, G., & Liton, A. M. (2020). Mobile edge computing for the internet of things (IoT): security and privacy issues. *Indonesian Journal of Electrical Engineering and Computer Science (IJEECS), 18*(3), 1486–1493. Retrieved From: https://pdfs.semanticscholar.org/356b/3229a6063e9333a8ca125fe23df4da30da8d.pdf [Retrieved On: 05.05.2022]

Awin, F. A., Alginahi, Y. M., Abdel-Raheem, E., & Tepe, K. (2019). Technical issues on cognitive radio-based Internet of Things systems: A survey. *IEEE Access, 7,* 97887–97908. Retrieved From: https://ieeexplore.ieee.org/abstract/document/8766798 [Retrieved On: 05.05.2022]

Awoyemi, B. S., Alfa, A. S., & Maharaj, B. T. (2020). Resource optimization in 5G and internet-of-things networking. *Wireless personal communications, 111*(4), 2671–2702. Retrieved From: https://link.springer.com/article/https://doi.org/10.1007/s11277-019-07010-9 [Retrieved On: 05.05.2022]

Bana, A. S., De Carvalho, E., Soret, B., Abrao, T., Marinello, J. C., Larsson, E. G., & Popovski, P. (2019). Massive MIMO for the internet of things (IoT) connectivity. *Physical Communication, 37*, 100859. Retrieved From: https://www.sciencedirect.com/science/article/abs/pii/S18 74490719303891 [Retrieved On: 05.05.2022]

Bektas, C., Monhof, S., Kurtz, F., & Wietfeld, C. (2018, December). Towards 5G: An empirical evaluation of software-defined end-to-end network slicing. In *2018 IEEE Globecom Workshops (GC Wkshps)* (pp. 1–6). IEEE. Retrieved From: https://ieeexplore.ieee.org/abstract/document/ 8644145 [Retrieved On: 05.05.2022]

Boursianis, A. D., Papadopoulou, M. S., Diamantoulakis, P., Liopa-Tsakalidi, A., Barouchas, P., Salahas, G., ... & Goudos, S. K. (2022). Internet of things (IoT) and agricultural unmanned aerial vehicles (UAVs) in smart farming: a comprehensive review. *Internet of Things, 18*, 100187. Retrieved From: https://www.sciencedirect.com/science/article/abs/pii/S25426605203 00238 [Retrieved On: 05.05.2022]

Chen, N., & Okada, M. (2020). Toward 6G Internet of Things and the Convergence with RoF System. *IEEE Internet of Things Journal, 8*(11), 8719–8733. Retrieved From: https://ieeexp lore.ieee.org/abstract/document/9309383 [Retrieved On: 05.05.2022]

Chettri, L., & Bera, R. (2019). A comprehensive survey on the Internet of Things (IoT) toward 5G wireless systems. *IEEE Internet of Things Journal, 7*(1), 16–32. Retrieved From: https://ieeexp lore.ieee.org/abstract/document/8879484 [Retrieved On: 05.05.2022]

Del Peral-Rosado, J. A., Granados, G. S., Raulefs, R., Leitinger, E., Grebien, S., Wilding, T., ... & Sackenreuter, B. (2018). Whitepaper on new localization methods for 5G wireless systems and the Internet-of-Things. Retrieved From: https://re.public.polimi.it/handle/11311/1069386 [Retrieved On: 05.05.2022]

Farhan, L., Kharel, R., Kaiwartya, O., Quiroz-Castellanos, M., Alissa, A., & Abdulsalam, M. (2018, July). A concise review of the Internet of Things (IoT)-problems, challenges, and opportunities. In *2018 11th International Symposium on Communication Systems, Networks & Digital Signal Processing (CSNDSP)* (pp. 1–6). IEEE. Retrieved From: https://ieeexplore.ieee.org/abstract/doc ument/8471762 [Retrieved On: 05.05.2022]

French, A. M., & Shim, J. P. (2016). The digital revolution: internet of things, 5G, and beyond. *Communications of the Association for Information Systems, 38*(1), 40. Retrieved From: https://aisel.aisnet.org/cais/vol38/iss1/40/ [Retrieved On: 05.05.2022]

Ghendir, S., Sbaa, S., Al-Sherbaz, A., Ajgou, R., & Chemsa, A. (2019). Towards 5G wireless systems: A modified Rake receiver for UWB indoor multipath channels. *Physical Communication, 35*, 100715. Retrieved From: https://www.sciencedirect.com/science/article/abs/pii/S18 74490718303471 [Retrieved On: 05.05.2022]

Goyal, S., Sharma, N., Kaushik, I., Bhushan, B., Kumar, N. (2021a). A Green 6G Network Era: Architecture and Propitious Technologies. In: Khanna, A., Gupta, D., Pólkowski, Z., Bhattacharyya, S., Castillo, O. (eds) Data Analytics and Management. Lecture Notes on Data Engineering and Communications Technologies, vol 54. Springer, Singapore. https://doi.org/ 10.1007/978-981-15-8335-3_7

Goyal, P., Sahoo, A. K., Sharma, T. K., & Singh, P. K. (2021b). Internet of Things: Applications, security, and privacy: A survey. *Materials Today: Proceedings, 34*, 752–759. Retrieved From: https://www.sciencedirect.com/science/article/pii/S221478532033385X [Retrieved On: 05.05.2022]

Habibi, M. A., Nasimi, M., Han, B., & Schotten, H. D. (2019). A comprehensive survey of RAN architectures toward 5G mobile communication system. *IEEE Access, 7*, 70371– 70421. Retrieved From: https://ieeexplore.ieee.org/abstract/document/8723481 [Retrieved On: 05.05.2022]

Hassan, R., Qamar, F., Hasan, M. K., Aman, A. H. M., & Ahmed, A. S. (2020). Internet of Things and its applications: A comprehensive survey. *Symmetry, 12*(10), 1674. Retrieved From: https://www.mdpi.com/2073-8994/12/10/1674 [Retrieved On: 05.05.2022]

He, Y., Khan, H. U., Zhang, K., Wang, W., Choi, B. J., Aly, A. A., ... & Baz, M. (2021). D2D-V2X-SDN: Taxonomy and architecture towards 5G mobile communication system. *IEEE*

Access. Retrieved From: https://ieeexplore.ieee.org/abstract/document/9610025 [Retrieved On: 05.05.2022]

Hossein Motlagh, N., Mohammadrezaei, M., Hunt, J., & Zakeri, B. (2020). Internet of Things (IoT) and the energy sector. *Energies, 13*(2), 494. Retrieved From: https://www.mdpi.com/1996-1073/13/2/494 [Retrieved On: 05.05.2022]

Hu, F., Chen, B., & Zhu, K. (2018). Full-spectrum sharing in cognitive radio networks toward 5G: A survey. *IEEE Access, 6*, 15754–15776. Retrieved From: https://ieeexplore.ieee.org/abstract/document/8281460/ [Retrieved On: 05.05.2022]

Hui, H., Ding, Y., Shi, Q., Li, F., Song, Y., & Yan, J. (2020). 5G network-based Internet of Things for demand response in smart grid: A survey on application potential. *Applied Energy, 257*, 113972. Retrieved From: https://www.sciencedirect.com/science/article/abs/pii/S03062619193 16599 [Retrieved On: 05.05.2022]

Husain, S., Kunz, A., Prasad, A., Samdanis, K., & Song, J. (2018, February). Mobile edge computing with network resource slicing for Internet-of-Things. In *2018 IEEE 4th World Forum on Internet of Things (WF-IoT)* (pp. 1–6). IEEE. Retrieved From: https://ieeexplore.ieee.org/abstract/document/8355232 [Retrieved On: 05.05.2022]

Hussein, A. F., Elgala, H., & Little, T. D. (2018, December). Evolution of multi-tier transmission towards 5G Li-Fi networks. In *2018 IEEE Global Communications Conference (GLOBECOM)* (pp. 1–7). IEEE. Retrieved From: https://ieeexplore.ieee.org/abstract/document/8647846 [Retrieved On: 05.05.2022]

Iannacci, J. (2018). Internet of things (IoT); internet of everything (IoE); tactile internet; 5G–A (not so evanescent) unifying vision empowered by EH-MEMS (energy harvesting MEMS) and RF-MEMS (radio frequency MEMS). *Sensors and actuators a: physical, 272*, 187–198. Retrieved From: https://www.sciencedirect.com/science/article/abs/pii/S0924424717320757 [Retrieved On: 05.05.2022]

Islam, N., Rashid, M. M., Pasandideh, F., Ray, B., Moore, S., & Kadel, R. (2021). A review of applications and communication technologies for the internet of things (IoT) and unmanned aerial vehicle (UAV) based sustainable smart farming. *Sustainability, 13*(4), 1821. Retrieved From: https://www.mdpi.com/2071-1050/13/4/1821 [Retrieved On: 05.05.2022]

Jungnickel, V., Hinrichs, M., Bober, K. L., Kottke, C., Corici, A. A., Emmelmann, M., … & Koonen, A. M. J. (2019, June). Enhance lighting for the internet of things. In *2019 Global LIFI Congress (GLC)* (pp. 1–6). IEEE. Retrieved From: https://ieeexplore.ieee.org/abstract/document/8864126 [Retrieved On: 05.05.2022]

Kadhim, K. T., Alsahlany, A. M., Wadi, S. M., & Kadhum, H. T. (2020). An overview of a patient's health status monitoring system based on the internet of things (IoT). *Wireless Personal Communications, 114*(3), 2235–2262. Retrieved From: https://link.springer.com/article/https://doi.org/10.1007/s11277-020-07474-0 [Retrieved On: 05.05.2022]

Kaur, C. (2020). Cloud computing and the internet of things (IoT). *International Journal of Scientific Research in Science, Engineering and Technology, 7*(1), 19–22. Retrieved From: https://d1wqtx ts1xzle7.cloudfront.net/65387970/6167-with-cover-page-v2.pdf? [Retrieved On: 05.05.2022]

Khan, I. H., & Javaid, M. (2021). Role of Internet of Things (IoT) in the adoption of Industry 4.0. *Journal of Industrial Integration and Management*, 2150006. Retrieved From: https://www.worldscientific.com/doi/abs/https://doi.org/10.1142/S2424862221500068 [Retrieved On: 05.05.2022]

Khanna, A., & Kaur, S. (2019). Evolution of Internet of Things (IoT) and its significant impact in the field of Precision Agriculture. *Computers and electronics in agriculture, 157*, 218–231. Retrieved From: https://www.sciencedirect.com/science/article/abs/pii/S01681699 18316417 [Retrieved On: 05.05.2022]

Khanna, A., & Kaur, S. (2020). Internet of things (IoT), applications and challenges: a comprehensive review. *Wireless Personal Communications, 114*(2), 1687–1762. Retrieved From: https://link.springer.com/article/https://doi.org/10.1007/s11277-020-07446-4 [Retrieved On: 05.05.2022]

Khatua, P. K., Ramachandaramurthy, V. K., Kasinathan, P., Yong, J. Y., Pasupuleti, J., & Rajagopalan, A. (2020). Application and assessment of internet of things toward the sustainability of energy systems: Challenges and issues. *Sustainable Cities and Society, 53,* 101957. Retrieved From: https://www.sciencedirect.com/science/article/abs/pii/S22106707193 10455 [Retrieved On: 05.05.2022]

Khujamatov, K., Reypnazarov, E., Khasanov, D., & Akhmedov, N. (2020, October). Networking and computing in the internet of things and cyber-physical systems. In *2020 IEEE 14th International Conference on Application of Information and Communication Technologies (AICT)* (pp. 1–6). IEEE. Retrieved From: https://ieeexplore.ieee.org/abstract/document/9368793 [Retrieved On: 05.05.2022]

Kim, Y., Kim, Y., Oh, J., Ji, H., Yeo, J., Choi, S., ... & Lee, J. (2019). New radio (NR) and its evolution toward 5G-advanced. *IEEE Wireless Communications, 26*(3), 2–7. Retrieved From: https://ieeexplore.ieee.org/abstract/document/8752473 [Retrieved On: 05.05.2022]

Knieps, G. (2019). Internet of Things, big data, and the economics of networked vehicles. *Telecommunications Policy, 43*(2), 171–181. Retrieved From: https://www.sciencedirect.com/science/article/abs/pii/S0308596117304640 [Retrieved On: 05.05.2022]

Li, S., Da Xu, L., & Zhao, S. (2018). 5G Internet of Things: A survey. *Journal of Industrial Information Integration, 10,* 1–9. Retrieved From: https://www.sciencedirect.com/science/article/abs/pii/S2452414X18300037 [Retrieved On: 05.05.2022]

Liu, Y., Tong, K., Mao, F., & Yang, J. (2020a). Research on digital production technology for traditional manufacturing enterprises based on the industrial Internet of Things in the 5G era. *The International Journal of Advanced Manufacturing Technology, 107*(3), 1101–1114. Retrieved From: https://link.springer.com/article/https://doi.org/10.1007/s00170-019-04284-y [Retrieved On: 05.05.2022]

Liu, Y., Peng, M., Shou, G., Chen, Y., & Chen, S. (2020b). Toward edge intelligence: multiaccess edge computing for 5G and internet of things. *IEEE Internet of Things Journal, 7*(8), 6722–6747. Retrieved From: https://ieeexplore.ieee.org/abstract/document/9123504 [Retrieved On: 05.05.2022]

Liu, L., Guo, X., & Lee, C. (2021). Promoting smart cities into the 5G era with multi-field Internet of Things (IoT) applications powered by advanced mechanical energy harvesters. *Nano Energy, 88,* 106304. Retrieved From: https://www.sciencedirect.com/science/article/abs/pii/S22112855210 05590 [Retrieved On: 05.05.2022]

Lu, Y., & Da Xu, L. (2018). Internet of Things (IoT) cybersecurity research: A review of current research topics. *IEEE Internet of Things Journal, 6*(2), 2103–2115. Retrieved From: https://iee explore.ieee.org/abstract/document/8462745 [Retrieved On: 05.05.2022]

Malik A, Bhushan B (2022) Challenges, standards, and solutions for secure and intelligent 5g internet of things (IoT) scenarios. Smart and Sustainable Approaches for Optimizing Performance of Wireless Networks. https://doi.org/10.1002/9781119682554.ch7

Mavromoustakis, C. X., Mastorakis, G., & Batalla, J. M. (Eds.). (2016). *Internet of Things (IoT) in 5G mobile technologies* (Vol. 8). Springer. Retrieved From: https://link.springer.com/book/ https://doi.org/10.1007/978-3-319-30913-2?noAccess=true [Retrieved On: 05.05.2022]

Mei, G., Xu, N., Qin, J., Wang, B., & Qi, P. (2019). A survey of Internet of Things (IoT) for geohazard prevention: Applications, technologies, and challenges. *IEEE Internet of Things Journal, 7*(5), 4371–4386. Retrieved From: https://ieeexplore.ieee.org/abstract/document/8895751 [Retrieved On: 05.05.2022]

Migabo, E. M., Djouani, K. D., & Kurien, A. M. (2020). The narrowband Internet of Things (NB-IoT) resources management performance state of art, challenges, and opportunities. *IEEE Access, 8,* 97658–97675. Retrieved From: https://ieeexplore.ieee.org/abstract/document/909 7268 [Retrieved On: 05.05.2022]

Narayanan, A., De Sena, A. S., Gutierrez-Rojas, D., Melgarejo, D. C., Hussain, H. M., Ullah, M., ... & Nardelli, P. H. (2020). Key advances in pervasive edge computing for the industrial internet of things in 5g and beyond. *IEEE Access, 8,* 206734–206754. Retrieved From: https://ieeexp lore.ieee.org/abstract/document/9257390 [Retrieved On: 05.05.2022]

Nguyen, D. C., Ding, M., Pathirana, P. N., Seneviratne, A., Li, J., Niyato, D., … & Poor, H. V. (2021). 6G Internet of Things: A comprehensive survey. *IEEE Internet of Things Journal*. Retrieved From: https://ieeexplore.ieee.org/abstract/document/9509294 [Retrieved On: 05.05.2022]

Ni, J., Lin, X., & Shen, X. S. (2019). Toward edge-assisted Internet of Things: From security and efficiency perspectives. *IEEE Network*, *33*(2), 50–57. Retrieved From: https://ieeexplore.ieee. org/abstract/document/8675172 [Retrieved On: 05.05.2022]

Oughton, E., Frias, Z., Russell, T., Sicker, D., & Cleevely, D. D. (2018). Towards 5G: Scenario-based assessment of the future supply and demand for mobile telecommunications infrastructure. *Technological Forecasting and Social Change*, *133*, 141–155. Retrieved From: https://www.scienc edirect.com/science/article/pii/S0040162517313525 [Retrieved On: 05.05.2022]

Painuly, S., Kohli, P., Matta, P., & Sharma, S. (2020). Advance applications and future challenges of 5G IoT. In *2020 3rd International Conference on Intelligent Sustainable Systems (ICISS)* (pp. 1381–1384). IEEE. Retrieved From: https://ieeexplore.ieee.org/abstract/document/ 9316004 [Retrieved On: 05.05.2022]

Pedersen, T., Fleury, B. H., & COST Action CA15104, I. R. A. C. O. N. (2018). Whitepaper on new localization methods for 5G wireless systems and the Internet-of-Things. Retrieved From: https://vbn.aau.dk/en/publications/whitepaper-on-new-localization-methods-for-5g-wir eless-systems-an [Retrieved On: 05.05.2022]

Poncha, L. J., Abdelhamid, S., Alturjman, S., Ever, E., & Al-Turjman, F. (2018, May). 5G in a convergent Internet of Things era: An overview. In *2018 IEEE International Conference on Communications Workshops (ICC Workshops)* (pp. 1–6). IEEE. Retrieved From: https://ieeexp lore.ieee.org/abstract/document/8403748 [Retrieved On: 05.05.2022]

Porambage, P., Okwuibe, J., Liyanage, M., Ylianttila, M., & Taleb, T. (2018). Survey on multi-access edge computing for the internet of things realization. *IEEE Communications Surveys & Tutorials*, *20*(4), 2961–2991. Retrieved From: https://ieeexplore.ieee.org/abstract/document/ 8391395 [Retrieved On: 05.05.2022]

Qiu, T., Chi, J., Zhou, X., Ning, Z., Atiquzzaman, M., & Wu, D. O. (2020). Edge computing in the industrial internet of things: Architecture, advances, and challenges. *IEEE Communications Surveys & Tutorials*, *22*(4), 2462–2488. Retrieved From: https://ieeexplore.ieee.org/abstract/ document/9139976 [Retrieved On: 05.05.2022]

R. Rathi, N. Sharma, C. Manchanda, B. Bhushan, and M. Grover, "Security Challenges & Controls in Cyber-Physical System," 2020 IEEE 9th International Conference on Communication Systems and Network Technologies (CSNT), 2020, pp. 242–247, DOI: https://doi.org/10.1109/CSNT48 778.2020.9115778.

Sadique, K. M., Rahmani, R., & Johannesson, P. (2018). Towards security on internet of things: applications and challenges in technology. *Procedia Computer Science*, *141*, 199–206. Retrieved From: https://www.sciencedirect.com/science/article/pii/S1877050918318180 [Retrieved On: 05.05.2022]

Salam, A. (2020). Internet of things for sustainable community development: introduction and overview. In the *Internet of Things for Sustainable Community Development* (pp. 1–31). Springer, Cham. Retrieved From: https://link.springer.com/chapter/https://doi.org/10.1007/978-3-030-35291-2_1? [Retrieved On: 05.05.2022]

Santos, B., Feng, B., & van Do, T. (2018, October). Towards a standardized identity federation for the internet of things in 5g networks. In *2018 IEEE SmartWorld, Ubiquitous Intelligence & Computing, Advanced & Trusted Computing, Scalable Computing & Communications, Cloud & Big Data Computing, Internet of People, and Smart City Innovation (SmartWorld/SCALCOM/ UIC/ATC/CBDCom/IOP/SCI)* (pp. 2082–2088). IEEE. Retrieved From: https://ieeexplore.ieee. org/abstract/document/8560327 [Retrieved On: 05.05.2022]

Sethi, R., Bhushan, B., Sharma, N., Kumar, R., Kaushik, I. (2021). Applicability of Industrial IoT in Diversified Sectors: Evolution, Applications, and Challenges. In: Kumar, R., Sharma, R., Pattnaik, P.K. (eds) Multimedia Technologies in the Internet of Things Environment. Studies in Big Data, vol 79. Springer, Singapore. https://doi.org/10.1007/978-981-15-7965-3_4

Sharma, S. K., Woungang, I., Anpalagan, A., & Chatzinotas, S. (2020). Toward tactile internet in beyond 5G era: Recent advances, current issues, and future directions. *IEEE Access, 8*, 56948–56991. Retrieved From: https://ieeexplore.ieee.org/abstract/document/9034103 [Retrieved On: 05.05.2022]

Shi, Q., Dong, B., He, T., Sun, Z., Zhu, J., Zhang, Z., & Lee, C. (2020). Progress in wearable electronics/photonics—Moving toward the era of artificial intelligence and the internet of things. *in format, 2*(6), 1131–1162. Retrieved From: https://onlinelibrary.wiley.com/doi/full/https://doi.org/10.1002/inf2.12122 [Retrieved On: 05.05.2022]

Sicari, S., Rizzardi, A., & Coen-Porisini, A. (2020). 5G In the internet of things era: An overview on security and privacy challenges. *Computer Networks, 179*, 107345. Retrieved From: https://www.sciencedirect.com/science/article/abs/pii/S1389128620300827 [Retrieved On: 05.05.2022]

Sodhro, A. H., Pirbhulal, S., Sodhro, G. H., Muzammal, M., Zongwei, L., Gurtov, A., ... & de Albuquerque, V. H. C. (2020). Towards 5G-enabled self-adaptive green and reliable communication in the intelligent transportation system. *IEEE Transactions on Intelligent Transportation Systems, 22*(8), 5223–5231. Retrieved From: https://ieeexplore.ieee.org/abstract/document/919 9859 [Retrieved On: 05.05.2022]

Sun, Y., Tian, Z., Li, M., Zhu, C., & Guizani, N. (2020). Automated attack and defense framework toward 5G security. *IEEE Network, 34*(5), 247–253. Retrieved From: https://ieeexplore.ieee.org/abstract/document/9083674 [Retrieved On: 05.05.2022]

Teli, S. R., Zvanovec, S., & Ghassemlooy, Z. (2018, November). Optical internet of things within 5G: Applications and challenges. In *2018 IEEE International Conference on Internet of Things and Intelligence System (IOTAIS)* (pp. 40–45). IEEE. Retrieved From: https://ieeexplore.ieee.org/abstract/document/8600894 [Retrieved On: 05.05.2022]

Wijethilaka, S., & Liyanage, M. (2021a). Survey on network slicing for Internet of Things realization in 5G networks. *IEEE Communications Surveys & Tutorials, 23*(2), 957–994. Retrieved From: https://ieeexplore.ieee.org/abstract/document/9382385 [Retrieved On: 05.05.2022]

Wijethilaka, S., & Liyanage, M. (2021b, January). Realizing the Internet of Things with network slicing: Opportunities and challenges. In *2021b IEEE 18th Annual Consumer Communications & Networking Conference (CCNC)* (pp. 1–6). IEEE. Retrieved From: https://ieeexplore.ieee.org/abstract/document/9369637 [Retrieved On: 05.05.2022]

Worlu, C., Jamal, A. A., & Mahiddin, N. A. (2019). Wireless sensor networks, the internet of things, and their challenges. *International Journal of Innovative Technology and Exploring Engineering, 8*(12S2), 556–566. Retrieved From: https://www.researchgate.net/profile/Azrul-Amri-Jamal/publication/339014974_Wireless_Sensor_Networks_Internet_of_Things_and_Their_Challenges/links/5e38ce53a6fdccd9658472a4/Wireless-Sensor-Networks-Internet-of-Things-and-Their-Challenges.pdf [Retrieved On: 05.05.2022]

Xia, J., Xu, Y., Deng, D., Zhou, Q., & Fan, L. (2019). Intelligent secure communication for the Internet of Things with statistical channel state information of attacker. *IEEE Access, 7*, 144481–144488. Retrieved From: https://ieeexplore.ieee.org/abstract/document/8854816 [Retrieved On: 05.05.2022]

Xing, L. (2020). Reliability in the Internet of Things: Current status and future perspectives. *IEEE Internet of Things Journal, 7*(8), 6704–6721. Retrieved From: https://ieeexplore.ieee.org/abstract/document/9089244 [Retrieved On: 05.05.2022]

Yan, G. (2019). Simulation analysis of key technology optimization of 5G mobile communication network based on Internet of Things technology. *International Journal of Distributed Sensor Networks, 15*(6), 1550147719851454. Retrieved From: https://journals.sagepub.com/doi/full/https://doi.org/10.1177/1550147719851454 [Retrieved On: 05.05.2022]

Yang, H. C., & Alouini, M. S. (2019). Data-oriented transmission in future wireless systems: Toward trustworthy support of advanced Internet of Things. *IEEE Vehicular Technology Magazine, 14*(3), 78–83. Retrieved From: https://ieeexplore.ieee.org/abstract/document/876 0250 [Retrieved On: 05.05.2022]

Ye, N., Li, X., Yu, H., Wang, A., Liu, W., & Hou, X. (2019). Deep learning aided grant-free NOMA toward reliable low-latency access to tactile Internet of Things. *IEEE Transactions on Industrial Informatics*, *15*(5), 2995–3005. Retrieved From: https://ieeexplore.ieee.org/abstract/document/8625480 [Retrieved On: 05.05.2022]

Zhang, Y., Chen, M., Leung, V. C., Xing, T., & Fortino, G. (2018). Guest editorial special issue on cognitive internet of things. *IEEE Internet of Things Journal*, *5*(4), 2259–2262. Retrieved From: https://ieeexplore.ieee.org/abstract/document/8430685 [Retrieved On: 05.05.2022]

Zheng, W., Sun, K., Zhang, X., Zhang, Q., Israr, A., & Yang, Q. (2020, August). Cellular communication for a ubiquitous internet of things in smart grids: present and outlook. In *2020 Chinese Control And Decision Conference (CCDC)* (pp. 5592–5596). IEEE. Retrieved From: https://ieeexplore.ieee.org/abstract/document/9164273 [Retrieved On: 05.05.2022]

Zikria, Y. B., Kim, S. W., Afzal, M. K., Wang, H., & Rehmani, M. H. (2018). 5G mobile services and scenarios: Challenges and solutions. *Sustainability*, *10*(10), 3626. Retrieved From: https://www.mdpi.com/2071-1050/10/10/3626 [Retrieved On: 05.05.2022]

Zikria, Y. B., Kim, S. W., Hahm, O., Afzal, M. K., & Aalsalem, M. Y. (2019). Internet of Things (IoT) operating systems management: opportunities, challenges, and solution. *Sensors*, *19*(8), 1793. Retrieved From: https://www.mdpi.com/1424-8220/19/8/1793 [Retrieved On: 05.05.2022]

Chapter 3
Network Architectures and Protocols for Efficient Exploitation of Spectrum Resources in 5G

Kande Archana, V. Kamakshi Prasad, M. Ashok, and G. R. Anantha Raman

Abstract With the emergence of 5G technology with its enhanced architecture, there are unprecedented possibilities in terms of Quality of Service (QoS), data rate, latency, and capacity. This chapter throws light on different aspects of 5G technology and associated technological innovations. It has support for user diversity, devise diversity, and other dimensions. It focuses on the architecture of 5G technology, its underlying mechanisms, support for Device-to-Device (D2D) communication, Multiple-Input Multiple-Output (MIMO) enhancements, advanced interference management, enhanced utility of ultra-dense networks, spectrum sharing, and cloud technologies associated with 5G networks. It also proposes a methodology with spectrum broker with underlying components with delay-aware and energy-efficient approach to leverage 5G base stations leading to the reduction of energy consumption. The concept of queueing delays is considered while proposing the scheme for delay-sensitive communications with energy efficiency. The spectrum broker has mechanisms to achieve this. Simulation study with the proposed scheme shows the energy efficiency of the proposed scheme when compared with the state of the art. This chapter not only provides the required know-how on different technologies but also provides the simulation study that may trigger further investigation into the resource management in 5G networks.

K. Archana (✉)
Research Scholar, Department of Computer Science and Engineering, Jawaharlal Nehru Technological University, Hyderabad, Telangana 500057, India
e-mail: kande.archana@gmail.com

Assistant Professor, Department of Computer Science and Engineering, Malla Reddy Institute of Engineering and Technology Maisammaguda, Hyderabad, Telangana 500100, India

V. Kamakshi Prasad
Department of Computer Science and Engineering, Jawaharlal Nehru Technological University, Hyderabad, Telangana 500057, India
e-mail: kamakshiprasad@jntuh.ac.in

M. Ashok · G. R. Anantha Raman
Department of Computer Science and Engineering, Malla Reddy Institute of Engineering and Technology Maisammaguda, Hyderabad, Telangana 500100, India

© The Author(s) 2023

B. Bhushan et al. (eds.), *5G and Beyond*, Springer Tracts in Electrical and Electronics Engineering, https://doi.org/10.1007/978-981-99-3668-7_3

45

Keywords D2D · 5G radio access networks · 5G mobile cognitive radio · Networks · Spectrum broker · 5G mobile communication system

Introduction

Enabler Technologies

Mobile technology and communications have undergone a sea change in the recent past. Mobile technology has been evolving at a rapid pace, and mobile devices, since the current millennium, have witnessed many features including gaming, GPS navigation, messaging, and so on. The mobiles are even able to participate in cloud computing in the form of Mobile Cloud Computing (MCC). Largely computer technology also depends on small hand-held smart devices. Tablet computers associated with mobile computing have become popular phenomenon. Mobile technology plays a vital role in digital infrastructure used for computing. The 5G is fifth-generation technology which is the successor of 4G. The 5G technology provides wider frequency bands besides increased spectral bandwidth. It is better than its predecessor in terms of peak bit rate, spectral efficiency, increased device connectivity, concurrency, speed, low battery consumption, assured connectivity in all geographical regions, increased device support, low-cost infrastructure, and reliability.

As presented in Fig. 3.1, the enabler technologies of 5G have their role in making it useful in real-world applications. The technologies are associated with network management, massive MIMO, millimeter wave, heterogeneous networks, convergence of access and backhaul, massive machine-type communication, communication with low latency, ultra-lean design, spectrum sharing, and flexibility. The sections below in this book provide more details on the architecture of 5G technology, its underlying mechanisms, support for Device-to-Device (D2D) communication, Multiple-Input Multiple-Output (MIMO) enhancements, advanced interference management, enhanced utility of ultra-dense networks, spectrum sharing, and cloud technologies associated with 5G networks. It also proposes a methodology with spectrum broker with underlying components with delay aware and energy-efficient approach to leverage 5G base stations leading to reduction of energy consumption. The concept of queueing delays is considered while proposing the scheme for delay-sensitive communications with energy efficiency.

The 5G network is based on various design considerations as explored in Agyapong et al. (2014), Marsch et al. (2016), and Gupta and Jha (2015). It has inherent support considerations for the use cases of IoT (Khalfi et al. 2017). There is provision to have better utilization of spectrum with 5G in an energy-efficient manner (Mavromoustakis et al. 2015). The notion of Software-defined radios (SDRs) with 5G technology can improve its controlling (Lin et al. 2015). With 5G networks, spectrum harvesting and energy are to be integrated for better results (Liu et al.

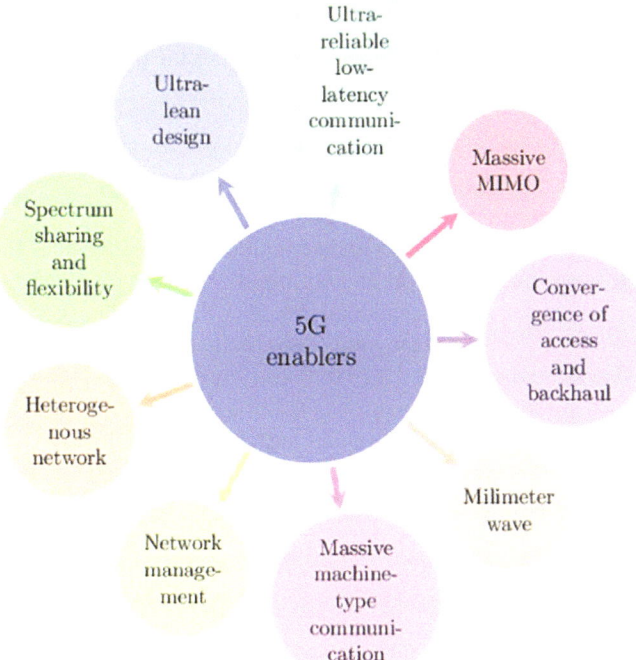

Fig. 3.1 5G enabler technologies

2015). There are some techniques discussed in Zhang et al. (2017) and Yang et al. (2016) for advanced spectrum sharing in 5G technology. Content-centric orientation and spectrum sharing are explored for 5G networks (Gur 2019). With 5G network in place, spectrum sharing among secondary users is investigated in Papageorgiou et al. (2020). SDN architecture for such networks is studied in Akyildiz et al. (2015). The contributions of this paper are as follows:

1. Investigation is made into architectures of 5G technology, MIMO usage in 5G, and D2D communication in 5G.
2. Focused on interference management, spectrum sharing, and ultra-dense networks with 5G technology.
3. Other areas explored with 5G include cloud computing technologies, design of energy efficient, and cooperative schemes with an algorithm and empirical study.

The remainder of the paper is structured as follows. Section 3.2 covers the architecture details of 5G. Section 3.3 throws light on the MIMO technology in 5G. Section 3.4 explores the D2D communication in 5G. Section 3.5 deals with the interference management for 5G. Section 3.6 covers the spectrum sharing with cognitive radio in 5G. Section 3.7 discusses the ultra-dense networks in multi-radio access technology association in 5G. Section 3.8 throws light on the cloud technologies for 5G. Section 3.9 covers the proposed methodology for the design of energy-efficient delay-aware cooperative scheme for 5G. Section 3.10 provides challenges and directions for future work. Section 3.11 concludes the paper.

Architecture of 5G

The 5G technology is the new global standard for mobile networks. It has the potential to have connectivity to more devices and machines with unprecedented possibilities. It is aimed at delivering multi-Gbps data speed with reliability, low latency, and massive capabilities. There are many important use cases of 5G. One important use case is the 5G network can be used for broad spectrum due to its reliability and high speed. It enables mission-critical communications and supports IoT technology. The 5G technology offers more flexibility and supports devices that need large-scale connectivity in future. It leverages mobile broadband, augmented reality, and virtual reality. The 5G technology bestows numerous benefits such as unprecedented speed, reliability, supporting future connectivity, and so on.

The 5G technology is equipped with advanced architecture with improved terminals and network elements. It also enables service providers to adopt it with advanced technologies to render value-added services to their customers. The upgradeability depends on cognitive radio technology with its features like location, temperature, and weather. Transceiver is made up of cognitive radio technology to enhance the operating efficiency. The technology can also distinguish the subtle environmental changes in its location and provide response to ensure high-quality and uninterrupted service.

As presented in Fig. 3.2, the 5G architecture is designed based on IP and the model is meant for both mobile and wireless networks. The 5G system consists of user terminal and is associated with many autonomous and independent radio access technologies. Each technology is treated as an IP link in the eyes of the Internet world. The IP-based approach is to have full control and routing IP packets in different application scenarios that involve sessions over the Internet between servers and client applications. It has flexibility provided to user in making decisions pertaining to routing of packets.

As presented in Fig. 3.3, the policy router layer facilitates communication between applications and servers. It has provision for different kinds of communications over IP model. The 5G technology has a master core to have flexible convergence point for other technologies.

As presented in Fig. 3.4, the master core technology associated with 5G has provision to have convergence point for different technologies such as photonic routers, beam transceivers, and nanotechnologies. It has support for concurrent approaches considering either 5G network mode or IP network mode. It is capable of controlling technologies and supports 5G-based deployments. It offers more flexibility, efficiency, power, and less complicated phenomena. There are many researchers proposed different architectures based on 5G. For instance, Integrated Access and

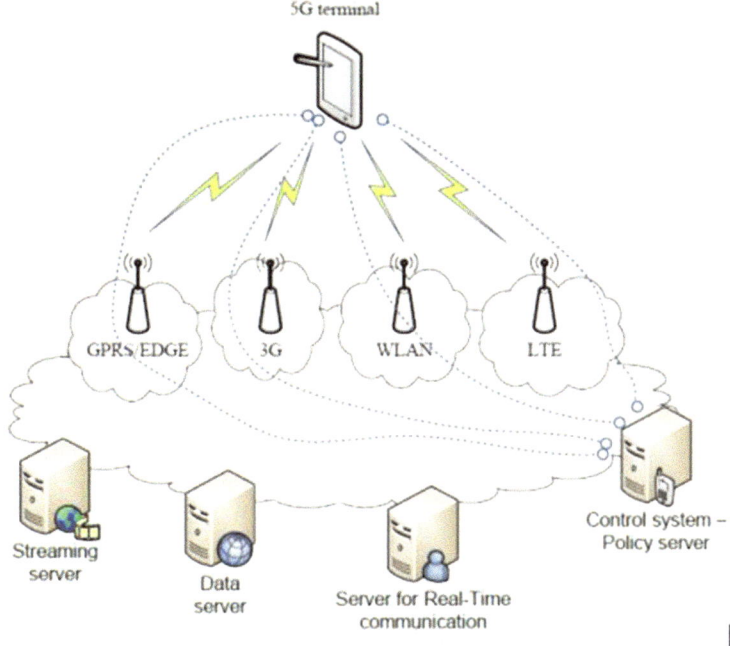

Fig. 3.2 Shows architectural overview of 5G

Backhaul (IAB) architecture is proposed in Ranjan et al. (2022) for exploiting 5G technology. The 5G mobile architecture is discussed in Tudzarov and Janevski (2011), while the key enabling technologies of the 5G are discussed in Idowu-Bismark et al. (2019). Evolution of wireless technologies including 5G is explored in Gupta and Jha 2015 architectural overview of 5G technology.

Arora et al. (2020) explored the evolution of 5G network in terms of Internet of Things (IoT) technology. They discussed the importance of 5G networks in order to solve the problems of IoT in the area of speed and convergence. Ghosh et al. (2019) also focused on the evolution of 5G technology. It has covered time-sensitive communication, integrated access, and backhaul. Ahmad et al. (2020b) discussed the evolution of IoT in order to use 5G technology. They found that 5G brings several benefits to IoT use cases. Goyal et al. (2021) explored about architecture and underlying aspects of a green 6G network. Arsh et al. (2021) explored design requirements and further developments needed to have IoT that makes use of IoT. Chetri and Bera (2019) studied the possibilities of using 5G in IoT use cases in order to reap its benefits in terms of speed and fulfillment of the user needs.

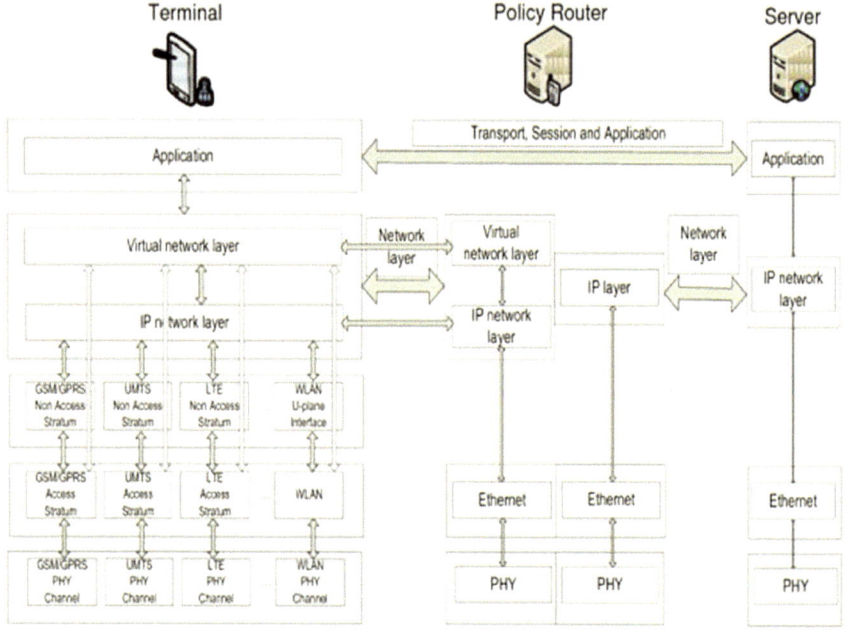

Fig. 3.3 Shows communication between client applications and server

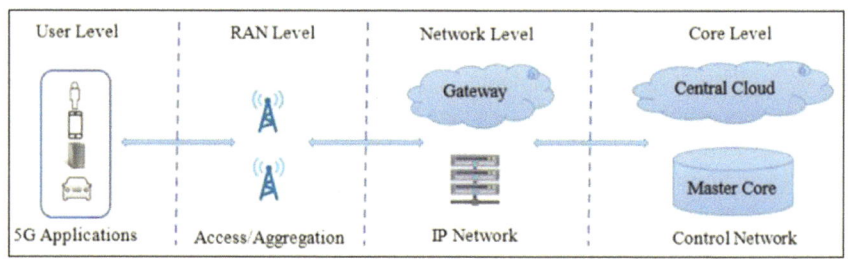

Fig. 3.4 Master core technology associated with 5G

Multiple-Input Multiple-Output Technology in 5G

Antenna Array for MIMO in 5G

In communications with wireless technology, MIMO is an improvement over its predecessor models for improving efficiency. MIMO enables a configuration that supports receiving and sending multiple signals concurrently. In fact, it is an improved and smart antenna technology that leverages performance in wireless communications. It is suitable for 5G technologies to reap its benefits. For 5G terminal, Abdullah

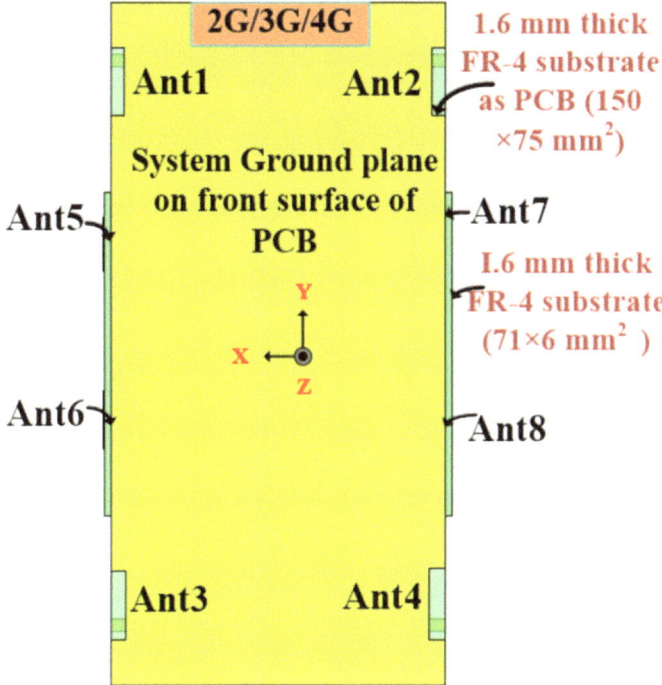

Fig. 3.5   Antenna array for MIMO in 5G

et al. (2019b) designed multiple antennas in order to realize MIMO technology. It is presented in Fig. 3.5.

It is designed for 5G mobile terminal that can operate at 2.6/3.5 GHz. Single antenna is used in traditional communication systems. It has many issues such as obstructions in the propagation line. It is overcome with MOMO technology and that can be used to exploit 5G benefits. MIMO technology has many applications like home networks, mobile communication networks, wireless local area networks (WLANs), and digital television (DTV). More on the MIMO technology usage, its future possibilities are found in Sharawi (2017). For 5G mobile terminals, the MIMO system is explored in Abdullah et al. 2019a.

Device-to-Device Communication in 5G

D2D communication refers to the communication that is directly between two devices without having conventional infrastructure such as access point. There are technologies like Wi-Fi Direct and Bluetooth that support D2D communication as explored in Shen (2015). Cellular networks have no support for direct communication between devices. Among different possibilities with 5G networks, D2D is one of the kinds of communication expected. With the advanced standard of Long-Term Evolution (LTE), D2D is expected to be realized in 5G cellular networks. With 5G cellular networks, mmWave technology is becoming popular. In such networks also D2D is expected. This technology is known for high propagation loss and short wavelengths and inability to penetrate into solids. These conditions provide suitability to realize D2D communications.

As presented in Fig. 3.6, there is a visual representation showing the approach in which cellular communication and D2D communication takes place. For service providers, D2D communication has been not viable. However, with the emergence of 5G, the subject of D2D became important once again. There are many use cases for D2D such as local data services, coverage extension, and Machine to Machine (M2M) communication. With respect to efficient spectrum usage, secondary users associated with Cognitive Radio Network (CRN) can exploit D2D communication that has a potential to get rid of interference to primary users of the spectrum. D2D is also expected to complement MIMO-enabled networks and HetNets with high data rates and efficiency. There is possibility of noise resilience and enhanced system capability with MIMO and D2D technologies. System capacity is significantly enhanced with D2D with MIMO devices.

As presented in Fig. 3.7, there is direct communication between two mobile devices. It is possible with "operator controlled link establishment." In this kind of D2D communication, both devices can exchange information without the involvement of base station. However, they are to be supported by the operator in order to have link established. As discussed in Adnan and Ahmad Zukarnain (2020), there are

Fig. 3.6 Illustrates device-to-device communication

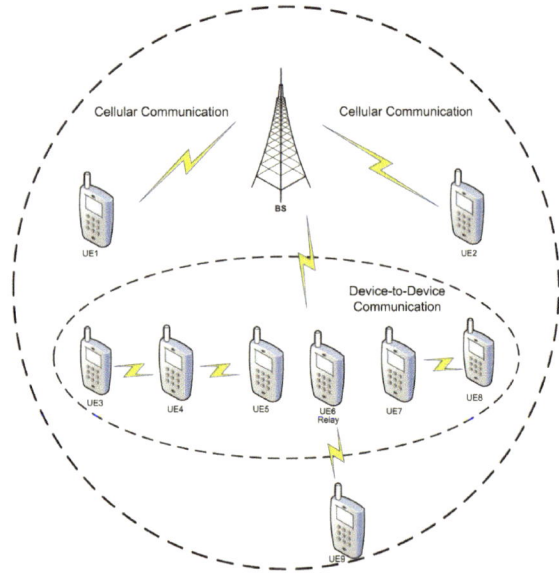

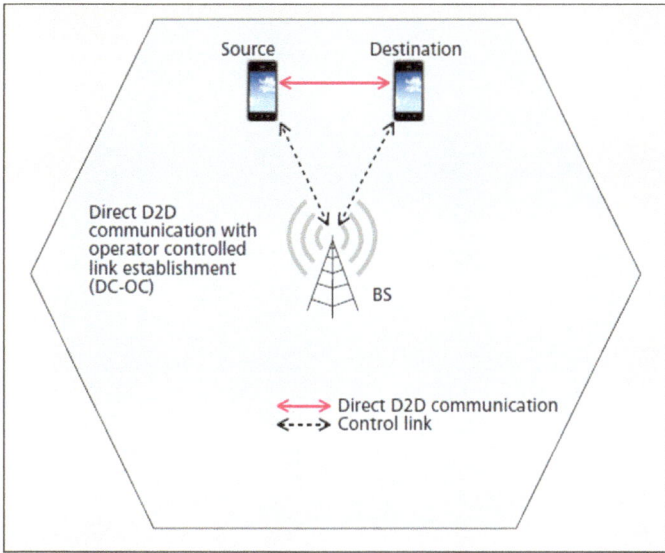

Fig. 3.7   Overview of D2D communication

many challenges associated with D2D communication. The challenges are in selection, mode, control, power, security, privacy, management, interference, discovery, and device.

Interference Management for 5G

Multi-tier 5G Architecture

Interference is one of the issues in 4G technology. Due to density in cell deployment, it caused interference. Especially co-channel interference is the main problem with traditional cellular networks. As discussed in Nam et al. (2014) interference management can be done with many techniques such as advanced receiver and joint scheduling. They proposed two kinds of interference management techniques with 5G. They are categorized into UE-side and network-side management of interference. In Hossain et al. (2014), it is opined that multi-tier architecture of 5G has mechanisms to have advanced interference management. 5G also promotes D2D communication, Quality of Service (QoS), and energy efficiency in spectrum usage. It is made possible with multi-tier architecture of 5G technology as presented in Fig. 3.8.

The 5G technology with its multi-tier architecture enables tier-based and traffic-based priorities. It has provision for energy efficiency, latency, and reliability. It

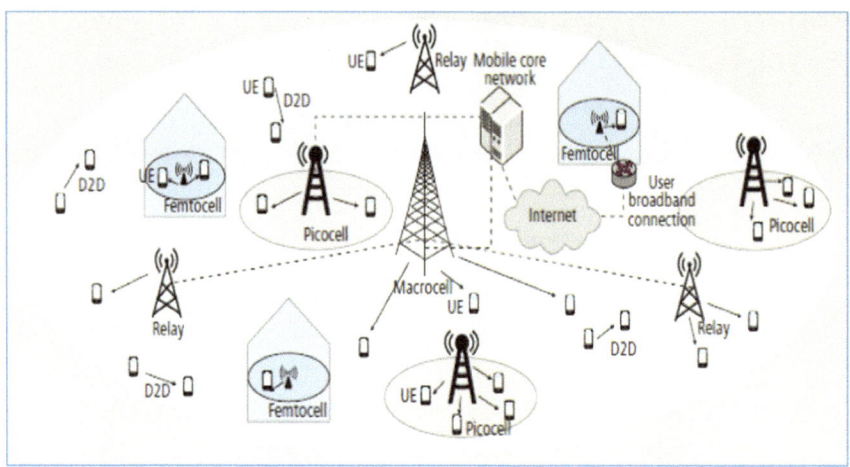

Fig. 3.8 Multi-tier 5G architecture

also minimizes interference in the system. There are certain key challenges associated with interference management. The challenges include heterogeneity, balancing traffic load and coverage, restrictions of public and private usage, and management of priorities. Interference management in 5G is explored with system model and analysis can be found in Sanguinetti et al. (2015).

Spectrum Sharing with Cognitive Radio in 5G

Spectrum sharing is an important aspect of wireless communications. With respect to CRNs, there are users like primary users and secondary users. Wireless network usage is rapidly increasing due to its advantages and conveniences. However, design of a wireless communication system is challenging as it needs to consider many factors such as spectrum utilization efficiency, high Quality of Service (QoS), larger channel capacity, and higher data rate in order to meet the requirements of wireless users. Cognitive Radio (CR) is the technology innovation in which wireless communications are more flexible. It enables users to modify transmitting and receiving parameters without causing interference to primary and secondary users in order to achieve more efficient performance in wireless communications. There are two types of users known as Primary Users (PUs) and Secondary Users (SUs) based on the priority in the assigned radio frequency spectrum. Under a specified policy, both the users can coexist in the system. The licensed users of the band are known as PUs. And PUs can least it to SUs in such a way that SUs use the spectrum without causing harmful interference to PUs in either underlay or overlay approach. Spectrum sensing is an important function of CR system in an overlay fashion. It is meant for finding the presence of PUs in the given frequency band. Cooperative spectrum sensing

(CSS) is "a solution to enhance the detection performance, in which secondary users collaborate with each other to sense the spectrum to find the spectrum holes." With 5G technology, spectrum sharing has improved possibilities while serving different enabling networks. The concept of dynamic spectrum sharing is thoroughly discussed in Ahmad et al. (2020a), while how full spectrum sharing takes place with different approaches is found in Feng and Bing (2016). In Briones-Reyes et al. (2020), a Markov process is discussed in detail to have spectrum sharing efficiently for 5G systems with a mathematical model.

Ultra-dense Networks in Multi-radio Access Technology Association in 5G

Ultra-Dense Networks (UDNs) are the networks where there are more cells present than the users. Such networks are found to be useful to handle explosive data traffic in 5G networks in future.

UDN Moving Cells with 5G+ Network

It has heterogeneous cells even that can be moving in 5G environments. Instead of static RANs, it is possible to have dynamic and moving RANs in future with the emergence of 5G technology. This concept is explored with different possibilities in the world beyond 5G capabilities in Andreev et al. (2019).

Figure 3.9 illustrates the possibilities of moving ultra-dense networks with 5G+ design. There are different unprecedented possibilities with moving cells of UDN with 5G networks. The possibilities include predictive handover, multi-connectivity, dynamic blockage, fleet coordination, swarm formation control, on-the-fly association, space–time methodology, proactive cell discovery, dual mobility, computation offloading, and dynamic association. As discussed in Habbal et al. (2015), it is possible to have a selection of context-aware radio access technology when 5G networks are used with UDNs. They opined that extreme densification has improved capabilities with 5G technology.

Cloud Technologies for 5G

Cloud computing has enabled many services that can be rendered in on-demand fashion. With the emergence of 5G technology, it is possible to have related services. As explored in Sabella et al. (2015), it is possible to have cloud technologies that

Fig. 3.9 UDN moving cells with 5G+ network

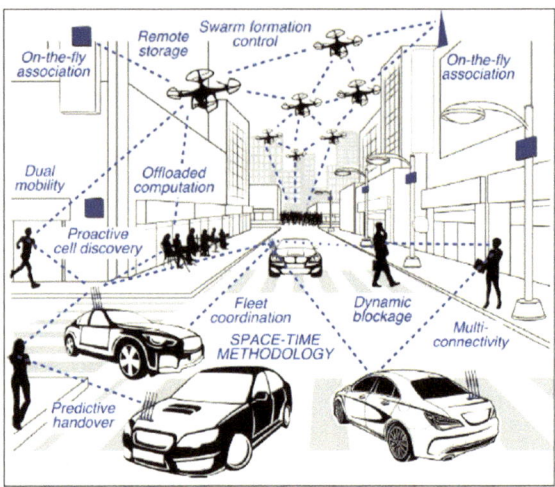

provide benefits associated with 5G. They proposed an architecture known as Cloud-RAN that provides RAN services in pay per used fashion. In other words, they proposed a novel service known as RAN as a Service (RANaaS) that consists of the required cloud-based infrastructure to render RAN services. The proposed virtualization infrastructure is associated with the RAN service. Similarly, in Rost et al. (2014), cloud technologies are leveraged to support flexibility in 5G RANs. Provision of RAN services is provided through cloud offering based on the runtime needs of systems. RAN services are centralized and appropriate software functionality is provided. As discussed in Al-Falahy and Alani (2017), 5G technology-related services can be rendered with cloud services. Different 5G enabler technologies aforementioned can be used in cloud in order to provide scalable, on-demand, and location-transparent services.

Methodology

The Methodology of the Proposed Scheme

This section provides a methodology for the design of energy-efficient delay-aware cooperative scheme for 5G. The proposed scheme is energy-efficient and delay-aware that is used to ensure that the 5G technology usage is made with energy efficiency. The proposed scheme is built based on the architecture shown in Fig. 3.10. It has an important component known as spectrum broker. Spectrum broker is responsible to exploit TV White Spaces (TVWS) by managing the process associated with energy consumption. TVWS consists of spectrum available to support the design of energy-efficient systems. The proposed methodology works based on queueing delays.

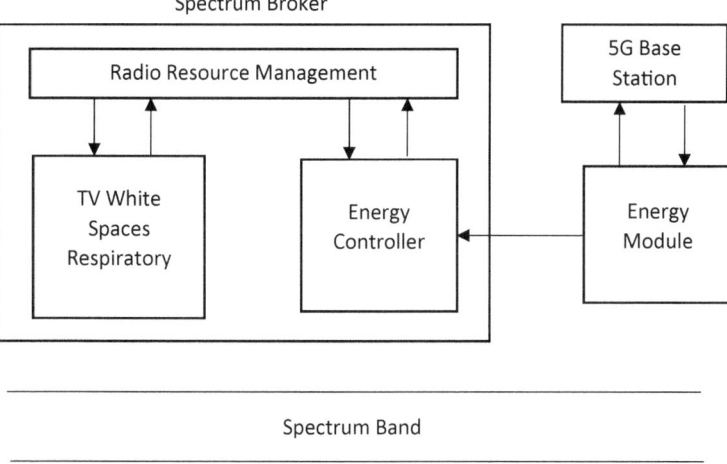

Fig. 3.10 The methodology of the proposed scheme

As presented in Fig. 3.10, the proposed methodology has spectrum broker which takes care of energy efficiency by coordinating other components. The concept of queueing delays is considered while proposing the scheme for delay-sensitive communications with energy efficiency. The spectrum broker has mechanisms to achieve this. The broker coordinates with energy module in order to ensure energy-efficient communications. Simulation study with the proposed scheme shows the energy efficiency of the proposed scheme when compared with the state of the art.

Spectrum Band

The 5G technology offers diversified spectrum bands that meet different requirements. For instance, a low-band spectrum with less than 1 GHz is desirable to many customers through it can provide latency incrementally.

5G Base Station

The 5G base station is a very powerful device that can help in supporting latency and also connectivity beyond the existing standards.

Energy Module

Energy module is responsible to help energy controller in order to achieve energy-efficient approaches with 5G technology.

Energy Controller

Energy controller works in tandem with energy module and radio resource management module for achieving energy efficiency.

Algorithm 1 Pseudocode of the proposed scheme

1. Start
2. Set 5G base stations
3. Activate radio resource management
4. While(true)
5. For each space in tv white space repository
6. Compute energy consumption and delay
7. Save it to map
8. End For
9. Find energy efficient spaces with least delay
10. Notify energy controller
11. Energy controller assists energy module
12. Energy module assists base station
13. Results in energy efficiency
14. End while
15. End

The proposed scheme is delay-aware and energy efficient. It continuously monitors TVWS in order to have opportunities to help 5G base stations to have energy-efficient functionality.

As presented in Fig. 3.11, the energy efficiency of the proposed system is compared with the scheme of Mekikis et al. (2014). The proposed scheme is found to be energy-efficient.

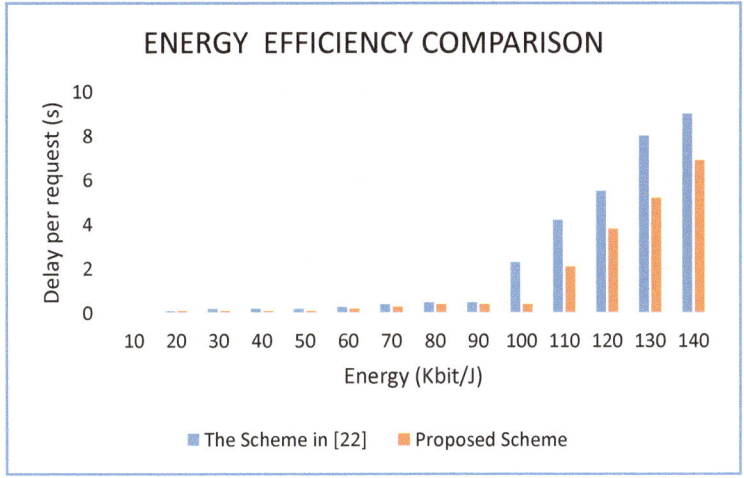

Fig. 3.11 Performance comparison in terms of energy efficiency

Challenges and Future Research Directions

Exploration of spectrum resources in 5G has its advantages in order to have better optimization of resource usage. Though 5G technology is found to be a game changer, it has certain challenges that are to be understood and addressed. The first challenge is to have 5G infrastructure or network development which needs small-cell technology in order to increase network capacity. The second challenge is that of the cost involved in the 5G infrastructure. Microcell costs are more and the cost is going to play its role in making obstructions in its adoption. The third problem is that 5G technology has issues with backhaul and coordination with current cellular technology. The fourth problem is that there is a challenge related to the provision of necessary bandwidth with wave spectrum in the presence of trees, buildings, and other objects acting as obstacles. The fifth challenge is that 5G has certain security challenges in terms of vulnerabilities in the existing infrastructure and new infrastructure. In the presence of these challenges, it is essential to have further research to acquire ways and means to handle these challenges.

Conclusion

This chapter has focused on different aspects of 5G technology and its impact on various communication technologies. It presents the architecture of 5G technology, its underlying mechanisms, support for D2D communication, MIMO enhancements, advanced interference management, enhanced utility of ultra-dense networks, spectrum sharing, and cloud technologies associated with 5G networks. It also proposes a methodology with spectrum broker with underlying components with delay-aware and energy-efficient approach to leverage 5G base stations leading to reduction of energy consumption. The concept of queueing delays is considered while proposing the scheme for delay-sensitive communications with energy efficiency. The spectrum broker has mechanisms to achieve this. Simulation study with the proposed scheme shows the energy efficiency of the proposed scheme when compared with the state of the art.

References

Abdullah M, Kiani SH, Iqbal A (2019a) Eight element multiple-input multiple-output (MIMO) antenna for 5G mobile applications. IEEE Access 1–9

Abdullah M, Kiani SH, Abdulrazak LF, Iqbal A, Bashir MA, Khan S, Kim S (2019b) High-performance multiple-input multiple-output antenna system for 5G mobile terminals. Electronics 8(10):1–16

Adnan MH, Ahmad Zukarnain Z (2020) Device-to-device communication in 5G environment: issues, solutions, and challenges. Symmetry 12(11):1–22

Agyapong P, Iwamura M, Staehle D, Kiess W, Benjebbour A (2014) Design considerations for a 5G network architecture. IEEE Commun Mag 52(11):65–75

Ahmad WSHMW, Radzi NAM, Samidi FS, Ismail A, Abdullah F, Jamaludin MZ, Zakaria MN (2020a) 5G technology: towards dynamic spectrum sharing using cognitive radio networks. IEEE Access 1–30

Ahmad A, Bhushan B, Sharma N, Kaushik I, Arora S (2020b) 2020b IEEE 9th international conference on communication systems and network technologies (CSNT)—importunity & evolution of IoT for 5G, Gwalior, India (2020b.4.10–2020b.4.12), pp 102–107

Akyildiz IF, Wang P, Lin S-C (2015) SoftAir: a software defined networking architecture for 5G wireless systems. Comput Netw 85:1–18

Al-Falahy N, Alani OY (2017) Technologies for 5G networks: challenges and opportunities. IEEE Comput Soc 12–20

Andreev S, Petrov V, Dohler M, Yanikomeroglu H (2019) Future of ultra-dense networks beyond 5G: harnessing heterogeneous moving cells. IEEE Commun Mag 1–7. https://doi.org/10.1109/mcom.2019.1800056

Arora S, Sharma N, Bhushan B, Kaushik I, Ahmad A (2020) 2020 IEEE 9th international conference on communication systems and network technologies (CSNT)—evolution of 5G wireless network in IoT, Gwalior, India (2020.4.10–2020.4.12), pp 108–113

Arsh M, Bhushan B, Uppal M (2021) Internet of things (IoT) toward 5G network: design requirements, integration trends, and future research directions, vol 1286.Springer, pp 887–899

Briones-Reyes et al., 2020Briones-Reyes A, Vásquez-Toledo LA, Prieto-Guerrero A, Aguilar-Gonzalez R (2020) Mathematical evaluation of spectrum sharing in cognitive radio networks for 5G systems using Markov processes. Comput Netw 182:1–14

Chettri L, Bera R (2019) A comprehensive survey on internet of things (IoT) towards 5G wireless systems. IEEE Internet Things J 1–18

Feng HU, Bing ZHU (2016) Full spectrum sharing in cognitive radio networks toward 5G: a survey. IEEE Access 4:1–24

Ghosh A, Maeder A, Baker M, Chandramouli D (2019) 5G evolution: a view on 5G cellular technology beyond 3GPP release 15. IEEE Access 1–13

Goyal et al. 2021Goyal S, Sharma N, Kaushik I, Bhushan B, Kumar N (2021) A green 6G network era: architecture and propitious technologies, vol 54. Springer, pp 59–75

Gupta A, Jha R (2015) A survey of 5G network: architecture and emerging technologies. IEEE Access 1–26

Gur G (2019) Spectrum sharing and content-centric operation for 5G hybrid satellite networks: prospects and challenges for space-terrestrial system integration. IEEE Veh Technol Mag 1–11

Habbal A, Goudar SI, Hassan S (2015) Context-aware radio access technology selection in 5G ultra dense networks. J Latex Class Files IEEE 14(8):1–13

Hossain E, Rasti M, Tabassum H, Abdelnasser A (2014) Evolution toward 5G multi-tier cellular wireless networks: an interference management perspective. IEEE Wirel Commun 21(3):118–127

Idowu-Bismark O, Kennedy O, Husbands R, Adedokun M (2019) 5G wireless communication network architecture and its key enabling technologies. Int Rev Aerosp Eng 12:70–82

Khalfi B, Hamdaoui B, Guizani M (2017) Extracting and exploiting inherent sparsity for efficient IoT support in 5G: challenges and potential solutions. IEEE Wirel Commun 24(5):68–73

Lin K, Wang W, Wang X, Ji W, Wan J (2015) QoE-driven spectrum assignment for 5G wireless networks using SDR. IEEE Wirel Commun 22(6):48–55

Liu Y, Zhang Y, Yu R, Xie S (2015) Integrated energy and spectrum harvesting for 5G wireless communications. IEEE Netw 29(3):75–81

Marsch P, Da Silva I, Bulakci O, Tesanovic M, El Ayoubi SE, Rosowski T, Kaloxylos A, Boldi M (2016) 5G radio access network architecture: design guidelines and key considerations. IEEE Commun Mag 54(11):24–32

Mavromoustakis CX, Bourdena A, Mastorakis G, Pallis E, Kormentzas G (2015) An energy-aware scheme for efficient spectrum utilization in a 5G mobile cognitive radio network architecture. Telecommun Syst 59(1):63–75

Mekikis PV et al (2014) Two-tier cellular random network planning for minimum deployment cost. In: IEEE ICC, Sydney, Australia

Nam W, Bai D, Lee J, Kang I (2014) Advanced interference management for 5G cellular networks. IEEE Commun Mag 52(5):52–60

Papageorgiou GK, Voulgaris K, Ntougias K, Ntaikos DK, Butt MM, Galiotto C, Marchetti N, Frascolla V, Annouar H, Gomes A, Morgado AJ, Pesavento M, Ratnarajah T, Gopala K, Kaltenberger F, Slock DTM, Khan FA, Papadias CB (2020) Advanced dynamic spectrum 5G mobile networks employing licensed shared access. IEEE Commun Mag 58(7):21–27

Ranjan S, Jha P, Karandikary A, Chaporkar P (2022) A flexible IAB architecture for beyond 5G network. pp 1–7

Rost P, Bernardos CJ, De Domenico A, Di Girolamo M, Lalam M, Maeder A, Sabella D, Wübben D (2014) Cloud technologies for flexible 5G radio access networks. IEEE Commun Mag 52(5):68–76

Sabella D, Rost P, Banchs A, Savin V, Consonni M, Di Girolamo M, Lalam M, Maeder A, Berberana I (2015) Benefits and challenges of cloud technologies for 5G architecture. IEEE Access 1–5

Sanguinetti L, Moustakas AL, Debbah M (2015) Interference management in 5G reverse TDD HetNets with wireless backhaul: a large system analysis. IEEE J Sel Areas Commun 33(6):1187–1200

Sharawi MS (2017) Current misuses and future prospects for printed multiple-input, multiple-output antenna systems. IEEE Antennas 162–170

Shen X (2015) Device-to-device communication in 5G cellular networks. IEEE Netw 29(2):2–3

Tudzarov A, Janevski T (2011) Design for 5G mobile network architecture. Int J Commun Netw Inf Secur 3:112–123

Yang C, Li J, Guizani M, Anpalagan A, Elkashlan M (2016) Advanced spectrum sharing in 5G cognitive heterogeneous networks. IEEE Wirel Commun 23(2):94–101

Zhang L, Xiao M, Wu G, Alam M, Liang Y-C, Li S (2017) A survey of advanced techniques for spectrum sharing in 5G networks. IEEE Wirel Commun 24(5):44–51

Chapter 4
Wireless Backhaul Optimization Algorithm in 5G Communication

Astha Sharma, Mukesh Soni, Abhaya Nand, Suryabhan Pratap Singh, and Sumit Kumar

Abstract A wireless backhaul optimization approach using a delay jitter is suggested to handle the wireless backhaul issue for 5G dynamic heterogeneous situations. First, the delay and delay jitter issues in 5G dynamic heterogeneous situations are carefully evaluated, optimization indicators are defined, and the fundamental backhaul model is further built. Then, considering the optimization action needs, include delay constraints to create better model 1; examine network overload, relax channel number allocation variables to construct improved model 2, and present a matching hierarchical method for a quick solution. The simulation results reveal that the suggested approach has improved delay jitter performance compared with three kinds of current wireless backhaul optimization algorithms.

Keywords 5G communication · Wireless network · Heterogeneous scenarios · Hierarchical algorithm · Backhaul algorithm

The original version of this chapter was revised: The name of the author Astha Sharma and the affiliation have been updated. The correction to this chapter is available at https://doi.org/10.1007/978-981-99-3668-7_14

A. Sharma
GL Bajaj Institute of Technology and Management, Greater Noida, India

M. Soni (✉)
Department of CSE, University Centre for Research & Development Chandigarh University, Mohali, Punjab, India
e-mail: soni.mukesh15@gmail.com

A. Nand
IIMT College of Management, Greater Noida, India

S. P. Singh
Institute of Engineering and Technology, Deen Dayal Upadhyaya Gorakhpur University, Gorakhpur, India

S. Kumar
Indian Institute of Management, Kozhikode, India
e-mail: sumit01phdpt@iimk.ac.in

© The Author(s) 2023, corrected publication 2023
B. Bhushan et al. (eds.), *5G and Beyond*, Springer Tracts in Electrical and Electronics Engineering, https://doi.org/10.1007/978-981-99-3668-7_4

Introduction

With the rapid development of science and technology and the continuous improve-ment of people's living standards, various types of mobile terminal equipment and mobile Internet services are widely used. It is estimated that by 2030, the number of mobile terminals will be close to 100 billion, and mobile service traffic will increase by nearly 20,000 times (Ge et al. 2019; Wang et al. May 2016). It will be difficult for the existing communication system to meet the access requirements of massive terminals and services in the future. In view of this, the fifth generation mobile communication system (5G) (Zhang et al. 2020; Madapatha et al. 2021) emerges as the times require. On the other hand, with the gradual rise of various interactive multimedia services (such as video conferencing and online games, etc.), delay and delay jitter have increasingly become the most important QoS indicators, and become the key to user service experience (Chaudhry et al. 2020a; Ahmad 2015). The current research mainly focuses on the analysis of the delay characteristics of data packets, and there are relatively few studies on delay jitter. Therefore, it is imperative to study the delay jitter for the 5G environment.

The control of service delay and delay jitter is mainly implemented at the network link layer with the packet as the granularity, and the forwarding scheduling and queue management of the data packets are carried out through the network nodes. At present, the results of studying delay jitter performance with packet as granularity are rare. References (Rezaabad et al. 2018; Hore et al. 2021) have explored the network delay jitter characteristics. Reference (Chaudhry et al. 2020b) uses delay mean square error to approximate the delay jitter for multi-hop wireless Mesh networks and obtains the delay jitter by solving the end-to-end delay distribution. It should be noted that most of the literature on delay jitter research is based on single-point systems, such as only studying the delay jitter of wireless access network or wired core network (Tran and Le 2018; Pham et al. 2019). However, the 5G network must be a complex and diverse heterogeneous network, and the backhaul network is an important part. The unified consideration of the access network and the backhaul network can more truly describe the future 5G network.

Therefore, in the assessment of the existing problems, this paper suggests an optimal channel resource allocation algorithm with delay jitter as the optimization index for the 5G hybrid backhaul scenario. Considering the dynamic characteristics of the channel, the delay jitter problem is comprehensively analyzed, the delay jitter index is obtained, and then various backhaul optimization models are constructed. Finally, the hierarchical algorithm is used to solve the problem quickly.

The rest of the chapter is organized as follows: section "Network Scenarios and System Assumptions" covers the 5G network scenario, section "Analysis of Delay Jitter Indicators" includes the analysis of delay jitter indicators, section "Opti-mization Model Establishments" includes the proposed model description, section "Improved Model Solutions" includes the improved model, section "Simulation Analysis" includes the simulation detail and result analysis, and at last, section "Conclusions" includes the conclusion and future work.

Network Scenarios and System Assumptions

As shown in Fig. 4.1, for a two-layer heterogeneous network scenario, the upper layer is a Backhaul Aggregator Node (BAN). The lower layer is multiple Small-Cell Base stations (SCBS) covered by the BAN. Dedicated optical fibers link the BAN to the rest of the network, and millimeter waves are used for communication with SCBS or end customers (Iradukunda et al. 2021).

While SCBS uses frequency bands below 6 GHz (Liu et al. 2020) to communicate with users, here, the BAN combines the functions of an aggregator and base station access. Considering the hybrid backhaul scenario, there are two backhaul methods to choose from: the first one, the user can attach to the BAN via a one-hop wireless connection, and the BAN will handle the backup via an Ethernet cable; the second one, the user can access the SCBS to which he belongs, and then owned by.

The SCBS accesses the BAN and the core network through a two-hop wireless link.

Define SCBS set $SC = \{C_1, C_2,..., C_1,..., C_L\}$, define the user set covered by cell C_1 as $UE_1 = \{UE_{11}, UE_{12},..., UE_{1n},..., UE_{1N1}\}$, where UE_{1n} represents the nth user, N_1 is the number of users in the cell SC_1. It is assumed that all SCBSs have a common channel. Now, the seamless coverage of the entire network is disjoint, that is, $UE_i \cap UE_j = \emptyset$ $(i \neq j)$. Therefore, the set of users is $UE = \{UE_1, UE_2,..., UE_1,..., UE_L\}$, then the number of users $N = \sum_{l=1}^{L} N_l$.

The access selection vector is defined as

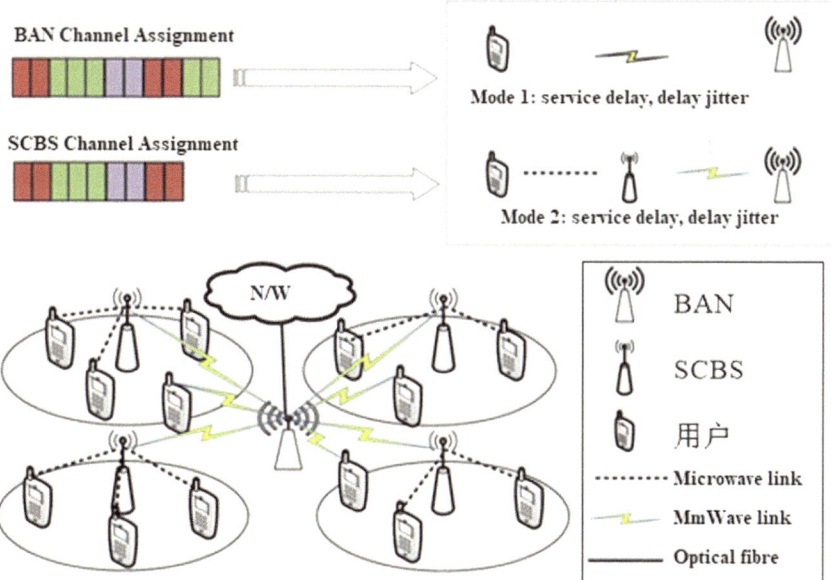

Fig. 4.1 5G two-layer heterogeneous network scenario

$$A = \{a_{11}, a_{12}, \ldots, a_{1N_2}, a_{21}, a_{22}, \ldots, a_{2N_2}, \ldots, a_{L1}, a_{L2}, \ldots, a_{LN_L}\}$$

Among them, $a_{ln} = 1$ $(n \leq N_1)$ means that user n in cell C_1 accesses the BAN, and the BAN performs the backhaul; $a_{ln} = 0$ $(n \leq N_1)$ means that the user n in cell C_1 accesses the C_1, and the C_1 performs the backhaul.

The channel allocation matrix that defines SCBS is

$$B = \begin{bmatrix} b_{11,1} & b_{11,2} & \ldots & b_{11,M_1} \\ b_{12,1} & b_{12,2} & \ldots & b_{12,M_1} \\ & \vdots & \vdots & b_{ln,m} & \vdots \\ b_{1N_1,1} & b_{1N_1,2} & \ldots & b_{1N_1,M_1} \\ b_{21,1} & b_{21,2} & \ldots & b_{21,M_1} \\ b_{22,1} & b_{22,2} & \ldots & b_{22,M_1} \\ & \vdots & \vdots & \ldots & \vdots \\ b_{1N_2,1} & b_{2N_2,2} & \ldots & b_{2N_2,M_1} \\ & \vdots & \vdots & \ldots & \vdots \\ b_{LN_L,1} & b_{LN_L,2} & \ldots & b_{LN_L,M_1} \end{bmatrix}$$

Among them, $b_{ln,m} = 1$ shows that the bandwidth W_m is allocated to the user UE_{ln}, and $b_{ln,m} = 0$ suggests that it is not given. Here, $W = \{W_1, W_2, \ldots, W_{M_1}\}$ denotes the channel vector that SCBS can allocate, and M_1 is the total number of SCBS channels.

The BAN channel distribution vector is defined as below, where M_2 is the total number of channels in BAN

$$C = \{c_{11}, c_{12}, \ldots, c_{1N_1}, \ldots, a_{11}, a_{12}, \ldots, c_{1n}, \ldots, c_{lN_l}, \ldots, c_{L1}, c_{L2}, \ldots, c_{LN_L}\}$$

The number of BAN channels assigned to UE_{ln} is denoted by c_{ln} and c_{ln} $1(l,n)$. It's important to remember that users require a certain quantity of BAN capacity whether they use the BAN backhaul or the SCBS backhaul. cln displays the number of channels assigned to the Cl BAN link in the UEln Cl BAN route for UEln packet transmission if aln is set to 1; otherwise, cln displays the number of channels assigned to the UEln BAN link for UEln data packet transmission if aln is set to 0.

The paper's method has been based on the following principles:

(1) Assuming that the channels are discrete, the BAN can assign different tracks to any SCBS or a particular user. The SCBS can also set other numbers of media to a specific user.
(2) For the service uplink transmission scenario, study the delay jitter index based on packet granularity.

(3) To simplify interference analysis, we will presume that the BAN and SCBS channels have the same total bandwidth, and the BAN's total bandwidth should be larger than the SCBS's total bandwidth.
(4) If the link's transmission rate is higher than its channel capacity, wireless packet loss will occur; each SCBS has an unlimited wait buffer to prevent packet loss from accumulating (or overcrowding); and if the link experiences packet loss, the transmission mechanism will resume to transmit the packet again.

Analysis of Delay Jitter Indicators

In the wireless communication environment, delay jitter is a physical quantity that measures the delay change experienced by data packets during network transmission—packet delay jitter. Therefore, to analyze the delay jitter problem, it is necessary to examine the delay problem first. Propagation delay and transmission delay are components of link delay. Propagation delays are affected by a number of factors, including the distance over which a signal must travel and the pace at which electromagnetic waves travel. (i.e., the speed of light). Due to the short distances over which 5G signals must travel, transmission delays are negligible. However, data fragment transfer causes a delay, which is known as transmission delay. Therefore, the analysis of the service delay and delay jitter system is analyzed separately below.

Delay Analysis

In this scenario, there are three types of wireless links for UE_{ln}: the link l_{ln}^1 for UE_{ln} to access BAN, the link for UE_{ln} to access SC_1 l_{ln}^2, and the link for SC_1 to access BAN correspondingly l_{ln}^3. All link channels are assumed to be subject to small-scale Rayleigh fading. For links l_{ln}^1 and l_{ln}^3, since the same track is not repeatedly allocated in the BAN, there is no interference in the BAN;

Both use millimeter wave communication, and the distance between BANs is relatively long. The above analysis denotes the signal-to-interference noise ratio of the link l_{ln}^i as $SINR_{ln}^i$. If $SINR_{ln}^i \geq SINR_{th}$, the transmission is considered successful, and the bit error rate BER_{ln}^i of the link l_{ln}^i is calculated.

Channel error correction coding is linked to the data loss rate. It is believed that the capacity of all packages is PL (PL $\geq$ 3). According to this report, the data package has three mistakes, or bits, and the box is presumed gone. If there is a loss of packets and the data has to be reissued, the packet loss rate is

$$PER_{ln}^i = 1 - (1 - BER_{ln}^i)^{PL} - C_{PL}^1(1 - BER_{ln}^i)^{PL-1}$$
$$BER_{ln}^i - C_{PL}^2(1 - BER_{ln}^i)^{PL-2}(BER_{ln}^i)^2 \qquad (4.1)$$

C_θ^ϑ represents the number of combinations to select ϑ elements from θ elements. Link l_{ln}^i is derived from the packet loss rate average transmission delay for

$$\tau_{ln}^i = \left(1 - PER_{ln}^i\right)T + PER_{ln}^i\left(1 - PER_{ln}^i\right)2T + \cdots + \left(PER_{ln}^i\right)^{RT}$$
$$\left(1 - PER_{ln}^i\right)(RT + 1)T + \left(PER_{ln}^i\right)^{RT+1}(RT + 1)T \tag{4.2}$$

T represents a transmission delay, while RT is the greatest number of retransmissions.

The latency study of the UE $\rightarrow$ BAN route is straightforward once we move beyond the connection level. With a one-hop wireless link, the latency introduced by the cable central network is negligible and not worth considering. Therefore, for the UE $\rightarrow$ SC $\rightarrow$ BAN path, the time for Packet1, Packet2, Packet3, ..., Packet k to reach SC_1 is τ_{ln}^2, $2\tau_{ln}^2$, $3\tau_{ln}^2$, $\cdots$, $k\,\tau_{ln}^2$ in sequence. If $\tau_{ln}^2 \geq \tau_{ln}^3$, there is no queue congestion at SC_1, so the time to reach BAN is $\tau_{ln}^2 + \tau_{ln}^3$, $2\,\tau_{ln}^2 + \tau_{ln}^3$, $3\,\tau_{ln}^2 + \tau_{ln}^3$, $\cdots$, $k\,\tau_{ln}^2 + \tau_{ln}^3$. If $\tau_{ln}^2 < \tau_{ln}^3$, there is a queuing delay in SC_1, so the queuing delay in SC_1 is 0 in sequence, $\tau_{ln}^3 - \tau_{ln}^2$, $2(\tau_{ln}^3 - \tau_{ln}^2)$, $\cdots$, $(k-1)(\tau_{ln}^3 - \tau_{ln}^2)$, the time for each data packet to arrive at the BAN is is $\tau_{ln}^2 + \tau_{ln}^3$, $\tau_{ln}^2 + 2\tau_{ln}^3$, $\tau_{ln}^2 + 3\tau_{ln}^3$, $\cdots$, $\tau_{ln}^2 + k\tau_{ln}^3$. Combining the above two situations, the path delay of UE $\rightarrow$ SC $\rightarrow$ BAN can be analyzed from the perspective of the expected average; the packet arrival delay interval is $\max\{\tau_{ln}^2, \tau_{ln}^3\}$, and the initial delay is $\tau_{ln}^2 + \tau_{ln}^3$.

Delay Jitter Analysis

Further, the delay jitter of the link l_{ln}^i can be calculated as

$$\sigma_{ln}^i = \left(1 - PER_{ln}^i\right)(T - \tau_{ln}^i)^2 + PER_{ln}^i\left(1 - PER_{ln}^i\right)(2T - \tau_{ln}^i)^2 + \cdots + (PER_{ln}^i)^{RT}$$
$$\left(1 - PER_{ln}^i\right)[(RT + 1)T - \tau_{ln}^i]^2 + (PER_{ln}^i)^{RT+1}[(RT + 1)T - \tau_{ln}^i]^2 \tag{4.3}$$

Among them, τ_{ln}^i represents the mean transmission delay of the link l_{ln}^i, PER_{ln}^i represents the packet loss rate, and T represents the one-time transmission delay. Furthermore, Eq. (4.3) describes the fluctuation degree of the transmission delay of the link l_{ln}^i relative to the average delay in the form of mathematical variance (Chaudhry et al. 2020b).

Spreading from the link to the path, for the transmission of the first data packet, since there is no waiting delay, each component link can be regarded as independent of each other, so the initial delay jitter of the path is the sum of the delay jitter of each component link. For the transmission of subsequent data packets, there are two situations of queuing and non-queuing at intermediate nodes. The packet retransmission mechanism makes the calculation of waiting delay more complicated. At this time, each component link can be regarded as interrelated, so the path. The delay jitter should be less than the sum of the delay jitters of each component link.

For simplicity, this paper uses the maximum delay jitter of each component link to approximate the delay jitter of the entire path.

Analyzing the latency and disturbance on the UE $\rightarrow$ BAN route (where the BAN immediately backhauls UE_{ln}) is straightforward. The cable network's backbone can be reached via a single wireless step. It assumes that the wired connection has no delay jitter, the path wireless. The initial delay jitter and the average delay jitter of the incoming side are both σ_{ln}^i. For the UE $\rightarrow$ SC $\rightarrow$ BAN path (i.e., the backhaul path where UE_{ln} accesses BAN via SC_1), its initial delay jitter is $\sigma_{ln}^2 + \sigma_{ln}^3$, and the average delay jitter of subsequent packets is max $(\sigma_{ln}^2, \sigma_{ln}^3)$.

Optimization Model Establishments

Basic Backhaul Model

The optimization goal can be written as, based on the delay jitter study above:

$$[X^*, Y^*, Z^*] = \underset{\forall a_{ln}, b_{ln,m}, c_{ln}}{\arg\min} \left\{ U = \sum_{\forall l} \sum_{\forall l} (1 - a_{ln}) \right.$$
$$\left. \left[r. \max\left(\sigma_{ln}^2, \sigma_{ln}^3\right) \right] + a_{ln} \cdot \sigma_{ln}^1 \right\} \tag{4.4}$$

Among them, U represents the optimization objective function of the basic backhaul model, which can be regarded as the sum of the average delay jitter of all user backhaul paths, X^*, Y^*, and Z^* represent the optimal solutions of X, Y, and Z, respectively, and $r \geq 1$ is the initial delay jitter compensation factor to reflect the effect of the initial delay jitter. In particular, when max$(\sigma_{ln}^2, \sigma_{ln}^3)$, the average delay jitter of the two types of paths is the same. In this case, UE $\rightarrow$ BAN with a more negligible initial delay jitter will be selected as the optimal path.

Then, the basic optimization model can be established:

$$\underset{\forall a_{ln}, b_{ln,m}, c_{ln}}{\arg\min} \left\{ \sum_{\forall l} \sum_{\forall l} (1 - a_{ln}) \left[r. \max\left(\sigma_{ln}^2, \sigma_{ln}^1\right) \right] + a_{ln}, \sigma_{ln}^1 \right\} \tag{4.5}$$

$$\text{s. t. } \min\left(\sum_{m=1}^{M_1} b_{ln,m}, a_{ln} \right) = 0, \forall 1, n \tag{4.6}$$

$$\sum_{n=1}^{N_l} b_{ln,m} \leq 1, \forall l, m \tag{4.7}$$

$$\sum_{\forall l, n} c_{ln} \leq M_2 \tag{4.8}$$

$$a_{ln} \in \{0, 1\}, \forall l, n \tag{4.9}$$

$$b_{\ln,m} \in \{0, 1\}, \forall l, n, m \tag{4.10}$$

$$c_{\ln} \geq 1, \forall l, n \tag{4.11}$$

Based on Eq. (4.6), UEln uses either the SCBS backhaul (for which the SCBS allocates multiple bandwidths) or the BAN backhaul (for which the SCBS does not assign a channel). According to Eq. (4.7), in order to prevent intra-cell crosstalk, each track within a given cell is only shared by a single user. In Eq. (4.8), M2 is the highest number of channels that can be provided by BAN, so the number of channels allotted by BAN must be less than or equal to M2. Equations (4.9) and (4.10) define the value space of $a_{\ln}$ and $b_{\ln,m}$. Finally, Eq. (4.11) indicates that the BAN bandwidth needs to be allocated no matter how the user is backhauled.

Improved Model 1

The basic model only optimizes the delay jitter and ignores the optimization of a service delay, so the delay interval constraint of the backhaul path of user $UE_{\ln}$ is added:

$$(1 - a_{\ln})\left[r.\max\left(\sigma_{\ln}^2, \sigma_{\ln}^3\right)\right] + a_{\ln} \cdot \tau_{\ln}^1 \leq \varepsilon_{\ln}, \forall l, n \tag{4.12}$$

Among them, $\varepsilon_{\ln}$ represents the maximum delay constraint to ensure the basic delay requirements of the business. Therefore, an improved model 1 is constructed:

$$\underset{\forall a_{\ln}, b_{\ln,m}, c_{\ln}}{\arg\min} \left\{ \sum_{\forall l} \sum_{\forall l} (1 - a_{\ln})\left[r.\max\left(\sigma_{\ln}^2, \sigma_{\ln}^3\right)\right] + a_{\ln} \cdot \sigma_{\ln}^1 \right\} \tag{4.13}$$

$$\text{s.t. } \min\left(\sum_{m=1}^{M_1} b_{\ln,m}, a_{\ln}\right) = 0, \forall 1, n \tag{4.14}$$

$$\sum_{n=1}^{N_1} b_{\ln,m} \leq 1, \forall l, m \tag{4.15}$$

$$\sum_{\forall l,n} c_{\ln} \leq M_2 \tag{4.16}$$

$$(1 - a_{\ln})\left[r.\max\left(\sigma_{\ln,}^2, \sigma_{\ln,}^3\right)\right] + a_{\ln}. \tau_{\ln}^1 \leq \varepsilon_{\ln}, \forall l, n \tag{4.17}$$

$$a_{\ln} \in \{0, 1\}, \forall l, n \tag{4.18}$$

$$b_{\ln,m} \in \{0, 1\}, \forall l, n, m \tag{4.19}$$

$$c_{\ln} \geq 1, \forall l, n \tag{4.20}$$

Improved Model 2

When the number of users exceeds the number of available channels on the network (that is, the network is overloaded), the above model will face an unsolvable problem, namely:

$$N = \sum_{l=1}^{L} N_l > M_2 \tag{4.21}$$

At this time, even if $c_{\ln} = 1(\forall l, n)$, there is still $\sum_{\forall 1,n} c_{\ln} = N > M_2$.

Since there is insufficient BAN route to satisfy all demands, the admittance control system must be triggered, disappointing some users. As for which users are dismissed and how resources are allocated to admit users, an improvement model 2 needs to be established:

$$\arg \min \left\{ U + w \left(M_2 - \sum_{\forall l,n} c_{\ln} \right) \right\} \tag{4.22}$$

$$\text{s.t. } \min \left(\sum_{m=1}^{M_1} b_{\ln,m}, a_{\ln} \right) = 0, \forall 1, n \tag{4.23}$$

$$\sum_{n=1}^{N_l} b_{\ln,m} \leq 1, \forall l, m \tag{4.24}$$

$$\sum_{\forall l,n} c_{\ln} \leq M_2 \tag{4.25}$$

$$(1 - a_{\ln}) \left[r. \max \left(\sigma_{\ln}^2, \sigma_{\ln}^3 \right) \right] + a_{\ln}. \tau_{\ln}^1 \leq \varepsilon_{\ln}, \forall l, n \tag{4.26}$$

$$a_{\ln} \in \{0, 1\}, \forall l, n \tag{4.27}$$

$$b_{\ln,m} \in \{0, 1\}, \forall l, n, m \tag{4.28}$$

$$c_{\ln} \geq 0, \forall l, n \tag{4.29}$$

where ω is the adjustment factor, if ω is small, U is inclined to be optimized; if ω is large, $M_2 - \sum_{\forall l,n} c_{ln}$ is inclined to be optimized to allocate resources to users as much as possible under the premise of ensuring a solution. Also, if $c_{ln} = 0$, the user request is denied.

Improved Model Solutions

Modified Model 1 and Modified Model 2 are fundamentally integer programming. To solve the above models quickly, according to their mathematical characteristics, corresponding solving algorithms are designed based on stratification and branch and bound.

Improved Model 1 Solution

Modified Model 1 consists of the Boolean vector X, the Boolean matrix Solution Y, and the number vector Z. This leads to the optimization problem being decomposed into its component parts. First, the initialization process is given as follows:

Initialization:

1. Generate $X = $ zeros$(1, N)$, $Y = $ zeros(N, M_1),
2. $Z = $ ones$(1, N)$
3. Input: W, C, L, N_1, M_1
4. for $b_{ln} \in Y, W_m \in W$
5. $if \left(m = \left(\sum_{l=1}^{l-1} N_l + n \right) \mathrm{mod} M_1 \right) || ((\sum_{l=1}^{l-1} N_l + n)/M_1 \in N^*)$
6. then $b_{ln,m} \leftarrow 1$ end if
7. end for
8. for $W_m \in W, C_1 \in C$
9. if $\sum_{n=1}^{N_l} b_{ln,m} \geq 2$
10. then $b_{ln,m'} = $ find$(b_{ln,m} = 1)$; $b_{ln,m'} \leftarrow 0$; $a_{ln} \leftarrow 1$ end if
11. end for
12. Output: Initial solution X,Y,Z

In the first layer, solve for the vector X. Vector X according to Eq. (4.18) to construct multiple subproblems. Each iteration can obtain an optimal solution to the current optimization problem. Among them, ff_{ln} represents the value of the formula on the left side of Eq. (4.17), and ff_0 represents its minimum value. In the algorithm, f_0 and $\{ f_{ln} \}$ represent the objective function value of the current feasible solution. The first layer of explanation that is, Algorithm 1, is as follows:

Algorithm 1 Algorithm for Computing X Vector
Input: X, Y, Z, M_2, R, RT, PL
    Calculate f_0;

1. for $a_{\ln} \in X$
2. if $a_{\ln} = 0$ then
3. $a_{\ln} \leftarrow 1$; $b_{\ln,m'} = \text{find}(b_{\ln,m} = 1)$; $b_{\ln,m'} \leftarrow 0$;
4. Calculate $ff_{\ln}$, $f_{\ln}$;
5. if $(ff_{\ln} \leq \varepsilon_{\ln})$&&$(f_{\ln} < f_0)$ then
6. $f_0 \leftarrow f_{\ln}$;
7. else $b_{\ln,m'} \leftarrow 1$; $a_{\ln} \leftarrow 0$; end if
8. end if
9. end for
10. Output: optimized value $X*Y$, f_0

Matrix Y must be solved in the second layer. Using the first layer's answer, the remaining channels in the cell have been allocated so as to fulfill the algorithm (17). Users who access SCBS get more channels, which will reduce the packet loss rate accordingly. The second layer of the solution that is, Algorithm 2, is as follows:

Algorithm 2 Algorithm for Computing Y Matrix
Input: $X*$, Y, Z, W, f_0, N_1, M_2, $\varepsilon_{\ln}$, R, RT, PL

1. for $a_{\ln} \in X$
2. if $a_{\ln} = 0$ then
3. $b_{\ln,m'} = \text{find}(b_{\ln,m} = 1)$;
4. for $W_m \in W$
5. if $(W_m \neq W_{m'})$&&$(\sum_{n'=1}^{N_l} b_{\ln' m} = 0)$ then
6. $b_{\ln,m} \leftarrow 1$; $b_{\ln,m'} \leftarrow 0$; Compute f_m;
7. if $f_m < f_0$ then $f_0 \leftarrow f_m$;
8. else $b_{\ln,m} \leftarrow 0$, $b_{\ln,m'} \leftarrow 1$ end if
9. end if
10. end for
11. end if
12. end for
13. for $a_{\ln} \in X$
14. if $a_{\ln} = 0$ then
15. while $ff_{\ln} > \varepsilon_{\ln}$
16. $b_{\ln,m'} = \text{find}(b_{\ln,m} = 1)$;
17. for $W_m \in W$
18. if $(W_m \neq W_{m'})$&&$(\sum_{n'=1}^{N_l} b_{\ln' m} = 0)$ then
19. $b_{\ln,m} \leftarrow 1$; Calculate $ff_{\ln,m}$; $b_{\ln,m} \leftarrow 0$ end if
20. end for
21. $ff_{\ln,m^*} = \arg \min(ff_{\ln,m})$; $b_{\ln,m^*} \leftarrow 1$;
22. end while
23. end if

24. end for
25. for $W_m \in W$, $a_{\ln} \in X$
26. if $a_{\ln} = 0$ then
27. Set B_temp = find($b_{\ln,m} = 1$);
28. if ($b_{\ln,m} \notin$ B_temp)&&($\sum_{n'=1}^{N_l} b_{\ln'm} = 0$) then
29. $b_{\ln,m} \leftarrow 1$; Calculate $f_{\ln,m}$, ff$_{\ln,m}$; $b_{\ln,m} \leftarrow 0$; end if
30. end if
31. $[f_{\ln,m*}] = \arg\min(f_{\ln,m})$;
32. if ($ff_{\ln,m*} \leq \varepsilon_{\ln}$)&&($f_{\ln,m*} < f_0$) then
33. $f_0 = f_{\ln,m*}$, $b_{\ln,m*} \leftarrow 1$ end if
34. end for
35. Output: optimized value of $Y*$

In this step, solve the matrix Z. After the first and second layers are solved, the current optimization problem becomes a pure integer programming problem. The remaining channels of the BAN are allocated, and one remaining channel is allocated for each traversal. The third layer of the solution that is, Algorithm 3 is as follows:

Algorithm 3 Algorithm for Computing Z Vector
Input: $X*$, $Y*$, Z, f_0, L, N_1, M_2, $\varepsilon_{\ln}$, r_2, RT, PL

1. while $\sum_{l=1}^{L} \sum_{n=1}^{N_l} c_{ln} < M_2$
2. for $c_{\ln} \in Z$
3. $Z' \leftarrow Z$; $c_{\ln} = c_{\ln} + 1$; compute $f_{\ln}$, $ff_{\ln}$; $Z \leftarrow Z'$;
4. end for
5. $[f_{1*n*}] = \arg\min(f_{\ln})$;
6. if ($ff_{1*n*} \leq \varepsilon_{\ln}$)&&($f_{1*n*} < f_0$) then
7. $f_0 \leftarrow f_{1*n*}$; $c_{1*n*} = c_{1*n*} + 1$ end if
8. end while
9. Output: optimized value of $Z*$

The above method for finding a solution can be seen to break down the initial issue into a set of three progressively more difficult challenges iteratively solved according to the characteristics of the solution space. The related problem's optimal solution can be obtained using the branch and bound method, and the iterative solution can gradually converge to the optimal solution of the original problem.

Improved Model 2 Solution

Based on the solution technique of the improved model 1, based on the mathematical characteristics of the enhanced model 2, a three-layer solution technique of the improved model 2 is proposed:

In the first layer, solve the matrix X. Relax (25) and set $\omega = 0$. The solution method is the same as algorithm 1 of the enhanced model 1.

In the second layer, solve matrix Y. The solution process is the same as algorithm 2 of the enhanced model 1.

In the third level, the vector Z must be resolved. It is required to reevaluate the limits of Eq. (4.25) based on the answers found in the first and second layers and to conduct $N - M2$ repetitions. At each step, the person with the largest delay fluctuation in the present optimization outcome is dropped. Algorithm 4 describes the method used to find the answer.

Algorithm 4 Algorithm for Solving Z Vector
Input: X^*, Y^*, Z, L, N_1, M_1, M_2, ω, $\varepsilon_{\ln}$, r, RT, PL

1. While $\sum_{l=1}^{L} \sum_{n=1}^{N_l} a_{\ln} + \sum_{l=1}^{L} \sum_{n=1}^{N_l} \sum_{m=1}^{M_1} b_{\ln,m} > M_2$ do
2.   for $a_{\ln} \in X$
3.   if $a_{\ln} = 1$ then $a_{\ln} \leftarrow 0$;
4.   Algorithm 1; Algorithm 2;
5.   compute $f_{\ln}$; $a_{\ln} \leftarrow 1$;
6.   else for $b_{\ln,m} \in Y$
7.   Set B_temp $=$ find($b_{\ln,m} = 1$);
8.   for $b_{\ln,m} \in$ B_temp then $b_{\ln,m} \leftarrow 0$ end for
9.   Algorithm 1; Algorithm 2; Calculate $f_{\ln}$;
10.   for $b_{\ln,m} \in$ B_temp then $b_{\ln,m} \leftarrow 0$ end for
11.   end for
12.   end if
13.   end for
14. $[f_{l*n*}] = \arg\max(f_{\ln})$; $a_{l*n*} \leftarrow 0$;
15. $b_{l*n*,1} \leftarrow 0$, $b_{l*n*,2} \leftarrow 0$, ..., $b_{l*n*,M1} \leftarrow 0$;
16.   end while
17.   for $a_{\ln} \in X$
18.   if $(a_{\ln} = 0)$&&$(\sum_{m=1}^{M_1} b_{\ln,m} = 0)$ then $c_{\ln} \leftarrow 0$ end if
19.   end for
20.   Output: optimized solution Z^*

Simulation Analysis

The BAN is positioned in the middle of a 500 m * 500 m, and all four are placed evenly at regular intervals. To ensure complete coverage, the SCBS has a typical contact radius of 175 m. Each of the N customers corresponds to one of the N SCBSs, and the N transmission channels are modeled as Rayleigh fading channels. Furthermore, the network's performance will be directly affected by the value of the number of retransmissions RT: a large RT will increase the transmission delay but decrease the packet loss rate, while a small RT will decrease the transmission delay but increase the packet loss rate. To find a happy medium between transmission lag and packet loss in a real-world system, try RT $= 5$. Table 4.1 shows additional simulation parameters.

Table 4.1 Parameters used in simulation

Parameter	Parameter	Parameter value
α	Free space transfer coefficient	3
n_0^{BAN}	Thermal noise of BAN/(dBm $\cdot$ Hz^{-1})	-174
n_0^{SCBS}	Thermal noise of SCBS/(dBm $\cdot$ Hz^{-1})	-174
Δ_{BAN}	Channel bandwidth allocated by BAN/MHz	20
Δ_{SC}	Channel bandwidth allocated by SCBS/MHz	20
M_2	Number of channels that can be given by BAN	250
P_{UE}	User's transmit power/dBm	7
P_{SC}	Transmit power of SCBS/dBm	27
H	Packet sending rate /(Mb $\cdot$ s^{-1})	100
PL	Packet size/bit	512
r	Initial delay compensation factor	1.1
Ω	Regulator	1000

In this paper, we propose the Improved Model 1-Based Solved Algorithm (IM1SA) and the Improved Model 2-Based Solved Algorithm (IM2SA) as two versions of a model-based solved algorithm for wireless backhaul optimization, and we model their performance using three distinct algorithms to evaluate their effectiveness. WBOASBS (Gu et al. 2018) is the wireless backhaul optimization algorithm for single backhaul scenarios, and it belongs to the first class of algorithms. This approach was developed for a particular kind of communication scenario. (that is, all users access the BAN through SCBS for backhaul). The second category of method is the Wireless Backhaul Optimization method for Static Channel Scenarios (WBOASCS) (Zhang et al. 2020). To optimize channel allocation, the third type of algorithm, WBOABIDJ, uses an average delay jitter index that does not take into account channel dynamics. This algorithm is considered for hybrid backhaul scenarios (which include two types of backhaul methods). All three categories of algorithms are very similar to the one presented in this article, and each one is developed with the help of MATLAB. Delay jitter is used as an efficiency metric to evaluate the aforementioned methods. Jitter in this context refers to the time it takes to implement the outcomes of channel distribution derived from various methods into a working network. The typical user delay fluctuation number is determined.

In the situation of minimal service demand, Fig. 4.2 compares the delay jitter of IM1SA, the other three kinds of algorithms, and the number of users (the number of users is less than M2). Increasing the number of users N causes the delay jitter of the four distinct methods to rise steadily, as shown below for varying values of M1; when the number of users is small, the network load is low, and the network performance is at a better level (that is, the link packet loss rate is standard), at this time, the delay jitter of the same algorithm under different M1 is very close; when the number of users is large, the network load is high, and the network performance deteriorates (that is, the link packet loss rate is high), At this time, the higher M1, the smaller

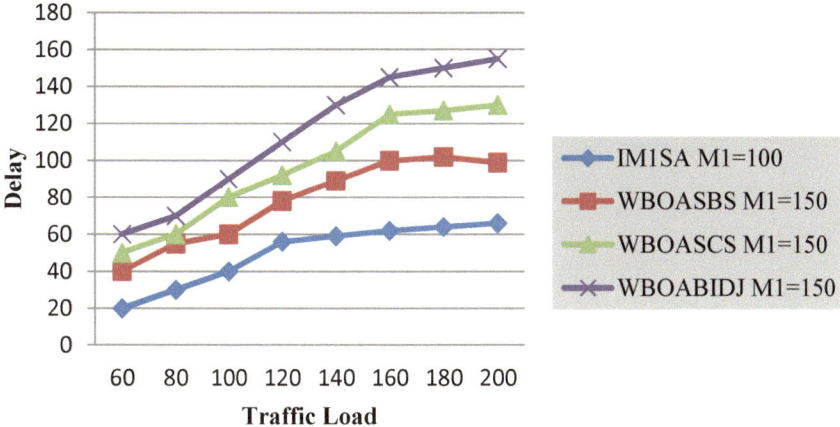

Fig. 4.2 Comparison of delay and jitter of four types of algorithms (low business load)

the delay jitter of the same algorithm; in any case, the delay jitter performance of IM1SA is always optimal (Table 4.2).

The modeling findings for the light traffic burden scenario are not shown in Fig. 4.2. Once the number of users rises above M2, all four kinds of algorithms will be unable to optimize the system; in other words, they will be useless in an overloaded system. As a result, IM2SA is proposed in this article. In the event of a heavy service demand, the delay fluctuation of IM2SA varies as shown in Fig. 4.3. (that is, the number of users is more significant than M2). As the number of users on IM2SA grows, the delay fluctuation will only vary within a narrow range. The IM2SA delay fluctuation decreases as the number of users, N, increases. Admission management and channel allocation can be efficiently executed by IM2SA.

In the case of IM2SA, the delay jitter performance based on IM2SA is always maintained at a certain level without deterioration (Tables 4.3 and 4.4).

Figure 4.4 shows the different results of the network parameter of IM2SA with the number of users in the case of a high service load (that is, the number of users is more significant than M2). It can be seen that the packet delivery ratio, end-to-end delay, and load of IM2SA fluctuates within a specific range with the increase in the number of users N. Given the number of users, the delay jitter of IM2SA shows a downward trend with the rise in M1. IM2SA can effectively carry out admission control and allocate channels reasonably.

Table 4.2 Comparison of delay and jitter of four types of algorithms (low business load)

	60	80	100	120	140	160	180	200
IM1SA M1 = 100	20	30	40	56	59	62	64	66
WBOASBS M1 = 150	40	55	60	78	89	100	102	99
WBOASCS M1 = 150	50	60	80	92	105	125	127	130
WBOABIDJ M1 = 150	60	70	90	110	130	145	150	155

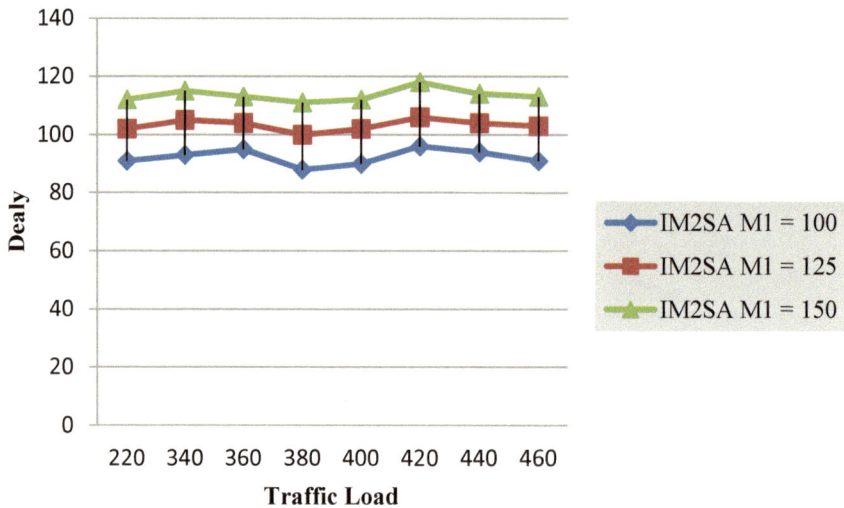

Fig. 4.3 IM2SA delay jitter results (high traffic load)

Table 4.3 IM2SA delay jitter results (high traffic load)

	220	340	360	380	400	420	440	460
IM2SA M1 = 100	91	93	95	88	90	96	94	91
IM2SA M1 = 125	102	105	104	100	102	106	104	103
IM2SA M1 = 150	112	115	113	111	112	118	114	113

Table 4.4 IM2SA evaluation over network parameter

Network scenario	Packet delivery ratio (PDR)	End-to-end delay	Load
IM1SA M1 = 100	78	56	58
WBOASBS M1 = 150	72	68	62
WBOASCS M1 = 150	68	72	74
WBOABIDJ M1 = 150	62	75	80

Conclusions

For 5G dynamic diverse situations, this article suggests an optimization method for wireless backhaul that takes delay uncertainty into account. Backhaul optimization metrics are created, and a fundamental backhaul model is built, based on a methodical

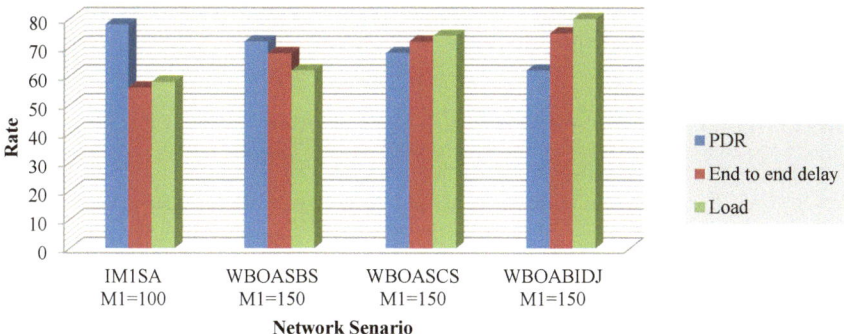

Fig. 4.4 IM2SA evaluation over network parameter

study of service delay and delay instability in active diverse techniques. Additionally, an optimized model 1 and an optimized model 2 are developed from the vantage points of delay optimization and network overflow, respectively, and a hierarchy method is suggested to handle them efficiently. This article uses a program to demonstrate the efficacy of the method.

References

Ahmad SA (2015) A waterfilling algorithm for multiple access point connectivity with constrained backhaul network. IEEE Wireless Commun Lett 4(5):517–520. https://doi.org/10.1109/LWC.2015.2451076

Chaudhry AU, Jacob N, George D, Hafez RHM (2020a) On the interference range of small cells in the wireless backhaul of 5G ultra-dense networks. Wireless Telecommun Symp (WTS) 2020:1–6. https://doi.org/10.1109/WTS48268.2020.9198725

Chaudhry AU, Jacob N, George D, Hafez RHM (2020b) On evaluating independent set heuristics for wireless backhaul network capacity of 5G ultra-dense networks. In: 2020b international symposium on networks, computers and communications (ISNCC), pp 1–6. https://doi.org/10.1109/ISNCC49221.2020.9297231

Ge X, Tu S, Mao G, Lau VKN, Pan L (2019) Cost efficiency optimization of 5G wireless backhaul networks. IEEE Trans Mobile Comput 18(12):2796–2810. https://doi.org/10.1109/TMC.2018.2886897

Gu Z, Zhang J, Ji Y (2018) Topology optimization for FSO-based fronthaul/backhaul in 5G+ wireless networks. In: 2018 IEEE international conference on communications workshops (ICC Workshops), pp 1–6. https://doi.org/10.1109/ICCW.2018.8403736

Hore A, Paul A, Maitra M (2021) Cost-effective policy for deployment of dense 5G RAN with fiber and wireless backhaul link. In: 2021 22nd asia-pacific network operations and management symposium (APNOMS), pp 142–147. https://doi.org/10.23919/APNOMS52696.2021.9562687

Iradukunda N, Pham Q-V, Zeng M, Kim H-C, Hwang W-J (2021) UAV-enabled wireless backhaul networks using non-orthogonal multiple access. IEEE Access 9:36689–36698. https://doi.org/10.1109/ACCESS.2021.3062627

Liu Y, Tang A, Wang X (2020) Joint incentive and resource allocation design for user provided network under 5G integrated access and backhaul networks. IEEE Trans Netw Sci Eng 7(2):673–685. https://doi.org/10.1109/TNSE.2019.2910867

Madapatha C, Makki B, Muhammad A, Dahlman E, Alouini M-S, Svensson T (2021) On topology optimization and routing in integrated access and backhaul networks: a genetic algorithm-based approach. IEEE Open J Commun Soc 2:2273–2291. https://doi.org/10.1109/OJCOMS.2021.3114669

Pham Q-V, Le LB, Chung S-H, Hwang W-J (2019) Mobile edge computing with wireless backhaul: joint task offloading and resource allocation. IEEE Access 7:16444–16459. https://doi.org/10.1109/ACCESS.2018.2883692

Rezaabad AL, Beyranvand H, Salehi JA, Maier M (2018) Ultra-dense 5G small cell deployment for fiber and wireless backhaul-aware infrastructures. IEEE Trans Veh Technol 67(12):12231–12243. https://doi.org/10.1109/TVT.2018.2875114

Tran TD, Le LB (2018) Joint wireless access-backhaul network slicing and content caching optimization. In: 2018 IEEE international conference on communications workshops (ICC Workshops), pp 1–6. https://doi.org/10.1109/ICCW.2018.8403710

Wang N, Hossain E, Bhargava VK (2016) Joint downlink cell association and bandwidth allocation for wireless backhauling in two-tier HetNets with large-scale antenna arrays. IEEE Trans Wireless Commun 15(5):3251–3268. https://doi.org/10.1109/TWC.2016.2519401

Zhang H, Huang C, Zhou J, Chen L (2020) QoS-aware virtualization resource management mechanism in 5G backhaul heterogeneous networks. IEEE Access 8:19479–19489. https://doi.org/10.1109/ACCESS.2020.2967101

Zhang Q, Luo K, Wang W, Jiang T (2020) Joint C-OMA and C-NOMA wireless backhaul scheduling in heterogeneous ultra dense networks. IEEE Trans Wireless Commun 19(2):874–887. https://doi.org/10.1109/TWC.2019.2949791

Chapter 5
Security Attacks and Vulnerability Analysis in Mobile Wireless Networking

Ayasha Malik, Bharat Bhushan, Surbhi Bhatia Khan, Rekha Kashyap, Rajasekhar Chaganti, and Nitin Rakesh

Abstract Security of data is very important while providing communication either by the wired or wireless medium. It is a very challenging issue in the world and the wireless mobile network makes it more challenging. In a wireless mobile network, there is a cluster of self-contained, self-organized networks that form a temporarily multi-hop peer-to-peer radio network, lacking any use of the pre-determined organization. As these networks are mobile and wireless connection links are used to connect these networks through each other, many of the times these kinds of networks are accomplished of self-manage, self-define, and self-configure. Due to their dynamic nature, wireless mobile networks/systems do not have a fixed infrastructure and, due to this, it is more vulnerable to many types of hostile attacks. Different kinds of security attacks that are present in wireless mobile networks are stated in the paper with their spotting and precaution techniques. Furthermore, the paper deliberates on the various types of mobile networks along with their numerous challenges and issues. Moreover, the paper defines the need and goals of security in wireless mobile networks as well as many security attacks along with their detection or prevention methods.

Keywords Wireless mobile networks · Attacks · Security · Sender · Receiver · Cryptography · Detection · Prevention · Vulnerability · Node · Transmission · Data

A. Malik
Delhi Technical Campus (DTC), GGSIPU, Greater Noida, India

B. Bhushan (✉) · N. Rakesh
Department of Computer Science and Engineering, School of Engineering and Technology, Sharda University, Greater Noida, India
e-mail: bharat_bhushan1989@yahoo.com

S. Bhatia Khan
Department of Data Science, School of Science, Engineering and Environment, University of Salford, Manchester, United Kingdom

R. Kashyap
Noida Institute of Engineering and Technology (NIET), Greater Noida, India

R. Chaganti
University of Texas, San Antonio, USA

© The Author(s) 2023
B. Bhushan et al. (eds.), *5G and Beyond*, Springer Tracts in Electrical and Electronics Engineering, https://doi.org/10.1007/978-981-99-3668-7_5

Introduction

For several years now, there has been a clear example seen in daily life as the transference from the fixed network to wireless mobile networks for easiness in communication as there is no need for a licensed frequency band to act and the wireless mobile network does not require any investment in infrastructure as it can able to form a dynamic structure (Lohachab and Jangra 2019). These properties have an important role to make them appealing for some commercial implementation in various fields and most important in the military field. As there are many good things in a wireless mobile network, there is another side too that says, in wireless mobile network many problems occur; among them, network security is the most important concern (Nurlan et al. 2022). Mobile technology is rapidly growing, wireless mobile networks have shown in many forms such as laptops, PDAs, etc. There is a very high chance for attackers, as in a wireless mobile network, a node can be operated as a source, destination, and router (intermediate node). Communication in a wireless mobile network is done through messages, a network can send the data to its adjacent network via messages. And these networks do not contain any information about any other nodes/networks, whether the network is prone to attack or safe. They do not know each other (Bhushan and Sahoo 2017, 2018). Securing a wireless mobile network is tough because there are many reasons such as no boundaries, attack from an unfriendly node into the network, no facility of central management, the power supply is limited, extension ability, no protection of channels, changes in topology, etc.

The wireless mobile network usually has small devices which are more memory-constrained and more susceptible to failures. Although energy is a scarce resource for both kinds of networks, these networks have tighter requirements on network lifetime, and recharging or replacing the node's batteries is much less of an option (Cuka et al. 2018). The basic purpose is focused on providing distributed computing and information gathering. Wireless mobile networks are used in environments like forests, mountains, rivers, etc. In order to be counterproductive and try to predict natural calamities such as forest fires, quakes, floods, cloudbursts, etc. (Bhushan and Sahoo 2020a; Han et al. 2019). Wireless mobile networks can be used for monitoring outgoing services, equipment, and nodes. It can also be used in surveillance of battlefield, atomic, biotic, and chemical attack detection. Wireless mobile networks can be used in health applications for telemonitoring of human physiological data, in telecare medicine information systems, for drug administration in hospitals, and for tracking and monitoring patients and doctors inside a hospital (Kibria et al. 2018). Figure 5.1 shows the transmission procedure of data in the form of packets in the wireless mobile network.

Wireless mobile systems are susceptible to safety outbreaks due to the broadcast behavior of the communication medium and the sensitive nature of collected information. Security effects by some parameters of wireless mobile networks that must be addressed are resource limitations, processing limitation, limited memory

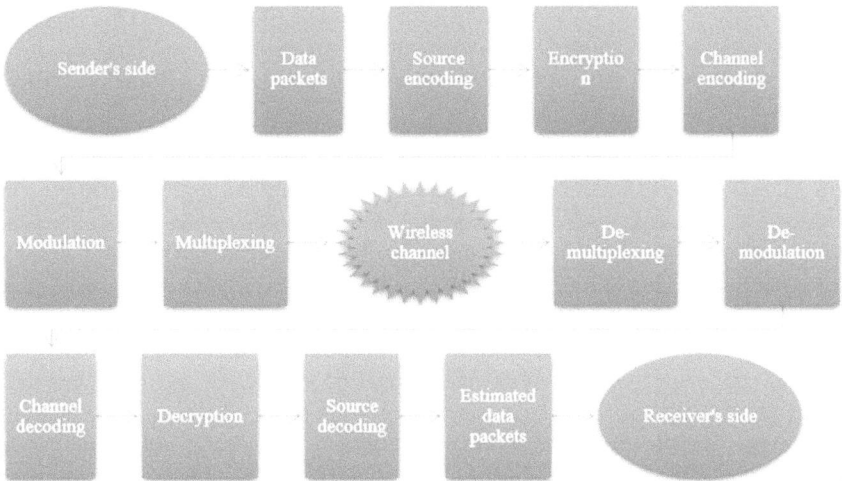

Fig. 5.1 Transmission procedure

and storage space, power limitation, etc. (Bhushan and Sahoo 2019a). Microcontrollers in nodes of wireless mobile networks range between 4 and 400 machine instructions per second which implement communication functions but are not sufficient to support security mechanisms. A small memory of nodes necessitates limiting the code size of security algorithms named encryption, decryption, verification, etc., (Bhushan and Sahoo 2019b) that employed in security algorithms need more processing, i.e., power consumption. And also more energy is mandatory to convey the safety-related data or overhead. Connectionless routing implies unreliable exchanges. Due to channel errors and congestion, packets may get damaged, resulting in lost or missing packets. Packets broadcasted on radio links may collide causing loss of information (Moin et al. 2021). The multi-hop routing in wireless mobile network nodes can lead to greater latency and makes it difficult to achieve synchronization. So, this causes a problem in detecting and reporting the events on time (Liu et al. 2020). Remote management makes it difficult to notice the physical interfering and physical caring concerns. Maybe a disseminated system without the management of central point makes network organization difficult (Zhao et al. 2019). Furthermore, the key inspiration of this study is as follows.

- The work discusses the background as well as different types of wireless mobile networks and the need of securing data in the network through wireless transmission/connections.
- The work deliberates the various challenges and issues of wireless mobile network, which comes during the transference of data.
- The work highlights the security goals and categories of attacks, and also discussed how to protect the wireless mobile network from attacks in detail.

- The work explores some recently proposed data related to networking and elaborates some methods to prevent an attack on a system.
- The work redefines the inspiration for protecting data with Java to form a securing application for Securing data.

The remainder of the paper is planned as follows, section "Types of Wireless Mobile Network" defines the different types of wireless mobile networks that construct on the basis of wireless connection. Furthermore, section "Challenges and Issues in Wireless Mobile Network" discusses the various challenges and issues that occur during the formation of wireless mobile networks or when the transmission takes place. Additionally, section "Security Goals of Wireless Mobile Network" elaborates the goals of security, where confidentiality, availability, authentication, integrity, and non-repudiation have been discussed. Moreover, section "Classification of Security Attacks" describes the classification of security attacks for securing applications or systems or wireless mobile networks. This section also defines some kinds of attacks and illustrates how the attacks can affect the wireless mobile network. Furthermore, this section also deliberates the information about various detection and prevention mechanisms for protecting the network from different kinds of attacks. Lastly, section "Conclusion" brings the paper to a conclusion.

Types of Wireless Mobile Network

The wireless links are used for connection between the devices by using the medium such as microwaves, communiqué satellites, radio waves, spread spectrum technologies, free-space optical transmissions, or numerous technologies that are used in mobile networks (Lyu et al. 2019). The different types of wireless mobile networks are shown in Fig. 5.2.

Wireless PAN

The wireless Personal Area Networks (PAN) connect all the network and end-to-end devices to a fairly small region, usually accessible to a person. For illustration, Bluetooth radio, as well as undistinguishable infrared rays, delivers wireless PAN headset connected to a portable computer. ZigBee also supports Wireless PAN applications that include sensors and many related devices (Awais et al. 2020).

Fig. 5.2 Types of wireless
network

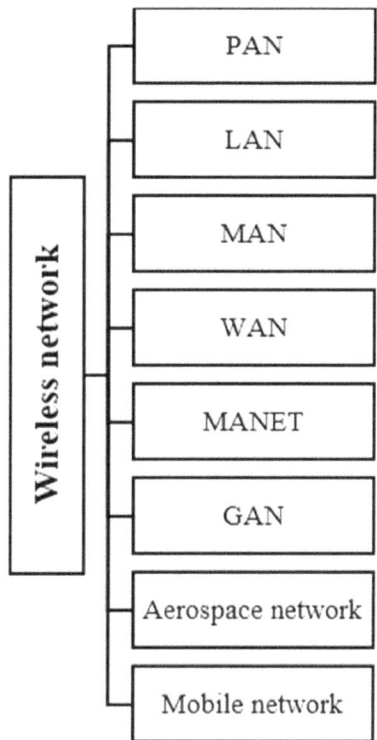

Wireless LAN

The wireless Local Area Network (LAN) joins two or more nodes as well as network devices to a short distance using a wireless dissemination technique, typically giving network access points over the internet. The use of spread-spectrum or wireless transmission technology could permit the users to navigate within the limited area and stay associated with the system. Immovable wireless technology uses point-to-point associations among computers or networks in two remote positions, usually using a devoted microwave or another ray converted into a line of sight. It is frequently used in capitals to attach the systems in two or more buildings without fixing a wireless connection (Fortino et al. 2018).

Manet

The wireless Mobile Ad hoc Network (MANET) is a wireless system that connects the node or devices based on the structure of mesh topology. Every device transmits the data packets instead of knowing other devices/nodes and every node forms a route.

Ad hoc networks can "support themselves", and automatically relocate to a depleted environment. Numerous network layer protocols are required to employ MANET, such as vector sequence tracking, associativity-based route, dynamic source route, and much more (Ayele et al. 2018).

Wireless MAN

The wireless Metropolitan Area Network (MAN) is a type of wireless network that links numerous wireless LANs to provide connectivity between two types of networks and covers the networks in a range of around thousands of kilometers or area covering two cities, for example, WiMAX (Malik et al. 2020).

Wireless WAN

The wireless Wide Area Network (WAN) often covers huge regions, like nearby villages and cities, or cities and suburbs. These systems can be used to join branch offices to companies or work as a public internet access system, for example, internet is a type of WAN, which connects people all over the world that says WAN is able to create or maintain the world largest networks easily. Wireless links among access points are generally point-to-point microwaves using parabolic vessels at 2.4 GHz band, at the place of omnidirectional horns, which are used with minor networks. The standard system consists of basic hubs, routers, gateways, access points, and relay wireless bridges. Another configuration system has spaces where each access point acts as a relay as well (Elhattab et al. 2017).

Mobile Network

It is a radio system dispersed over the world called cells, each of which is supplied with at least one immovable transceiver, known as a mobile site or base station. In this type of network, all cell uses a diverse group of radio frequencies across adjacent cells to escape any disruption. When they are put together, these cells provide radio broadcasts throughout the country (Tzanakaki and Anastasopoulos 2019).

Gan

The Global Area Network (GAN) is a network that is used to support mobile phones with a certain number of wireless LANs, satellite-covered environments, etc. It is a kind of network in which different networks or devices or transmission mediums are interconnected to cover an unlimited geographical space. The biggest challenge in mobile communication is the transfer of user communications from one location to another. The IEEE 802 project involves a series of groundless LANs (Zhao et al. 2021).

Aerospace Network

Aerospace networks are the networks that are used to communicate between spacecraft, usually in areas close to the universe. It has been giving instructional sustenance, software resolutions, and media invention facilities to both scholastic and commercial clients. An example of this type of network is the NASA space network (Liang et al. 2019).

Challenges and Issues in Wireless Mobile Network

In today's world, providing a reliable and dependable mode of communication, especially in emergencies or applications is one of the important research concerns and challenges. Some important issues and challenges in a wireless mobile network are buffer management, node discovery, forwarding of the message, security of network and data, etc., which are briefly explained below and also mentioned in Fig. 5.3.

Information Management

Wireless mobile network is a type of environment where most of the focus is on the delivery of information to achieve this most of the routing protocols use flooding-based mechanisms. This type of protocol has a habit to load the network by transferring a huge amount of information into the network. So, to handle these issues, the authors provide various other types of protocols that are based on the forwarding approach rather than the flooding approach (Ding et al. 2018).

Endless Communication

In wireless mobile networks, communication between two devices provides the basis for interaction. The communication issue is exacerbated by an absence of prior information about the position, time, and required bandwidth. Route agreements that use the context, profile, or history of mobile users as well as all connected devices

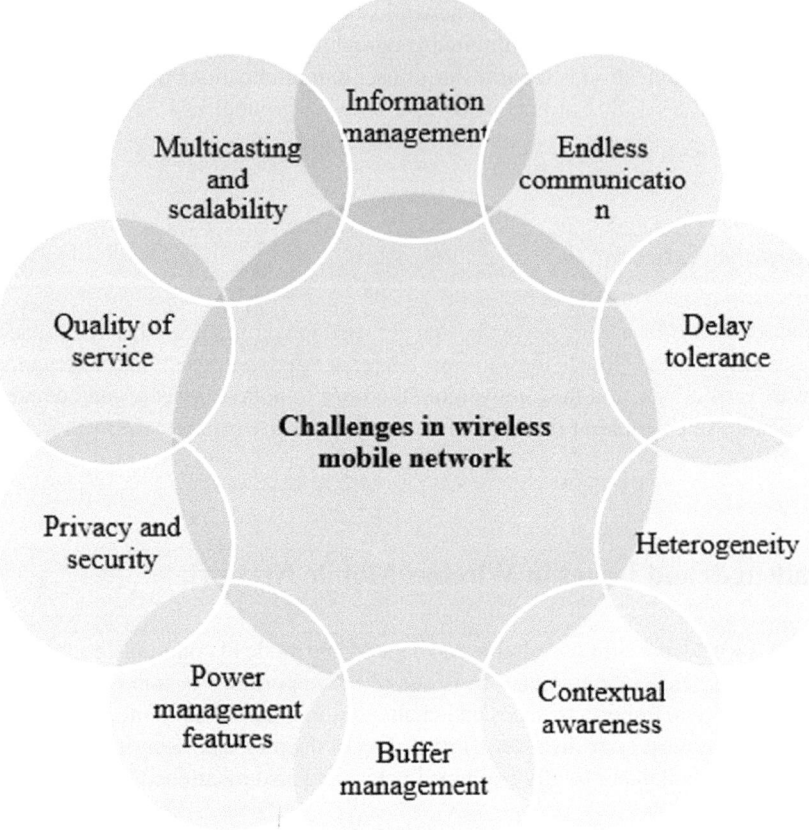

Fig. 5.3 Challenges in wireless mobile network

should be examined for use on mobile networks. It will be necessary to upgrade the middleware methods so that the mask can be delayed and hide the complexity of the flexible methods in the operating systems. The information obtained must be analyzed to archive, refine, and disseminate as the storage capacity and bandwidth are restricted (Shnaiwer et al. 2019).

Delay Tolerance

Effective use of Delay Tolerant Network (DTN)'s applications has been proven very useful for wireless mobile networks. Tolerance delay plays a significant role in the mobile computer as all individuals did not want to wait or waste their single minute hence, it is very essential to provide smooth communication without any kind of delay (Liu 2021).

Heterogeneity

Possibly, many types of nodes may come together automatically such as cell phones, PDs, laptops digital notebooks, sensors, cameras, and RFID devices. These devices can be maintained by a variety of communication abilities and radio signals. The interplay of communication between these pairs of different devices is the main experiment (Samanta and Misra 2018).

Contextual Awareness

It is an important key to searching/finding a secure system. Most of the content is important for people who are directly close to the source, creating a temporary, local public where they want to share. This needs of inaugurating strong and reliable momentary relationships among people and equipment. Content knowledge and profiles of devices, individuals, and applications as well as repository development strategies are needed to effectively manage the content repository. In order to share the information on social media, researchers have projected social networking sites. A social archive is a logical compilation of a device for each device that stores information that is useful to members of its social network. Given that participants are predictable to meet regularly, and data stored in a public repository can be used effectively by more members, temporary community retention can meaningfully improve system performance (Lee and Ke 2018; Kibria et al. 2018).

Buffer Management

The most important part of devices is storage or buffer space in themselves. On a mobile or movable computer, devices store the information of another person in their repository that should be carefully monitored by removing undesirable data and protecting the data usable from applications on the network's device, such as those devices by which our peers are expected to encounter next. The content repository can be processed through other applications, content, or other methods (Wadii et al. 2019).

Power Features

Power is another important feature of the portable device, where most devices are powered by a battery. Power management is a separate issue in terms of stowage and bandwidth management. Improved data transfer on a wireless optical connector causes more power, while local data storage may incur significant energy costs in memory control (Sinha et al. 2017).

Privacy and Security

Finding security and trust between anonymous nodes in this type of network is a challenge. However, social networking structures provide the basis for improving trust and providing protection through the use of "communities" of similar devices within themselves, physically or mentally. The idea of using social networking infrastructures to improve network security is not a novelty. Actually, the works cover a few suggestions based on the use of social networks to combat email spam and to protect the networks against various kind of attacks. Conversely, the use of social networking is the complete separation of networks is a new and challenging task as, in these surroundings, security resolutions based on a central server or trusted online specialists cannot be achieved. In this case, the natural direction of the pursuit of exploitation of electronic social networks and the relationship between trust and safety is deeply ingrained in human relationships (Petrov et al. 2018).

Security Goals of Wireless Mobile Network

Wireless network is more feasible as compared to a wired one but it is very essential to offer safe and secure communication or connection between the users. There are five security goals that needed to be accomplished to conserve smooth communication in a wireless mobile network as shown in Fig. 5.4.

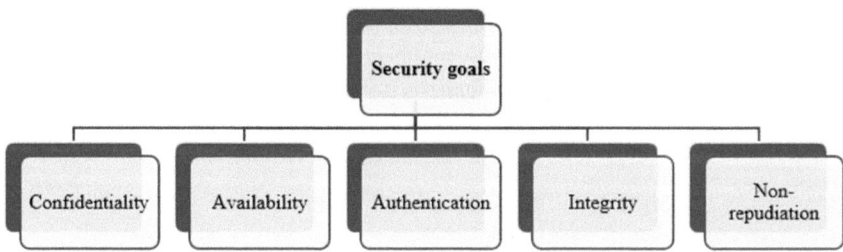

Fig. 5.4 Goals of security

- *Confidentiality*—It refers to the protection of data sent by a device so that it becomes unreadable by an unsanctioned person or access point. Because wireless networks are open, all networks are within direct transmission range, making data retrieval simple so it is very important to keep the data confidential from the unauthorized user or device.
- *Availability*—The "activeness of communication" is most essential in the network, network services should always be available when they are needed. The availability means the data, the transmission medium, as well as the node, will be available or reachable in the network for communication or connection if they are not busy in another network (Edirisinghe et al. 2021).
- *Authentication*—It identifies that a network or a client is genuine or not fake and prevents parody so that any devices carrying the virus with it cannot easily connect with the genuine network or perform illegal actions. As the fake node has the identity of a genuine network or device to access the sensitive information easily, it is much needed to authenticate the user.
- *Integrity*—Integrity relates to the fact that the sender's message should reach the receiver intact, with no changes or deletions. That means when the data packet which is sent by the sender to the receiver through an insecure and open transmission medium should be same without any alteration or change of a single bit in the data packet.
- *Non-repudiation*—It gave the assurances to the sender that a sent message cannot deny after sending that the message or that the message was received by the intended recipient will not be deniable. It's used to separate and identify infected access points. If Network X receives an infected message from Network Y, Network X should be able to blame Network Y and inform the other networks about it using non-repudiation (Fernando et al. 2019).

Classification of Security Attacks

The transmission medium of a wireless network is broadcast in nature. Due to this, wireless network is very sensitive to different kinds of security threats. In a wireless network, security attacks can be categorized as follows.

- *Passive versus active attacks*—In passive attacks, the data travels over the network without any disrupting operations applied to transmission. Although in an active attack, information disruption, alteration, deletion, construction, etc., can disturb the normal functioning of the wireless network (Lin et al. 2020).
- *Internal versus external attacks*—Internal attacks are performed inside the network by compromised networks that lie inside the network, while external attacks are performed by those networks that do not lie inside the network.
- *Stealthy versus non-stealthy attacks*—The attacker tries to hide his/her identity/ actions and operate quietly to disturb the network. In non-stealthy attacks, the attacker doesn't hide his/her action/identity (Guan and Ge 2018).

Table 5.1 Security attacks on different layers

Layers (Loreti and Bracciale 2020)	Attacks (Loreti and Bracciale 2020)
Physical layer	Eavesdropping, Interference
Datalink layer	Disturbing frames of MAC, Monitoring
Network layer	Location revelation, Rushing attack, Blackhole attack, Wormhole attack, Byzantine
Transport layer	Flooding, Hijacking
Application layer	Data exploitation, Cross-site scripting, Denial

- *Cryptographic versus non-cryptographic attacks*—Digital signature attacks, hash collision attacks, and many more are kinds of attacks that lie under the category of cryptographic attacks. Flooding attacks, blackhole attack, etc., are those attack that lies under the category of non-cryptographic attacks (Yang and Wen 2021).
- *Attacks on different layers*—Table 5.1 shows attacks, which are classified based on networking layers of the internet model. There are some attacks, which come under various layers like impersonation, replay, man-in-the-middle, etc., as shown below.

Most Prevalent Attacks in Wireless Systems

There is various kind of security attacks, which are performed by the attacker to gain access and admittance in the network and harm the network as well as data. In this subsection, various security attacks are classified or stated that how they perform malicious behavior.

Denial of Service (DoS) Attack

In this attack, the attacker familiarizes himself/herself with many fake or bogus data packets in the system to affect the system conflict in the wireless server. Sometimes, the infected system may pretend as a busy network and deny communicating with others (Ashfaq et al. 2019). In this attack, many bogus requests or other kind of requests floods over the system or server to keep the network busy and to make them not able to perform any genuine task. These impact network accessibility, furthermore, the detection, as well as prevention techniques of this attack are as follows.

- **Strengthening their security status**: This includes strengthening all internet-based resources to prevent compromising, installing and maintaining anti-virus software, setting up security walls designed to guard the network against DoS attacks, and following strict safety procedures to observe and control undesirable traffic (Okamura et al. 2019).

Flooding Attack

The purpose of an infected network is to deplete the resources in the network like consuming the power of the battery of the networks by flooding unnecessary requests. It is also termed a resource consumption attack or bogus information attack (Nundloll et al. 2020). Moreover, the prevention technique for flooding attacks is stated.

- **Blacklist the infected network**—Every network has a threshold value in a network that is priory defined. If the network sends the RREQ request more than its threshold value, then that network gets blacklisted from the network and any request that comes from the blacklist network is simply dropped by another network (Bhushan and Sahoo 2020b).

Jamming

The main purpose of this attack is to prevent sending and reception of legitimate packets from source to destination. Sometimes, it can be performed to capture the way and gain access. In this attack, unnecessary request and response messages are flooded to jam the routes so that the functionality of the network decreases. At last, all the possible routes between networks in the network get destroyed and no communication is done, it is also called an SYN flood attack (Liu and Labeau 2021). Moreover, some of the detection and prevention techniques of this attack are as follows.

- **Anti-Jamming reinforcement system**—It is used to see if there's any jamming going on. To lessen the jamming effects, it provides rate adaptation and power control measures such as ARES (A software package that allows the file to be immediately downloaded into the system) (Tsiota et al. 2019).
- **Uncoordinated Direct Sequence Spread Spectrum (UDSSS)**—The receiver has a certificate of the sender's public key in this broadcast situation, but they don't exchange the secret key. As a result, the receiver will be able to verify the request (Zhang et al. 2020).

- *Steiner Triple System and the Traversal Design (STS &TD)*—These two
  approaches, STS and TD, are proposed to provide jamming prevention (Gautam
  et al. 2019).

Intervention

Radio communication can be obstructed by the invader to harm or injure the data so
that it cannot reach the receiver. It happens when the user is able to access a little
solid information about the network without direct access to it. The purpose of this
unintentional attack is to combine the information on a single level of security in
order to determine the truth that should be protected at the highest level of security
(Malik and Gupta 2019).

Sleep Deprivation Attack

For the extra consumption of battery of networks, the sleep deprivation attack is
done. In this, networks are enforced to continue wakeful by the invader to reduce the
battery life and to shut down the networks. This attack is the most hazardous type
of attack at this stage, as the malicious node makes requests to the nodes only to
keep the victims awake. The victim's nodes are therefore reserved for the network
wakeful and not able to complete energy-based tasks (Bhushan et al. 2017).

Blackhole Attack

In this attack, an illusion is created by the infected network that it has the shortest
way from sender to receiver. Once it is done, then all the packets coming to the
infected network get to fall. If more than one infected network work in combination
and try to suffer the whole network, then it is called a collaborative blackhole attack
or packet dropping attack (Malik and Gautam 2019). Moreover, the detection and
prevention techniques that are used to protect the network from blackhole attack and
cooperative blackhole attack are as follows.

- *Prevention of a Cooperative Blackhole Attack (PCBHA)*—The concept of
  PCBHA is to use fidelity level in networks. Initially, each network has a defaulting
  fidelity level, and after distribution of an RREQ, a source network waits to
  receive return RREPs after the neighbor networks, only that neighbor network
  gets selected which has an advanced reliability level and surpasses the threshold
  value, for passing the data packets. It is mandatory for the destination network to
  return an ACK message after receiving data packets. When the source network
  receives an ACK message from the destination network, it adds 1 to the fidelity

level of the neighboring network. If no ACK response is received by the source network from the destination network, then 1 is subtracted from the fidelity level of the neighbor network which shows a possibility of a blackhole node in the network on the followed route (Heo et al. 2018).

- *Protection based on cryptography*—To protect blackhole and collaborative black-hole attacks, some cryptographic techniques are also used like fingerprint (hash or hash MAC), digital signature, and other protections (Zhou et al. 2020).
- *Protocol modification*—To protect blackhole and collaborative blackhole attacks, some modifications to the protocol can also be done like cross-check, degree of trust, and data routing information table (Nishanth and Mujeeb 2021).
- *Redundant route method*—To prevent blackhole attack, not only one but also many routes are found from sender to receiver. The minimum number of valid routes from sender to receiver, in any case, is three as per (Kafaie et al. 2018).

Rushing Attack

In this attack, the source network sends RREQ to the destination network via some networks in between. Concurrently, another RREQ is sent to the same destination network by the attacker's network. If the neighbor network of that destination network gets the attacker network's request first, then that infected route is selected (route having an infected network). After that, the original request which is conducted from the source network is received by the neighbor network and will be discarded. As a result, the communication between the source network and destination network is only done via the infected network or attacker network (Sivanesh et al. 2019). Moreover, the detection technique which is used to detect the rushing attack in the network is stated as follows.

- *Secure neighbor detection*—It confirms that a neighboring network falls in a maximum communication scale by introducing a delegation message which is based on the sign based on some routing table's entries (Zhou et al. 2020).

Sybil Attack

In this attack, a network is taking over the whole network and then claims numerous individualities. Generally, it disturbs the accessibility but, on the second hand, it also impacts the rest of the goals of security. Sybil attack is a type of computer network attack where an attacker overrides the reputation of the system by creating a large number of fake identities and using them to gain unparalleled influence (Baza et al. 2022). The detection and prevention techniques for this attack are stated below.

- *Trusted certification*—Every system in the network has a single identity certi-fication which is given by the centralized authority that cannot be alterable or multiple. It is a kind of certification which is a legal document given by central-ized authority to certify the individuality of both the trust and action of the trustee (Avoussoukpo et al. 2021).

- *Trusted devices*—In this approach, a network card is used to provide authenticity of entities and it is mandatory for all of the entities to have a card in a network, it is a kind of certification which is a legal document given by centralized authority (Yao et al. 2019).

Sinkhole Attack

This attack is done inside a system. An invader accommodates a network inside the whole network and inaugurates an attack like packet drop, fake routing update, and modification. To detect and prevent sinkhole attacks, a mechanism is developed that considers the operation of the AODV protocol as well as the behavior of sinkhole attacks. The mechanism is divided into four phases—the initialization phase, storage phase, Iivestigation, and resumption phase (Malik et al. 2022).

Gray Hole Attack

It is a superior case of the blackhole attack. It is very similar to the blackhole attack. The only variance is blackhole attack drops all the data packets while the gray hole attack can or cannot drop the data packets. It does not have a fixed behavior. It often switches its state from infected to normal networks and vice versa (Khan et al. 2021). To protect system from this attack, we can use all the discussed detection and prevention techniques of blackhole attacks as working of both attacks are very similar.

Byzantine Attack

In this attack creating routing loops, forwarding packets through non-existing paths, or dropping attacks are performed in a byzantine attack by an infected network or group of infected networks that work in collusion (Taggu and Marchang 2019).

Jellyfish Attack

The motive of this attack is to make unwanted delays while data packets are being sent. It introduces pauses in forwarding the data packets producing high end-to-end delays. It is known as a Jellyfish attack or GTS attack and Timing attack (Deepika and Saxena 2018). The various detection and prevention techniques for jellyfish attack are as follows.

- *2ACK*—The 2ACK basic technique is based on the idea that a specific two-hop acknowledgment called 2ACK to send by the destination network to the source network just to point out that the data packet was received successfully by the destination network.
- *Credit-based systems*—In that approach token or credit is used by the network, the moment it begins to send its packet in order to encourage successful transmissions (Thapar and Sharma 2020).
- *Reputation-based scheme*—In this system, single networks are capable to detect misbehaving networks (such as CONFIDANT) (Yang et al. 2021).

Wormhole Attack

The data packets are caught by the invader from one place and are tunneled to another place to disorder the routing. Sometimes this attack may also affect the accessibility of the network. If the tunneling mechanism is not applied properly, all the packets may be dropped by the attacker (Prasse and Mieghem 2020). Furthermore, the detection and prevention techniques for wormhole attack are stated.

- *Clustering*—The whole network is partitioned into small clusters (group of networks) containing a cluster head. In a cluster, the number of members (networks) is priory-defined. A cluster head is a leader of the cluster by has the power to transmit the information to the entire membership. There is no communication link between members, it is only done via cluster head (Yoshino et al. 2018).
- *Packet leash*—Two types of leashes are used to detect and prevent wormhole attacks. The first one is a geographical leash, and another is a temporal leash. In a geographical leash, the network sends its location and transmission time before sending the data packet. When the receiver receives the data packets, it calculates the traversal time of packets just to match the information which is sent by the sender. RTT (Round trip time) and time of flight are some methods that come under the geographical leash to detect the attack. On another side, in temporal leashes, the packet is sent with a sending timestamp added by the sender, and the traveling distance of that packet is calculated by the receiver (Gul 2021)
- *Other techniques*—Some other techniques can also be used to prevent this attack such as DS, Network monitoring, and GPS-based wormhole combating technique (directional antenna) (Thanuja et al. 2018).

Eavesdropping

It is the blocking and casting of an eye over data and conversations by the attacker. It disturbs the privacy of the network. It is also termed traffic analysis or sniffing attack. It is theft of data as it is transmitted to a network via a node, smartphone, or another connected device. It uses the opportunity of an unsecured network connection to access data as it is sent or received by its user. It occurs when cybercriminals steal information sent or received by a user through an insecure network. Additionally, by the usage of strong encryption techniques, we can able to mitigate this attack and can protect the system/network (Li et al. 2019).

Disclosure of Information Attack

In this type of attack, the attacker reveals information related to the network topology, confidential data, geographic position of networks, or ideal paths to actual networks in the network. It is also known as an Information leakage attack, it occurs when a website accidentally discloses the sensitive data to its users. Dependent on the framework, websites may reward all kinds of data/information for a potential invader, including data of other users, such as usernames or financial information. It occurs when the request does not adequately protect sensitive information that may eventually be disclosed to the parties who should not have access to it (Li et al. 2021).

Man-In-The-Middle Attack

The attacker sits between the sender and recipient and observes information, while transmission and theft of the essential data/information under this attack. This attack is a common name where the perpetrator puts himself in a conversation between the user and the request listener or pretends to be one of the parties, making it seem like they are exchanging common information. This attack also helps the vicious attacker, without any type of participant you see until it is too late, to break into another person's targeted data and should not be sent at all (Khatod and Manolova 2020).

Replay Attack

It is a passive attack, in this attack, the attacker stored a message or data packet of a network and used the stored message for further communication by controlling and resending them later to access the network and perform impersonate actions (Malik 2019).

ACK Attack

Under this attack, a fake acknowledgment is sent by the hacker to the receiver, intermediate node as well sender to eavesdrop on the network. When a request is sent by the sender then an attacker takes advantage of this to send a fake acknowledgment as a response at the place of receiver or intermediate nodes. Once the fake acknowledgment is received by the sender which is exactly the same as the acknowledgment of the actual receiver hence the sender gets trapped and sends the original data as a response to the fake received acknowledgment (Boche et al. 2021).

Spoofing Attack

When an infected network/node misprints its identification to a genuine node as multiple nodes, resulting in topology changes, delays, change in data, illegal actions, and data losses. An infected system may pretend to be a valid network member after retaining the same IP address of a network member. These are known as spoofing attacks, IP spoofing attacks, and session hijacking attacks (Wu et al. 2020).

Link Spoofing Attack

In This Attack- when a fake path to non-existent network/s is built or a fake updating in the routing table is performed, then routing protocol is directly affected. It's also called a Link spoofing attack, a fabrication attack, or a Global Positioning System (GPS) attack (Huan and Kim 2021). Hence, the detection technique that is used to protect the network from a link spoofing attack is stated.

- *Location information-based detection*—Each network has GPS and a timestamp attached with it by this technique. GPS works with cryptographic methods. In the network, each network must announce its current and actual location information with the help of GPS to other networks so that every network that is present in the network becomes to know the location details of other networks in the network. The distance between two networks that pretended to be neighbors can be verified and false links may be turned down (Wang et al. 2019).

Spear-Phishing Attack

Spear-phishing is also known as email spoofing, in this, the attacker forces the victim to open his or her email to acquire access and retrieve important information. It is a malicious email attack directed at an organization or individual, seeking unauthorized access to sensitive information. It is a direct attempt to steal sensitive information such as account information or financial information from the victim, usually for malicious reasons. This is achieved by obtaining personal information from the victim such as friends, hometown, places they frequently visit, and what they have recently purchased online (Swarnalatha et al. 2021).

Repudiation Attack

Repudiation assault refers to the denial in taking part in communication activity that is the node affected by repudiation attack will continuously deny to make a connection or take part in sharing the data packets by showing them busy (Zhang et al. 2021). The two techniques proposed in the literature for protecting the network against repudiation attacks are—Create Secure Audit Trails (CSAT) and Digital Signatures (DS) (Luo et al. 2018).

Infected Code Injection

Viruses, worms, logic bombs, spyware, adware, and Trojan horses are examples of harmful programming that can target both the operating system and user application as well as the network. It's also known as a malware attack (Bhardwaj et al. 2021). The detection or prevention technique for this attack is as follows.

- *Static code analysis*—It is the most effective method of preventing harmful malware from infecting business systems. Nowadays, leading scanners can rapidly expose infected code such as anti-debugging techniques, steady information, data leakage, time bombs, rootkits, etc. (Liu et al. 2019).

Colluding Mis-Relay Attack

In a colluding mis-relay attack, instead of a single attacker, a group of attackers works together quietly to modify data packets or drop-sending packages to disrupt the network's normal operation. When the attackers drop packets, it has an impact on the network's availability. To detect this attack, an acknowledgment-based approach is used at the receiver's end as well as the sender's end (Abdalzaher et al. 2019).

Selective-Forwarding Attack

The attacker gets all data packets originating from the source and then forwards some of the data packets to the destination node that select randomly, while the remaining data packets are stolen by the attacker so that a malicious action can be performed (Gonzalez and Jung 2019).

Database Hack Attack

In a network, all the activities or data stored in the database must be properly configured but when it is configured appropriately. It can, however, be hacked if it is configured incorrectly (Gonzalez and Jung 2019). Moreover, table 5.2 presents a tabular summary of the whole paper.

Conclusion

The paper discussed and presented the various security issues present in mobile or wireless networks which disturb the normal functions of the network. The mobile nature of the networks makes them even more vulnerable to security attacks like DoS attack, Blackhole attack, Jamming, Flooding attack Sybil attack, Gray hole attack, IP spoofing attack, Rushing attack, Sleep deprivation attack, Wormhole attacks, etc. The paper discussed the various categories of wireless mobile networks, different challenges of wireless mobile networks, security goals, and classification of attacks into different categories on various measures. At last, the paper presented some detection and prevention of attacks as proposed by different researchers. The paper presents a comprehensive survey of the attacks on wireless mobile networks and purposed solutions. In future work, we will design a technique that can able to secure the network from different kinds of attacks.

Table 5.2 Security attacks in wireless mobile network

Attack	Type	Action	Detection and prevention mechanisms	Affects
(Loreti and Bracciale 2020; Ashfaq et al. 2019; Okamura et al. 2019; Nundloll et al. 2020; Bhushan and Sahoo 2020b; Liu and Labeau 2021; Tsiota et al. 2019; Zhang et al. 2020, 2021; Gautam et al. 2019; Bhushan and Gupta 2019; Malik and Gautam et al. 2017; Malik and Gautam 2019; Heo et al. 2018; Zhou et al. 2019; Nishanth and Mujeeb 2020; Kafaie et al. 2018; Sivanesh et al. 2019; Zhou et al. 2020; Baza, et al. 2022; Avoussoukpo et al. 2021; Yao, et al. 2019; Malik et al. 2022; Khan et al. 2021; Taggu and Marchang 2019; Deepika and Saxena 2018; Thapar and Sharma 2020; Yang et al. 2021; Prasse and Mieghem 2020; Yoshino et al. 2018; Gul 2021; Thanuja et al. 2018; Li et al. 2019, 2021; Khatod and Manolova 2020; Malik 2019; Boche et al. 2021; Wu et al. 2020; Huan and Kim 2021; Wang et al. 2019; Swarnalatha et al. 2021; Luo et al. 2018; Bhardwaj et al. 2021; Liu et al. 2019; Abdalzaher et al. 2019; Gonzalez and Jung 2019)	(Loreti and Bracciale 2020; Ashfaq et al. 2019; Okamura et al. 2019; Nundloll et al. 2020; Bhushan and Sahoo 2020b; Liu and Labeau 2021; Tsiota et al. 2019; Zhang et al. 2020, 2021; Gautam et al. 2019; Bhushan et al. 2017; Malik and Gautam 2019; Heo et al. 2018; Zhou et al. 2020; Nishanth and Mujeeb 2021; Kafaie et al. 2018; Sivanesh et al. 2019; Zhou et al. 2020; Baza, et al. 2022; Avoussoukpo et al. 2021; Yao, et al. 2019; Malik et al. 2022; Khan et al. 2021; Taggu and Marchang 2019; Deepika and Saxena 2018; Thapar and Sharma 2020; Yang et al. 2021; Prasse and Mieghem 2020; Yoshino et al. 2018; Gul 2021; Thanuja et al. 2018; Li et al. 2019, 2021; Khatod and Manolova 2020; Malik 2019; Boche et al. 2021; Wu et al. 2020; Huan and Kim 2021; Wang et al. 2019; Swarnalatha et al. 2021; Luo et al. 2018; Liu et al. 2019; Abdalzaher et al. 2019; Gonzalez and Jung 2019)	(Loreti and Bracciale 2020; Ashfaq et al. 2019; Okamura et al. 2019; Nundloll et al. 2020; Bhushan and Sahoo 2020b; Liu and Labeau 2021; Tsiota et al. 2019; Zhang et al. 2020, 2021; Gautam et al. 2019; Bhushan and Gupta 2019; Malik and Gautam et al. 2017; Malik and Gautam 2019; Heo et al. 2018; Zhou et al. 2020; Nishanth and Mujeeb 2020; Kafaie et al. 2018; Sivanesh et al. 2019; Zhou et al. 2020; Baza, et al. 2022; Avoussoukpo et al. 2021; Yao, et al. 2019; Malik et al. 2022; Khan et al. 2021; Taggu and Marchang 2019; Deepika and Saxena 2018; Thapar and Sharma 2020; Yang et al. 2021; Prasse and Mieghem 2020; Yoshino et al. 2018; Gul 2021; Thanuja et al. 2018; Li et al. 2019, 2021; Khatod and Manolova 2020; Malik 2019; Boche et al. 2021; Wu et al. 2020; Huan and Kim 2021; Wang et al. 2019; Swarnalatha et al. 2021; Luo et al. 2018; Bhardwaj et al. 2021; Liu et al. 2019; Abdalzaher et al. 2019; Gonzalez and Jung 2019)	(Loreti and Bracciale 2020; Ashfaq et al. 2019; Okamura et al. 2019; Nundloll et al. 2020; Bhushan and Sahoo 2020b; Liu and Labeau 2021; Tsiota et al. 2019; Zhang et al. 2020, 2021; Gautam et al. 2019; Bhushan and Gupta et al. 2019; Malik and Gupta 2019; Bhushan et al. 2017; Malik and Gautam 2019; Heo et al. 2018; Zhou et al. 2020; Nishanth and Mujeeb 2021; Kafaie et al. 2018; Zhou et al. 2020; Sivanesh et al. 2019; Zhou et al. 2020; Baza, et al. 2022; Avoussoukpo et al. 2021; Yao, et al. 2019; Malik et al. 2022; Khan et al. 2021; Taggu and Marchang 2019; Deepika and Saxena 2018; Thapar and Sharma 2020; Yang et al. 2021; Prasse and Mieghem 2020; Yoshino et al. 2018; Gul 2021; Thanuja et al. 2018; Li et al. 2019, 2021; Khatod and Manolova 2020; Malik 2019; Boche et al. 2021; Wu et al. 2020; Huan and Kim 2021; Wang et al. 2019; Swarnalatha et al. 2021; Luo et al. 2018; Bhardwaj et al. 2021; Liu et al. 2019; Abdalzaher et al. 2019; Gonzalez and Jung 2019)	(Loreti and Bracciale 2020; Ashfaq et al. 2019; Okamura et al. 2019; Bhushan and Sahoo 2020b; Liu and Labeau 2021; Tsiota et al. 2019; Zhang et al. 2020, 2021; Gautam et al. 2019; Bhushan and Gupta 2019; Malik and Gautam et al. 2017; Malik and Gautam 2019; Heo et al. 2018; Zhou et al. 2020; Nishanth and Mujeeb 2020; Kafaie et al. 2018; Sivanesh et al. 2019; Zhou et al. 2020; Baza, et al. 2020; Avoussoukpo et al. 2021; Yao, et al. 2019; Malik et al. 2022; Khan et al. 2021; Taggu and Marchang 2019; Deepika and Saxena 2018; Thapar and Sharma 2020; Yang et al. 2021; Prasse and Mieghem 2020; Yoshino et al. 2018; Gul 2021; Thanuja et al. 2018; Li et al. 2019, 2021; Khatod and Manolova 2020; Malik 2019; Boche et al. 2021; Wu et al. 2020; Huan and Kim 2021; Wang et al. 2019; Swarnalatha et al. 2021; Luo et al. 2018; Bhardwaj et al. 2021; Liu et al. 2019; Abdalzaher et al. 2019; Gonzalez and Jung 2019)

(continued)

Table 5.2 (continued)

Sinkhole attack	Active attack	The attacker sends a fake routing detail to collect all network traffic. The data packages may also be modified by the attacker	SAR	All security goals
Gray hole attack	Active attack	Providing a sham request and then dropping (may or may not) the packets	Ignore the infected network	Availability
Wormhole Attack	Passive attack/Active attack	Infected networks use a high-speed link to connect to the source and act as the genuine neighbors	Clustering, Packet least, Digital signature (Round trip time), Directional antenna	Confidentiality
Jamming	Active attack	Block genuine packets from being sent or received	ARES, UDSSS, UFH	Availability
Blackhole attack and collaborative blackholes attack	Active Attack	The attacker creates routes upon receipt of the route request and the data packet received are not sent to the network	PCBHA, Cryptography-based protection, Protocol modification, and Redundant route method	Availability
Traffic analysis	Passive attack	The attacker closely monitors the route and transmission of data packets. To capture important personal information	Strong encoding methods	Confidentiality
Flooding attack	Active attack	Flood the network with needless requests and comeback messages, consuming network resources	Blacklist the infected network	Availability

(continued)

Table 5.2 (continued)

Rushing attack	Active attack	When an invader (attacker) receives a route request data package, it delivers it out to the entire network until genuine networks	Secure neighbor detection	Availability
Link spoofing attack	Active attack	To modify the routing, placed route to non-existent	Location information-based detection	Authenticity
Colluding mis-relay attack	Active attack	Some modifications were applied on packets and dropped the packets to destroy the normal function of the network	Acknowledgment based approach	Confidentiality and availability
Eavesdropping	Passive attack	Find the confidential information by sniffing the data packets	Strong encryption mechanisms	Confidentiality
Jellyfish attack	Active attack	Introduces unnecessary delay during delivery of packets	Credit-based systems, 2ACK, Reputation-based scheme	Availability
Denial of service	Active attack	The attacker acts as a busy network, denying or dropping packet forwarding	Enticement means based on repudiation	Availability
Infected code Attacks	Active attack	Infected code is inserted into data or over a network	SCA	Integrity
Sybil attack	Active attack	More than one individuality is generated by the attacker for a single network	Trusted certification, trusted devices, received signal strength	Availability
Repudiation Attack	Passive attack	Refusal of sharing	DS and CSAT	Non-repudiation

References

Abdalzaher MS, Samy L, Muta O (2019) Non-zero-sum game-based trust model to enhance wireless sensor networks security for IoT applications. IET Wirel Sens Syst 9(4):218–226. https://doi. org/10.1049/iet-wss.2018.5114

Ashfaq Q, Khan R, Farooq (2019) A comparative analysis of static code analysis tools that check Java Code Adherence to Java Coding Standards. In: 2019 2nd international conference on communication, computing and digital systems (C-CODE), pp 98–103. https://doi.org/10.1109/ C-CODE.2019.8681007

Avoussoukpo CB, Ogunseyi TB, Tchenagnon M (2021) Securing and facilitating communication within opportunistic networks: a holistic survey. IEEE Access 9:55009–55035. https://doi.org/ 10.1109/ACCESS.2021.3071309

Awais M et al (2020) Towards void hole alleviation enhanced geographic and opportunistic routing protocols in harsh underwater WSNs. IEEE Access 8:96592–96605. https://doi.org/10.1109/ ACCESS.2020.2996367

Ayele E, Meratnia N, Havinga PJM (2018) Towards a new opportunistic IoT network architecture for wildlife monitoring system. In: 2018 9th IFIP international conference on new technologies, mobility and security (NTMS), pp 1–5. https://doi.org/10.1109/NTMS.2018.8328721

Baza M et al (2022) Detecting sybil attacks using proofs of work and location in VANETs. IEEE Trans Dependable Secure Comput 19(1):39–53. https://doi.org/10.1109/TDSC.2020.2993769

Bhardwaj V, Kukreja V, Sharma C, Kansal I, Popali R (2021) Reverse engineering: a method for analyzing malicious code behavior. In: 2021 international conference on advances in computing, communication, and control (ICAC3), pp 1–5. https://doi.org/10.1109/ICAC353642.2021.969 7150

Bhushan B, Sahoo G (2020b) Requirements, protocols, and security challenges in wireless sensor networks: an industrial perspective. Handbook Comput Netw Cyber Secur 683–713. https://doi. org/10.1007/978-3-030-22277-2_27

Bhushan B, Sahoo G, Rai AK (2017) Man-in-the-middle attack in wireless and computer networking—a review. In: 2017 3rd international conference on advances in computing, communication & automation (ICACCA) (Fall). https://doi.org/10.1109/icaccaf.2017.8344724

Bhushan B, Sahoo G (2017) Recent advances in attacks, technical challenges, vulnerabilities and their countermeasures in wireless sensor networks. Wireless Pers Commun 98(2):2037–2077. https://doi.org/10.1007/s11277-017-4962-0

Bhushan B, Sahoo G (2018) Routing protocols in wireless sensor networks. Comput Intell Sens Netw Stud Comput Intell. https://doi.org/10.1007/978-3-662-57277-1_10

Bhushan B, Sahoo G (2019a) ISFC-BLS (intelligent and secured fuzzy clustering algorithm using balanced load sub-cluster formation) in WSN environment. Wireless Pers Commun. https://doi. org/10.1007/s11277-019-06948-0

Bhushan B, Sahoo G (2019b) $E^2 SR^2$: An acknowledgement-based mobile sink routing protocol with rechargeable sensors for wireless sensor networks. Wireless Netw 25(5):2697–2721. https:/ /doi.org/10.1007/s11276-019-01988-7

Bhushan B, Sahoo G (2020a) Requirements, protocols, and security challenges in wireless sensor networks: an industrial perspective. Handbook Comput Netw Cyber Secur. https://doi.org/10. 1007/978-3-030-22277-2_27

Boche H, Schaefer RF, Vincent Poor H (2021) Algorithmic detection of adversarial attacks on message transmission and ACK/NACK feedback. In: ICC 2021—IEEE international conference on communications, pp 1–6. https://doi.org/10.1109/ICC42927.2021.9500592

Cuka M, Elmazi D, Bylykbashi K, Spaho E, Ikeda M, Barolli L (2018) A fuzzy-based system for selection of IoT devices in opportunistic networks considering IoT device storage, waiting time and node centrality parameters. In: 2018 IEEE 32nd international conference on advanced information networking and applications (AINA), pp 710–716. https://doi.org/10.1109/AINA. 2018.00107

Deepika D, Saxena S (2018) Performance evaluation of AODV with self-cooperative trust scheme using jellyfish delay variance attack. In: 2018 second international conference on intelligent computing and control systems (ICICCS), pp 1191–1196. https://doi.org/10.1109/ICCONS.2018.8662962

Ding S, He X, Wang J, Liu J (2018) Pre-decoding recovery mechanism for network coding opportunistic routing in delay tolerant networks. IEEE Access 6:14130–14140. https://doi.org/10.1109/ACCESS.2018.2813382

Edirisinghe S, Ranaweera C, Lim C, Nirmalathas A, Wong E (2021) Universal optical network architecture for future wireless LANs [Invited]. J Opt Commun Netw 13(9):D93–D102. https://doi.org/10.1364/JOCN.426215

Elhattab M, Elmesalawy MM, Ibrahim II (2017) Opportunistic device association for heterogeneous cellular networks with H_2H/IoT co-existence under QoS guarantee. IEEE Internet Things J 4(5):1360–1369. https://doi.org/10.1109/JIOT.2017.2702199

Fernando N, Loke SW, Avazpour I, Chen F-F, Abkenar AB, Ibrahim A (2019) Opportunistic fog for IoT-challenges and opportunities. IEEE Internet Things J 6(5):8897–8910. https://doi.org/10.1109/JIOT.2019.2924182

Fortino G, Russo W, Savaglio C, Viroli M, Zhou M (2018) Opportunistic cyberphysical services—a novel paradigm for the future Internet of Things. In: 2018 IEEE 4th world forum on internet of things (WF-IoT), pp 488–492. https://doi.org/10.1109/WF-IoT.2018.8355174

Gautam S, Malik A, Singh N, Kumar S (2019) Recent advances and countermeasures against various attacks in IoT environment. In: 2019 2nd international conference on signal processing and communication (ICSPC). https://doi.org/10.1109/icspc46172.2019.8976527

Gonzalez C, Jung G (2019) Database SQL injection security problem handling with examples. In: 2019 international conference on computational science and computational intelligence (CSCI), pp 145–149. https://doi.org/10.1109/CSCI49370.2019.00031

Guan Y, Ge X (2018) Distributed attack detection and secure estimation of networked cyber-physical systems against false data injection attacks and jamming attacks. IEEE Trans Signal Inform Process Netw 4(1):48–59. https://doi.org/10.1109/TSIPN.2017.2749959

Gul O (2021) Near-optimal opportunistic spectrum access in cognitive radio networks in the 5G and IoT era. In: 2021 IEEE 46th conference on local computer networks (LCN), pp 403–406. https://doi.org/10.1109/LCN52139.2021.9525001

Han C, Chen L, Wang W (2019) Compressive sensing in wireless powered network: regarding transmission as measurement. IEEE Wireless Commun Lett 8(6):1709–1712. https://doi.org/10.1109/LWC.2019.2938517

Heo J, Kim J, Paek J, Bahk S (2018) Mitigating stealthy jamming attacks in low-power and lossy wireless networks. J Commun Netw 20(2):219–230. https://doi.org/10.1109/JCN.2018.000028

Huan X, Kim KS, Zhang J (2021) NISA: node identification and spoofing attack detection based on clock features and radio information for wireless sensor networks. IEEE Trans Commun 69(7):4691–4703. https://doi.org/10.1109/TCOMM.2021.3071448

Kafaie S, Chen Y, Dobre OA, Ahmed MH (2018) Joint inter-flow network coding and opportunistic routing in multi-hop wireless mesh networks: a comprehensive survey. IEEE Commun Surv Tutor 20(2):1014–1035. https://doi.org/10.1109/COMST.2018.2796101

Khan AU, Chawhan MD, Mushrif MM Neole B (2021) Performance analysis of Adhoc on-demand distance vector protocol under the influence of black-hole, gray-hole and worm-hole attacks in mobile adhoc network. In: 2021 5th international conference on intelligent computing and control systems (ICICCS), pp. 238–243. https://doi.org/10.1109/ICICCS51141.2021.9432072

Khatod V, Manolova A (2020) Effects of man in the middle (MITM) attack on bit error rate of bluetooth system. In: 2020 joint international conference on digital arts, media and technology with ECTI northern section conference on electrical, electronics, computer and telecommunications engineering (ECTI DAMT & NCON), pp 153–157. https://doi.org/10.1109/ECTIDAMTNCON48261.2020.9090721

Kibria M, Nguyen K, Villardi G, Zhao O, Ishizu K, Kojima F (2018) Big data analytics, machine learning, and artificial intelligence in next-generation wireless networks. IEEE Access 6:32328–32338. https://doi.org/10.1109/ACCESS.2018.2837692

Kibria MG, Nguyen K, Villardi GP, Liao W, Ishizu K, Kojima F (2018) A stochastic geometry analysis of multiconnectivity in heterogeneous wireless networks. In: IEEE Trans Veh Technol 67(10):9734–9746. https://doi.org/10.1109/TVT.2018.2863280

Lee H, Ke K (2018) Monitoring of large-area IoT sensors using a LoRa wireless mesh network system: design and evaluation. IEEE Trans Instrum Meas 67(9):2177–2187. https://doi.org/10.1109/TIM.2018.2814082

Li B, Yao Y, Zhang H, Lv Y (2019) Energy efficiency of proactive cooperative eavesdropping over multiple suspicious communication links. IEEE Trans Veh Technol 68(1):420–430. https://doi.org/10.1109/TVT.2018.2880768

Li C et al (2021) Mimosa: protecting private keys against memory disclosure attacks using hardware transactional memory. IEEE Trans Dependable Secure Comput 18(3):1196–1213. https://doi.org/10.1109/TDSC.2019.2897666

Liang W et al (2019) WIA-FA and its applications to digital factory: a wireless network solution for factory automation. Proc IEEE 107(6):1053–1073. https://doi.org/10.1109/JPROC.2019.2897627

Lin H, Kim KS, Shin W (2020) Interference-aware opportunistic random access in dense IoT networks. IEEE Access 8:93472–93486. https://doi.org/10.1109/ACCESS.2020.2996221

Liu W (2021) Harmonious wireless networks: perspective of interference management. Intell Converged Netw 2(3):198–204. https://doi.org/10.23919/ICN.2021.0013

Liu J, Labeau F (2021) Detection of false data injection attacks in industrial wireless sensor networks exploiting network numerical sparsity. IEEE Trans Signal Inform Process Netw 7:676–688. https://doi.org/10.1109/TSIPN.2021.3122289

Liu J, Zhao Z, Ji J, Hu M (2020) Research and application of wireless sensor network technology in power transmission and distribution system. Intell Converged Netw 1(2):199–220. https://doi.org/10.23919/ICN.2020.0016

Liu Z, Liu W, Ma Q, Liu G, Zhang L, Fang L, Sheng VS (2019) Security cooperation model based on topology control and time synchronization for wireless sensor networks. J Commun Netw 21(5):469–480. https://doi.org/10.1109/jcn.2019.000041

Lohachab A, Jangra A (2019) Opportunistic Internet of Things (IoT)—demystifying the effective possibilities of opportunisitc networks towards IoT. In: 2019 6th international conference on signal processing and integrated networks (SPIN), pp 1100–1105. https://doi.org/10.1109/SPIN.2019.8711621

Loreti P, Bracciale L (2020) Optimized neighbor discovery for opportunistic networks of energy constrained IoT devices. IEEE Trans Mobile Comput 19(6):1387–1400. https://doi.org/10.1109/TMC.2019.2908402

Luo H, Zhang G, Liu Y, Das SK (2018) Adaptive routing in wireless sensor networks. Adapt Cross Layer Design Wireless Netw 263–299. https://doi.org/10.1201/9781315219813-9

Lyu C, Zhang X, Liu X, Chi C-H (2019) Selective authentication based geographic opportunistic routing in wireless sensor networks for internet of things against DoS attacks. IEEE Access 7:31068–31082. https://doi.org/10.1109/ACCESS.2019.2902843

Malik KM, Malik H, Baumann R (2019) Towards vulnerability analysis of voice-driven interfaces and countermeasures for replay attacks. In: 2019 IEEE conference on multimedia information processing and retrieval (MIPR), pp 523–528. https://doi.org/10.1109/MIPR.2019.00106

Malik A, Gautam S (2019) Comparative analysis of AODV routing protocol vs. nodes in manet. J Emerg Technol Innov Res (JETIR) 6(3):53–61. www.jetir.org/papers/JETIR1903610.pdf

Malik A, Gupta V (2019) Comprehensive survey on blackhole attack with various detection/prevention techniques in Ad-hoc network. Int J Appl Eng Res 14(8):2009–2017. www.ripublication.com/ijaer19/ijaerv14n8_35.pdf

Malik A, Gautam S, Khatoon N, Sharma N, Kaushik I, Kumar S (2020) Analysis of black-hole attack with its mitigation techniques in ad-hoc network. In: Sagayam K, Bhushan B, Andrushia

A, Albuquerque VC (eds) Deep learning strategies for security enhancement in wireless sensor networks. IGI Global, pp 211–232. https://doi.org/10.4018/978-1-7998-5068-7.ch011

Malik A, Bhushan B, Kumar A (2022) Association rule-based routing protocol for opportunistic network. In: Saini HS, Singh RK, Tariq Beg M, Mulaveesala R, Mahmood MR (eds) Innovations in electronics and communication engineering. lecture notes in networks and systems, vol 355. Springer, Singapore. https://doi.org/10.1007/978-981-16-8512-5_42

Moin A, Thielens A, Araujo A, Sangiovanni-Vincentelli A, Rabaey JS (2021) Adaptive body area networks using kinematics and biosignals. IEEE J Biomed Health Inform 25(3):623–633. https://doi.org/10.1109/JBHI.2020.3003924

Nishanth N, Mujeeb A (2021) Modeling and detection of flooding-based denial-of-service attack in wireless ad hoc network using bayesian inference. IEEE Syst J 15(1):17–26. https://doi.org/10.1109/JSYST.2020.2984797

Nundloll V, Elkhatib Y, Elhabbash A, Blair GS (2020) An ontological framework for opportunistic composition of IoT systems. In: 2020 IEEE international conference on informatics, IoT, and enabling technologies (ICIoT), pp 614–621. https://doi.org/10.1109/ICIoT48696.2020.9089467

Nurlan Z, Zhukabayeva T, Othman M, Adamova A, Zhakiyev N (2022) Wireless sensor network as a mesh: vision and challenges. IEEE Access 10:46–67. https://doi.org/10.1109/ACCESS.2021.3137341

Okamura Y, Yamamoto R, Ohzahata S, Kato T (2019) Opportunistic routing for heterogeneous IoT networks. In: 2019 IEEE international conference on consumer electronics—Taiwan (ICCE-TW), pp 1–2. https://doi.org/10.1109/ICCE-TW46550.2019.8991759

Petrov V et al (2018) Vehicle-based relay assistance for opportunistic crowdsensing over narrowband IoT (NB-IoT). IEEE Internet Things J 5(5):3710–3723. https://doi.org/10.1109/JIOT.2017.2670363

Prasse B, Van Mieghem P (2020) Network reconstruction and prediction of epidemic outbreaks for general group-based compartmental epidemic models. IEEE Trans Netw Sci Eng 7(4):2755–2764. https://doi.org/10.1109/TNSE.2020.2987771

Samanta A, Misra S (2018) Energy-efficient and distributed network management cost minimization in opportunistic wireless body area networks. In: IEEE Trans Mob Comput 17(2):376–389. https://doi.org/10.1109/TMC.2017.2708713

Shnaiwer YN, Sorour S, Al-Naffouri TY, Al-Ghadhban DN (2019) Opportunistic network coding-assisted cloud offloading in heterogeneous fog radio access networks. IEEE Access 7:56147–56162. https://doi.org/10.1109/ACCESS.2019.2913860

Sinha P, Jha VK, Rai AK, Bhushan B (2017) Security vulnerabilities, attacks and countermeasures in wireless sensor networks at various layers of OSI reference model—a survey. In: 2017 international conference on signal processing and communication (ICSPC). https://doi.org/10.1109/cspc.2017.8305855

Sivanesh S, Sarma Dhulipala VR (2019) Comparitive analysis of blackhole and rushing attack in MANET. In: 2019 TEQIP III sponsored international conference on microwave integrated circuits, photonics and wireless networks (IMICPW), pp 495–499. https://doi.org/10.1109/IMICPW.2019.8933192

Swarnalatha K, Ramchandra K, Ansari K, Ojha L, Sharma S (2021) Real-time threat intelligence-block phising attacks. In: 2021 IEEE international conference on computation system and information technology for sustainable solutions (CSITSS), pp 1–6. https://doi.org/10.1109/CSITSS54238.2021.9683237

Taggu A, Marchang N (2019) Random-byzantine attack mitigation in cognitive radio networks using a multi-hidden markov model system. In: 2019 international conference on electrical and computing technologies and applications (ICECTA), pp 1–5. https://doi.org/10.1109/ICECTA48151.2019.8959766

Thanuja R, Ram ES, Umamakeswari A (2018) A linear time approach to detect wormhole tunnels in mobile adhoc networks using 3PAT and transmission radius (3PATw). In: 2018 2nd international conference on inventive systems and control (ICISC), pp 837–843. https://doi.org/10.1109/ICISC.2018.8398917

Thapar S, Sharma SK (2020) Direct trust-based detection algorithm for preventing jellyfish attack in MANET. In: 2020 4th international conference on electronics, communication and aerospace technology (ICECA), pp 749–753. https://doi.org/10.1109/ICECA49313.2020.9297601

Tsiota A, Xenakis D, Passas N, Merakos L (2019) On jamming and black hole attacks in heterogeneous wireless networks. IEEE Trans Veh Technol 68(11):10761–10774. https://doi.org/10.1109/TVT.2019.2938405

Tzanakaki A, Anastasopoulos M. P, Simeonidou D (2019) Converged optical, wireless, and data center network infrastructures for 5G services. J Opt Commun Netw 11(2):A111–A122. https://doi.org/10.1364/JOCN.11.00A111

Wadii J, Rim H, Ridha B (2019) Detecting and preventing Sybil attacks in wireless sensor networks. In: 2019 IEEE 19th mediterranean microwave symposium (MMS), pp 1–5. https://doi.org/10.1109/MMS48040.2019.9157321

Wang F, Xu G, Xu G (2019) A provably secure anonymous biometrics-based authentication scheme for wireless sensor networks using chaotic map. IEEE Access 7:101596–101608. https://doi.org/10.1109/access.2019.2930542

Wu Z, Zhang Y, Yang Y, Liang C, Liu R (2020) Spoofing and anti-spoofing technologies of global navigation satellite system: a survey. IEEE Access 8:165444–165496. https://doi.org/10.1109/ACCESS.2020.3022294

Yang M, Jia L, Xie W, Gao T (2021) Research on risk assessment model of epidemic diseases in a certain region based on markov chain and AHP. IEEE Access 9:75826–75839. https://doi.org/10.1109/ACCESS.2021.3081720

Yang L, Wen C (2021) Optimal jamming attack system against remote state estimation in wireless network control systems. IEEE Access 9:51679–51688. https://doi.org/10.1109/ACCESS.2020.3046483

Yao Y et al (2019) Multi-channel based sybil attack detection in vehicular ad hoc networks using RSSI. IEEE Trans Mob Comput 18(2):362–375. https://doi.org/10.1109/TMC.2018.2833849

Yoshino Y, Nakasaki S, Ikeda M, Barolli L (2018) An integrated message suppression controller with epidemic and MaxProp protocols—performance evaluation for VDTNs. In: 2018 32nd international conference on advanced information networking and applications workshops (WAINA), pp 159–163. https://doi.org/10.1109/WAINA.2018.00080

Zhang J, Gao K, Yang YR, Bi J (2020) Prophet-toward fast, error-tolerant model-based throughput prediction for reactive flows in DC networks. IEEE/ACM Trans Netw 28(6):2475–2488. https://doi.org/10.1109/TNET.2020.3016838

Zhang R, Zhang L, Choo K, Chen T (2021) Dynamic authenticated asymmetric group key agreement with sender non-repudiation and privacy for group-oriented applications. IEEE Trans Depend Secure Comput. https://doi.org/10.1109/TDSC.2021.3138445

Zhao S, Lu Z, Wang C (2019) Measurement integrity attacks against network tomography—feasibility and defense. IEEE Trans Depend Secure Comput. https://doi.org/10.1109/TDSC.2019.2958934

Zhao D, Lun G, Xue R (2021) Coding-aware opportunistic routing for sparse underwater wireless sensor networks. IEEE Access 9:50170–50187. https://doi.org/10.1109/ACCESS.2021.3069077

Zhou H, Chen X, He S, Zhu C, Leung VCM (2020) Freshness-aware seed selection for offloading cellular traffic through opportunistic wireless mobile networks. IEEE Trans Wireless Commun 19(4):2658–2669. https://doi.org/10.1109/TWC.2020.2967658

Zhou P, Yan Q, Wang K, Xu Z, Ji S, Bian K (2020) Jamsa: a utility optimal contextual online learning framework for anti-jamming wireless scheduling under reactive jamming attack. IEEE Trans Netw Sci Eng 7(3):1862–1878. https://doi.org/10.1109/TNSE.2019.2955464

Chapter 6
Utilities of 5G Communication Technologies for Promoting Advancement in Agriculture 4.0: Recent Trends, Research Issues and Review of Literature

Parijata Majumdar, Diptendu Bhattacharya, and Sanjoy Mitra

Abstract The ultrafast 5G network will play a significant role in the farming industry over the upcoming couple of years, serving to boost crop yield and quality while requiring minimal labour. Farmers will be more informed to make smart decisions regarding irrigation by using smart and precision farming. The introduction of 5G will significantly alter the farming characteristics and agriculture practices in this era of Agriculture 4.0. 5G network's IoT-based cloud computing service offers smart farming solutions that are both flexible and resourceful. This will permit the seamless operation of various unmanned agricultural devices during ploughing, sowing seed and managing phases of crop farming, resulting in secure, dependable, environment-friendly and energy-efficient operations, as well as the creation of unmanned farms. This paper examines the need for and role of smart and precision farming in the agricultural sector incorporating 5G applications in precision farming in the present era of Agriculture 4.0, such as real-time monitoring, data analytics, cloud repositories, virtual consultation and predictive maintenance and also discusses upcoming opportunities. 5G-based IoT solutions focusing towards Ultra-Reliable Low Latency Communication (URLLC) like automated control and self-driven vehicles to support rapid response times and higher dependability will diminish communication delays in time-sensitive agriculture applications and non-public networks to allocate part of frequency spectrum on demand, network slicing alternatives are also discussed here.

Keywords Smart agriculture · 5G · Precision farming · IoT · URLLC · Non-public networks · Agriculture 4.0

P. Majumdar · D. Bhattacharya
Department of Computer Science and Engineering, National Institute of Technology, Agartala, Jirania, Agartala, Tripura 799046, India

S. Mitra (✉)
Department of Computer Science and Engineering, Tripura Institute of Technology, Agartala, Narsingarh, Agartala, Tripura 799009, India
e-mail: mail.smitra@gmail.com

B. Bhushan et al. (eds.), *5G and Beyond*, Springer Tracts in Electrical and Electronics Engineering, https://doi.org/10.1007/978-981-99-3668-7_6

Introduction

Agriculture is most countries' principal factor of income and plays a critical role in their economic development. Different styles of agriculture are performed all over the world, with the goal of healthy food production to feed the world's population. Agriculture is a developing country's primary source of revenue. Modern farming began around the eighteenth century, during the British Agricultural Revolution, when numerous improvements to farming were accomplished in a little time, resulting in a productive increase in yield and more efficient ways. Food production must be raised swiftly to keep up with the rapid growth of the global population. Traditional farming practices result in irregular output, resource overuse and trash creation that is unchecked. Farmers will require more advanced technology to meet these demands, which will allow them to produce more while requiring manually less labour. This is when automation enters the picture. In this context, the introduction of 5G communications provides a potentially disruptive factor. In terms of communication, 5G's enhanced data rate, less end-to-end latency and wider coverage have the ability to meet even the ever-increasing demand of IoT end users (Heidari et al. 2021). Its ability to accommodate a massive number of devices allows for the creation of a truly global Internet of Things. Furthermore, as it focuses on the integration of access methods, 5G could serve as a unified interconnection framework, allowing "things" to connect to the Internet seamlessly. The purpose of this study is to examine in depth the potential of 5G for the Internet of Things for reaping full benefits of smart farming. With machine-to-machine services, however, the adoption of 5G will assist speed up the complete procedure. The real-time data transfer capabilities of 5G can aid in the efficient operation of these technologies, allowing for quick, reliable, data-driven and real-time decision-making. Applications of 5G in agriculture include AI-enhanced machinery, drone sprayers, accurate harvest estimation, effective irrigation and livestock tracking as well as management (Tang et al. 2021).

Challenges Faced by Existing Network Technologies

The 4G networking paradigm faces significant restrictions that may restrain the technology from reaching its full prospective in the agriculture industry. One of the most significant limits is the operating area. Remote locations are not covered by existing wireless networks. Due to variable data rates, resource allotment, handoff and channel state, issues between heterogeneous networks facilitating QoS (Tong et al. 2019) networks poses a considerable challenge. The large number of antennae and transmitters causes poor battery life in mobile nodes. Because many current agricultural gadgets, such as drones and agribots are powered by battery, these cannot be incorporated in remote crop fields for long periods of time. Several devices are continuously rising in number, requiring greater intelligence, scalability, processing capacity, secure communication, etc., to conduct profound computational operations. Ultralow latency along with high connection is essential for IoT devices to deliver quick performance and low prices. Because it only permits IP-based packet switching connectivity, the existing 4G network (LTE) cannot provide such functionalities (Martin et al. 2011). Transitioning to next-generation 5G will address these limitations of previous generations.

Motivation of the Work

Farmers can expect the following benefits in the near future as a result of 5G's accessible capabilities like faster communication where 5G will offer a data speed of up to 10 Gbps, which is 100 times faster than its predecessor, 4G. Real-time communication between stakeholders will be facilitated by faster speeds and much lower latency. Machine-to-machine data transfer will facilitate direct information transfer between 5G-enabled equipment without the need for human intervention that can improve agricultural processes' speed and efficiency. 5G will reduce the costs where farm owners can significantly boost revenue by requiring less agri-inputs, labour and other resources. It's possible that 5G will take longer to fully expand out and cover all distant locations. When it happens, though, this new agricultural technology will cut labour requirements while bringing automation. The 5G network minimizes the per unit time for data transfer, supports larger data rate, ensure secure and dependable connections required for efficiency of time-sensitive applications like irrigation scheduling which is dependent on real-time prediction of droughts and floods.

Existing Problems and Contribution

As 5G coverage spreads over the world, it will provide extensive coverage for numerous new applications. The 5G energy network uses LTE for both machines (LTE-M) and narrowband IoT technologies (Heidari et al. 2021). To meet service requirements, the 5G will form an extensively dense network. The high density creates a mobility organization difficulty and causes larger energy utilization. Energy harvesting methods help to deploy a large number of wireless sensors in both urban and rural locations. Intelligence is also added that will result in substantial energy savings during transmission. The ideal parameter settings to minimize energy loss can be attained via advanced analytics of network data (Tang et al. 2021). Furthermore, rather than the traditional reactive approach to energy management, a proactive method may be established. For a stable 5G future, we studied energy management and harvesting solutions for IoT devices. Based on energy harvesting schemes, IoT devices will be power conserving and management strategies at the circuit, system and device levels that will be implemented in the near future.

Scope of 5G-Enabled Networks in AgriIoT

5G provides a diverse set of capabilities to fulfil the needs of eMBB, mMTC and URLLC services. The notion of "network slicing" allows for the operation of many dedicated networks on a single platform. The flexibility of 5G specifications to dedicate a dedicated slice of the network for certain application areas will also enable new distant and mobile IoT applications, unlike earlier generations of mobile networks. In 5G, network slicing allows for various connection segments to be used to implement one or more use cases. In 5G, network slicing allows for various connection segments to be used to implement one or more use cases. The 5G network will be connected by a large number of IoT nodes. This will make ultra-reliable or ultra-low rate of time consumed for data transfer and communications possible (GSMA 2019). Edge computing and Artificial Intelligence at the edge, incorporating 5G, will perform novel augmented reality (AR), time-critical industrial IoT applications, virtual reality (VR), etc. The VR eye tracking interphase, that shows the user's focal point and supplies images of high resolution at the point of focal plane, is an application area of VR-IoT. To save energy, reduced resolution is used elsewhere. 5G can provide accommodation to millions of 5G devices in a square kilometres because of its capability for "large machine type communication." 5G technology is ideally suited to meet the reduced timing requirement to perform data transfer and dependability needs of crucial IoT equipments. The ability to deliver services for important and dependable systems, like agriculture monitoring based on real-time weather parameter sensing is critical to 5G with cellular networks. URLLC is a primary characteristic of 5G and one of its main foundation stones. URLLC IoT is utilized to better control traffic and prevent congestion while providing users with early warnings (Foerster et al. 2020). The majority of 5G IoT devices will be power-driven entirely by batteries. So, extending the network lifetime of IoT devices requires an energy-conserving plan like modifying the frequency of sensing and data collection. Low-cost ubiquitous computing is a major IoT enabler to ensure energy efficiency. The size of unit computing has shrunk over the last five decades, and this, together with new 5G network technologies like massive MIMO and millimetre-wave transmission, can help IoT realize its full potential. The reduction in accessible energy is the shortcoming of ubiquitous computing. Over time, the size of a single processing unit has shrunk dramatically. In the meantime, battery and energy storage technologies are progressing very slowly (Sen et al. 2018). As a result, the IoT nodes have a limited quantity of energy available. The sensors have a battery size which is modest. Because battery life is typically significantly shorter than electronic lifespan, developing an energy harvesting-based system that can achieve net-zero energy for sensor nodes will be a superior strategy.

Energy Consumption Management Methods in 5G Networks

Energy Harvesting technologies will be critical in extending network lifetime by providing a controlled manner of recharging batteries. It is a potential advancement which does not lessen the power requirement of devices but rather makes it easier to convert to self-powering in the event of a power outage. The 5G energy harvesting system can be divided into energy sources, energy conversion methods, energy harvesting phases, harvesting models, etc. Transducers can convert energy harvesting sources to usable electrical power that performs power conversion and energy storage capacity, such as supercapacitors or batteries, to store converted power. There are a variety of ambient energy sources available in the environment (Lee et al. 2018) like Solar, electromagnetic, mechanical vibrations or kinetic energy sources, thermal etc. are all examples of renewable energy sources. In general, electromagnetic radiation can be used to generate energy non-radiatively and radiative methods. The non-radiative method has a higher competence but less range, but the radiative method has an inferior efficiency but a comparatively better range. As a result, in the 5G and beyond future, effective utilization of these approaches is critical. Solar energy has already been shown to be a practicable source of significant power generation. Generators convert the mechanical energy into useful electrical power from there. Using a thermoelectric generator, energy is quickly extracted using thermal energy harvesting. Kinetic energy sources could be a critical enabler in the development of 5G-enabled components. With its speedy connectivity, 5G wireless capabilities may open up new options for agriculture-related data analytics (Yuan et al. 2020). If a generic power management framework can serve all 5G IoT devices with energy harvesting capabilities, the field's extendibility is limited in which intelligent power management is required. The ability to stop a system while it is not in use is arguably the most efficient way to save energy. Sensors, power management, communication transceivers and energy harvesters make up the sensing nodes of IoT devices. The most practical way for reducing the system's energy usage is to use the sleep mode to turn the BS on and off. This guarantees that the captured energy is used efficiently. Because their subsystems stay active in the idle state, 5G IoT devices may then also exhaust a large amount of energy when they are not transmitting data/sensing. Meanwhile, one of the aspects of next-generation networks is energy proportionality with traffic. Each BS sleep level is defined by a transition latency threshold. The sum of the part's reactivation and deactivation latencies is the subcomponent transition rate of data transferred per unit time. When the BS is in sleep state, a quick activation time is used to keep subcomponents with a long transition rate of data transfer per unit time always active. In most current systems, this strategy is employed to preserve QoS. Utilizing the IoT device hardware capability, the network's flexibility and the measurement periodicity acquired by the device, the sleep state can achieve noteworthy energy savings. The shortest BS sleep period is in sleep mode one (Debaillie et al. 2015). A BS that is in sleep mode 1 is still active and can receive data. When a BS is not transmitting data, it automatically switches to this sleep state. Sleep mode two denotes a middle-state sleep situation in which more subparts are deactivated, and it equates to 1 ms. Finally, the BS is in standby mode in sleep mode four, which lasts for a minimum of 1 s. In sleep mode 4, the BS is disabled, although it can be reactivated. The data transfer of IoT components can be scheduled with the four separate BS sleep modes to ensure that the energy harvested is used while still meeting QoS. For the 5G frame structure, five distinct numerologies have been established. Because of its high power density and efficiency, Lithium batteries can provide a long battery life. Massive IoT applications necessitate miniaturized and autonomous devices, restraining power management and energy storage ability. Non-rechargeable batteries will also be limited in their usage as a key energy source for vast IoT applications due to frequent replacement, environmental consequences and a shortage of energy sources. Another battery technology is solid-state thin film, which has a high energy concentration but low power density. Because of properties like bendability and manufacture in IC packages, these batteries enable significant size and cost reduction. Nowadays, supercapacitors are used in place of rechargeable batteries as it has an unconstrained charge–discharge cycle (Somov and Giaffreda 2015). Because batteries can be moulded into a variety of shapes and sizes, they remain a viable option for large-scale IoT deployments that require extremely low power consumption and a 10-year lifespan. Integrating rechargeable batteries with energy-conserving approaches is crucial to extend the lifespan of the 5G-enabled devices by recharging the batteries.

Application Areas of 5G in Smart Farming

Data Aggregation

For centralized data aggregation in large farming operations, 5G technology holds a lot of potential. To aggregate data from micro-monitored crop management systems, a large corporate farm may construct a private 5G network. These systems incorporate soil moisture sensor density that is hundreds of times greater than what is currently supported by available technologies. This network could allow for a more efficient real-time monitoring system, complete with triggers for limiting irrigation and other crop support systems (Xu et al. 2017).

Predictive Analytics

Large industrial farms can better utilize predictive analytics thanks to 5G technology, which enables data aggregation. Analytics software develops models and forecasts based on past and current data on circumstances (e.g. soil moisture and pesticide use) to assist farmers in making decisions. Analytics will become more exact as 5G enables denser real-time data, maximizing farm production and efficiency (Sevgican et al. 2020).

Drone Operations

Drones are increasingly being used by farmers to check their crops. Drones are less expensive than driving tractors through fields, and they provide more precise data on crop damage and other aspects. Drones will be able to collect higher-quality video data and transmit it faster thanks to 5G's high-bandwidth technology. These high-speed data transfer capabilities will allow for the development of AI drone technology and real-time reports (Tang et al. 2021).

Animal Tracking and Real-Time Monitoring

Animal monitoring sensors will most likely remain connected through Wi-Fi, Bluetooth or LTE LPWAN Until Rel 17 increases the practicality of 5G low-power and denser sensor networks. Large concentrated farms, where 5G infrastructure can be installed across a small area (e.g. a chicken farm) and individual animals may be tracked, are an exception. Herd management sensors, such as smart collars and ear tags, have been developed by agricultural technology developers to track an animal's position and health. An automated remedial action can be triggered based on any variation in these variables in order to preserve the typical circumstances for crop yield. Sensor data obtained for agriculture practices are in various forms depending on the precision and compatibility will necessitate the use of the relevant interfaces. To cover minimal or maximal distances, communication protocols are very important in IoT-based smart irrigation practices. Short ranges are covered by ZigBee or Wi-Fi, whereas to cover long ranges LoRaWAN, LPWAN protocols and Bluetooth are used. Narrowband IoT and long-term evolution of machine-type communications (GSMA 2019) are paving the way for 5G integration in the future and will have a significant impact on smart farming in the next years. The sensors must have maximal-range communication and should be energy-saving (Yao and Bian 2019). As a result, the sensors' lifetime is significantly extended by transferring data at reduced energy and eliminating data repetition. 5G NR improves network energy performance and decreases interference by allowing adaptive bandwidth switching from lower to higher bandwidth, while interworking and LTE coexistence allow existing cellular networks to be used while still accommodating future evolution.

Autonomous Agriculture Vehicles

Farm tools will benefit from the development of autonomous vehicle technology in other industries. Tractors with onboard computers already allow operators to regulate minute details of farming tasks. Farm equipment that is self-driving will improve, allowing farmers to have more flexibility and efficiency while also saving money on labour. IoT sensor benefits can also be reaped by trucks used for crop transportation. These sensors can monitor cargo temperature and inform you if it gets too hot or cold. High-latency technologies like LPWAN will likely continue to be used by small mobile sensors like asset trackers. 5G will allow autonomous vehicles to send and receive larger, ultra-low-latency data streams, including videos using more powerful onboard computers (Tang et al. 2021).

Weather Stations

Farming operations are at the mercy of the weather. Large sections of crops can be lost due to illnesses and damage that can be avoided. Farmers can tackle this problem by using connected weather stations in the field to provide data on agricultural conditions. The InField monitoring system, designed by AMA Instruments, is one example. InField monitors soil moisture and texture, air temperature, wind speed and exposure to the sun. Weather stations in remote locations will very certainly continue to use LPWAN connectivity in the near future. 5G will help them because it will allow for more data-dense observation and edge computing. Smart farming will continue to grow as the cellular-connected world switches to 5G. Farmers will be able to make better decisions based on data and predictive analytics, resulting in increased productivity and efficiency (Tang et al. 2021).

5G Enabled Components in Agriculture

To implement seamless farming practices 5G enabled the use of a number of components as summarized below.

Drones and Unmanned Aerial Vehicles (UAVs)

UAVs can boost crop yields, save time and maximize long-term performance. These drones can be utilized for a variety of purposes. Both aircraft and ground-based missions are possible. Drones are helpful for doing quick and effective livestock monitoring (Vayssade et al. 2019). Farmers may fly a drone across a long distance

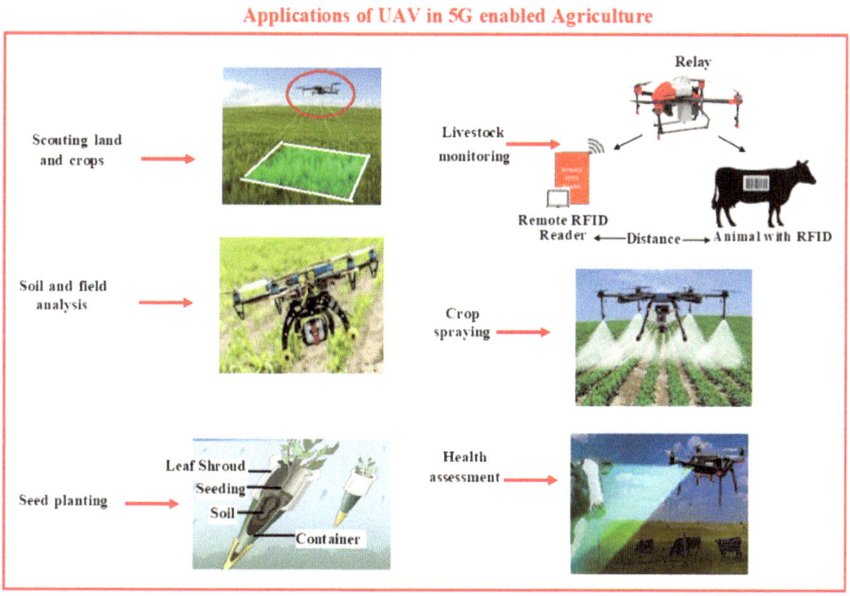

Fig. 6.1   Applications of unmanned aerial vehicles (UAVs) in the field of agriculture using 5G

using 5G technology (Faraci et al. 2018). In comparison to previous-generation mobile networks, the 5G network allows farmers to get real-time data as well as other critical sensory data faster. Drones do not utilize a lot of processing power, and all the data can be transferred to the cloud. Multiple drones (Razaak et al. 2019) can interact with one another to provide coordinated autonomous flight to perform several tasks with least energy expenditure, allowing for extended sensing time and economic operations. For decision-making, a large volume of data can be kept and processed. Drones can fly to supply agricultural products using 5G network's vast network coverage and steady connection. The 5G cellular network combines with drone traffic supervision technologies to improve operations' high-quality connectivity. Since, there is a large amount of data to be transferred, a data link with maximal rate of data transfer per unit time is needed, which is provided by robust 5G network coverage. Figure 6.1 shows applications of Unmanned Aerial Vehicles (UAVs) in Agriculture using 5G.

Virtual Consultation and Predictive Maintenance

Virtual consultation allows session services to attend to the requirements of farmers. Domain experts can straightforwardly derive a live streaming in real-time using sensors to obtain thorough views regarding condition and provide farmers solution for the improvement of agriculture and irrigation practice. Precision agriculture, soil sampling, disease management and animal health monitoring are some of the services provided by consultation services. Multiple machineries can be monitored in real time with 5G having fast transmission speed and low latency to monitor in advance and give repairs on time without any interruption. Using several sensors to monitor a huge number of weather conditions in real time, 5G will provide a new maintenance paradigm called advanced predictive maintenance. Based on feedback, the farmer is alerted of any forthcoming issues and any weakening parts, allowing repairs to be planned at suitable time rather than postponing any operations (Compare et al. 2020). This can drastically diminish unintended downtime caused by defective apparatus or machine malfunction.

Augmented and Virtual Reality

Farmers can benefit from augmented reality (AR) and virtual reality (VR) equipments in a diversity of ways. Through wearable glasses and smartphones, AR can provide diversities of information such as crop, animal and machinery statistics, soil and weather pattern changes, disease exposure for livestock, land examination and more (Garzón et al. 2020). The farmers can acquire important information like if the crops are unwell or when they can be reaped or sown using AR glasses. So, farmers can cultivate in a more efficient way to potentially lessen labour and ensure timely delivery while ensuring a premium quality harvest. Virtual reality can be utilized for immersive agriculture training and practice (Wang et al. 2014). An interactive VR experience increases the connectivity requirements even further. By offering realistic and powerful experiences for learners, 5G will enable online interactive learning taking maximum benefit from augmented reality and virtual reality over conventional offline education.

Agriculture Robots

The collaboration of AI and 5G exposes new advantages in live video monitoring, remote diagnostics, as well as the stabilization of drones and robots using precise parameter management. AI in agriculture is budding on a continuous rate to give modern solutions for enhancing crop yield and AI-powered robots are ready to revolutionize the industry. Recently, agricultural robots have been used to autonomously plant a variety of crops across large areas of land. Robots are designed to plot a route themselves all through the Machine vision is a core potential, allowing them to perceive, identify, confine and implement intelligent actions on plants. To avoid collisions, laser rangefinder is used to detect impediments in the robot's pathway (Ramin Shamshiri et al. 2018). The robot can plot a route to its surroundings and be directed from any location utilizing cloud computation and a noteworthy amount of data is sent via 5G. Transmission of real-time photos recorded from sensors with super low latency via 5G network (Aijaz et al. 2017).

Cloud-Based Data Analytics

Data is one of the most significant aspects that are developing the smart agricultural business forward. On numerous farms, all the data acquired from sources like sensors and drones, are saved in cloud. 5G and edge computing allows data transfers to the cloud, allowing real-time analytics to help automate the farming process. Larger data must be transported to the cloud and then returned to users. To reduce complexity, cloud-based edge computing is mostly employed in smart robots. The cloud can be utilized as a data centre or a host for storing robot navigation and data processing control services. In the precision agriculture scenario, intelligence analyses these data in real time to develop AI for protective drones or machineries. By placing the GPU on the edge server, the need for the robot's graphics processing unit (GPU) is eliminated. Because the data processing bandwidth is so high, only 5G can handle it. The robots' physical size, power expenditure and cost have all been decreased dramatically. Over existing cell networks, 5G will vastly improve the data transfer experience. A huge volume of data may be successfully sent across several devices while minimizing data loss, reducing connection downtime and avoiding retransmission of data that consumes much time while transmitting a large number of data unnecessarily. Cloud computing provides faster data acquisition, transmission and processing at the cloud with minimal round transfer latency between various 5G devices, enabling maximum efficiency for sustainable agriculture management (Song, et al. 2019). Sensors, drones, robotics and smart devices, are the usage of 5G. There are several 5G characteristics like device density, ultra-low rate of data transfer per unit time, ultra-reliability and security. Drones, robotics and IoT sensors work together to increase output and drastically decrease price. Figure 6.2 shows applications of data science and cloud repository.

Recent Development Scenarios in 5G-Based Agriculture

Relevant Literature based on recent development scenarios in 5G-based agriculture concerning UAVs, predictive maintenance, AR and VR, real-time monitoring and agribots are summarized as follows (Table 6.1).

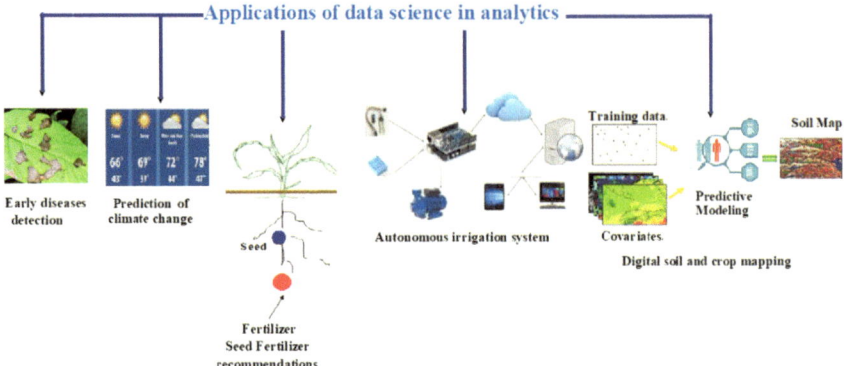

Fig. 6.2 Applications of data science in data analytics and cloud repository

Conclusion and Future Work

The 4G network although allows faster data transmission rates and adequate coverage, it is unable to transmit the massive amount of data between the number of devices. 5G comes into picture to meet the requirement of precision agriculture for improved output with a lesser amount of effort. In the forthcoming days, the 5G networks will be implemented in all industries; so, internet price will be much reduced and connectivity is going to be boosted. The utilization of 5G will drastically lower the implementation costs, which will be a godsend to farmers. Farmers will be well equipped for smart farming, with the capacity to use their mobile phones to forecast and prevent crop disease. By expanding their physical infrastructure, mobile carriers will make substantial contributions to precision agriculture. Data from the field will be collected by sensors and saved in the cloud. Sensors having a prolonged battery life will grow smaller and less expensive and networks will be more efficient, become smarter as well as secured. Although 5G has several applications and benefits in the agricultural industry, it will fundamentally alter the structure of jobs. There's a good chance that the number of agricultural jobs will decrease. Specific power supervision approaches, like sleep modes, have to be implemented to map the 5G network's no-load traffic allocation and maximize the use of harvested energy. To accomplish the ideal active periods of the 5G base stations while fulfilling the quality of service rendered, the active times of IoT devices should be efficiently coordinated.

Table 6.1 Recent development scenarios in 5G-based agriculture

Methods	Contribution
Unmanned aerial vehicles (Li et al. 2019)	Various drones can interact with one another to offer synchronized autonomous flying across a geographic area to execute numerous jobs with little data communication loss and battery expenditure, allowing comprehensive sensing time and gainful economic operation
Unmanned aerial vehicles (Aweiss et al. 2018)	A large number of obtained data can be kept and processed. A number of self-directed drones can soar across the air to deliver agricultural items to doorsteps using the 5G cellular network's stable connection. Unmanned aircraft system traffic management, flights outside line of sight and sensor data transmission are three major requirements for connecting drones utilizing 5G cellular connectivity
Real-time monitoring (GSMA 2019)	Machine-type communications with long-term evolution and narrowband IoT are paving the way for 5G integration in the future. It is used to connect numerous sensors to a single cellular tower, allowing farmers to set up IoT devices to perform tasks proficiently
Predictive maintenance (Compare et al. 2020)	Based on the response of 5G sensors, the end user is alerted regarding any upcoming issues and deteriorating parts, allowing maintenance to be scheduled in a timely way and without stopping any operations. It can significantly reduce unintended downtime caused by machine breakdown
Augmented reality and virtual reality (Garzón et al. 2020)	Wearable glasses and smartphones can provide useful information about crop, animal and machinery statistics, weather updates, AI-powered disease identification, land assessment, etc
Agribots (Skvortov et al. 2018)	AI-powered robots are programmed for using Global Positioning System to plot a route and recognize fruits and vegetables that are ripe to harvest
Augmented reality and virtual reality (Skvortsov et al. 2018)	Using a rapid and forceful mobile network, broadcast of inputs and audio output is reverted to the user with ultra-low data transfer time, allowing users without experiencing motion sickness

References

Aijaz A, Dohler M, Aghvami AH, Friderikos V, Frodigh M (2017) Realizing the tactile internet: haptic communications over next generation 5G cellular networks. IEEE Wirel Commun 24(2):82–89. https://doi.org/10.1109/MWC.2016.1500157RP

Aweiss AS, Owens BD, Rios JL, Homola JR, Mohlenbrink CP (2018) Unmanned aircraft systems (UAS) traffic management (UTM) National Campaign II. In: AIAA information systems-AIAA Infotech at aerospace, 2018, no 209989. https://doi.org/10.2514/6.2018-1727

Compare M, Baraldi P, Zio E (2020) Challenges to IoT-enabled predictive maintenance for industry 4.0. IEEE Internet Things J 7(5):4585–4597. https://doi.org/10.1109/JIOT.2019.2957029

Debaillie B, Desset C, Louagie F (2015) A flexible and future-proof power model for cellular base stations. In: IEEE vehicular technology conference, 2015, pp 1–7

Faraci G, Raciti A, Rizzo S, Schembra G (2018) A 5G platform for unmanned aerial monitoring in rural areas: design and performance issues. In: 2018 4th IEEE conference on network softwarization and workshops. NetSoft 2018, vol 1, no. NetSoft, pp 356–361. https://doi.org/10. 1109/NETSOFT.2018.8459960

Foerster JR, Costa-Perez X, Prasad RV (2020) Communications for iot: connectivity and networking. IEEE Internet Things Mag 3:6–7

Garzón J, Acevedo J, Pavón J, Baldiris S (2020) Promoting eco-agritourism using an augmented reality-based educational resource: a case study of aquaponics. Interact Learn Environ. https:// doi.org/10.1080/10494820.2020.1712429

GSMA (2019) NB-IoT deployment guide to basic feature set requirements, Gsma, vol Release 3, no June, pp 1–80 [Online]. https://www.gsma.com/iot/resources/nbiot-deployment-guide-v3/

GSMA (2019) Global system for mobile communication alliance: internet of things in the 5G era: opportunities and benefits for enterprises and consumers. https://www.gsma.com/iot/wpcontent/ uploads/2019/11/201911-GSMA-IoT-Report-IoT-inthe-5G-Era.pdf

Heidari H, Onireti O, Das R, Imran M (2021) Energy harvesting and power management for IoT devices in the 5G era. IEEE Commun Mag 59(9):91–97

Lee W-K, Schubert MJ, Ooi B-Y, Ho SJ-Q (2018) Multi-source energy harvesting and storage for floating wireless sensor network nodes with long range communication capability. IEEE Trans Ind Appl 54:2606–2615

Li B, Fei Z, Zhang Y (2019) UAV communications for 5G and beyond: recent advances and future trends. IEEE Internet Things J 6(2):2241–2263. https://doi.org/10.1109/JIOT.2018.2887086

Martin J, Amin R, Eltawil A, Hussien A (2011) Limitations of 4G wireless systems. In: Proceedings 2011 Virginia tech wireless symposium. (Blacksburg, VA), no. January, 2011, [Online]. http://www.researchgate.net/publication/228848043_Limitations_of_4G_Wir eless_Systems/file/e0b4951b0df4110bdd.pdf.

Ramin Shamshiri R et al (2018) Research and development in agricultural robotics: a perspective of digital farming. Int J Agric Biol Eng 11(4):1–11. https://doi.org/10.25165/j.ijabe.20181104. 4278

Razaak M et al (2019) An integrated precision farming application based on 5G, UAV and deep learning technologies. Commun Comput Inform Sci 1089:109–119. https://doi.org/10.1007/ 978-3-030-29930-9_11

Sen S, Koo J, Bagchi S (2018) TRIFECTA: security, energy efficiency, and communication capacity comparison for wireless IoT devices. IEEE Internet Comput 22:74–81

Sevgican S, Turan M, Gökarslan K, Yilmaz HB, Tugcu T (2020) Intelligent network data analytics function in 5G cellular networks using machine learning. J Commun Netw 22(3):269–280

Skvortsov EA, Skvortsova EG, Sandu IS, Iovlev GA (2018) Transition of agriculture to digital, intellectual and robotics technologies. EoR 14(3):1014–1028. https://doi.org/10.17059/2018- 3-23

Somov A, Giaffreda R (2015) Powering IoT devices: technologies and opportunities. IEEE IoT Newsletter

Song C et al (2019) Hierarchical edge cloud enabling network slicing for 5G optical fronthaul. J Opt Commun Netw 11(4):B60–B70. https://doi.org/10.1364/JOCN.11.000B60

Tang Y, Dananjayan S, Hou C, Guo Q, Luo S, He Y (2021) A survey on the 5G network and its impact on agriculture: challenges and opportunities. Comput Electron Agric 180:105895

Tong W, Feng X, Chen XJ (2019) Research on remote control and management based on '4G Network' in modern and high efficiency agriculture, pp 233–241

Torres Vega M et al (2020) Immersive interconnected virtual and augmented reality: a 5G and IoT perspective, no. 0123456789. Springer US, 2020

Vayssade J-A, Arquet R, Bonneau M (2019) Automatic activity tracking of goats using drone camera. Comput Electron Agric 162:767–772. https://doi.org/10.1016/j.compag.2019.05.021

Wang C-X, Haider F, Gao X, You X-H, Yang Y, Yuan D, Aggoune H, Haas H, Fletcher S, Hepsaydir E (2014) Cellular architecture and key technologies for 5G wireless communication networks. IEEE Commun Mag 52(2):122–130. https://doi.org/10.1109/MCOM.2014.6736752

Xu Y, Feng G, Liang L, Qin S, Chen Z (2017) MTC data aggregation for 5G network slicing. In: 2017 23rd Asia-Pacific conference on communications (APCC). IEEE, pp 1–6

Yao Z, Bian C (2019) Smart agriculture information system based on cloud computing and NB-IoT. DEStech Trans Comput Sci Eng no. cisnrc. https://doi.org/10.12783/dtcse/cisnrc2019/33340

Yuan M, Das R, Ghannam R, Wang Y, Reboud J, Fromme R et al (2020) Electronic contact lens: a platform for wireless health monitoring applications. Adv Intell Syst 2:1900190

Chapter 7
Security Attacks and Countermeasures in 5G Enabled Internet of Things

A. K. M. Bahalul Haque, Tasfia Nausheen, Abdullah Al Mahfuj Shaan, and Saydul Akbar Murad

Abstract The use of previous generation networks like 4G was vastly used in the Internet of Things (IoT) devices. The constant need to grow and develop just so the network can fulfill the requirement of IoT devices is still going on. The exponential growth of the data services substantially challenged the security and the networks of IoT because they were run by the mobile internet requiring high bit rate, low latency, high availability, and performances within various networks. The IoT integrates several sensors and data to provide services and a communication standard. Fifth Generation Communication System (5G) enabled IoT devices to allow the seamless connectivity of billions of interconnected devices. Cellular connections have become a central part of the society that powers our daily lives. Numerous security issues have come to light because of the exponential expansion of 5G technologies and the adaptation of the slow counterpart of IoT devices. Network services without security and privacy pose a threat to the infrastructure and sometimes endanger human lives. Analyzing security threats and mitigation is a crucial and fundamental part of the IoT ecosystem. Authorization of data, confidentiality, trust, and privacy of 5G enabled IoT devices are the most challenging parts of the system. And to provide a solution to these, we need a robust system to handle cyberattacks and prevent vulnerabilities by countermeasures. This paper includes a comprehensive discussion of 5G, IoT fundamentals, the layered architecture of 5G IoT, security attacks and their mitigation, current research, and future directions for 5G enabled IoT infrastructure.

Keywords IoT · 5G · Security and privacy · 5G IoT networks · SDN · NOMA · WSN

A. K. M. Bahalul Haque (✉) · T. Nausheen · A. Al Mahfuj Shaan
Department of ECE, North South University, Dhaka, Bangladesh
e-mail: bahalul.haque@lut.fi

T. Nausheen
e-mail: tasfia.nausheen@northsouth.edu

A. Al Mahfuj Shaan
e-mail: abdullah.mahfuj@northsouth.edu

S. A. Murad
Faculty of Computing, Universiti Malaysia Pahang, Pekan, Malaysia

B. Bhushan et al. (eds.), *5G and Beyond*, Springer Tracts in Electrical
and Electronics Engineering, https://doi.org/10.1007/978-981-99-3668-7_7

Introduction

Fifth Generation Communication (5G) networks are constantly developing, and it is becoming a significant catalyst for the growth of Internet of Things (IoT) devices (I-Scoop 2018). Future applications of IoT related application are going to be integration of various other technologies. IoT integration with other technologies can facilitate seamless connections between heterogeneous devices and promote ubiquitous computing infrastructure throughout our daily life. However, multiple complex integration challenges have emerged because of the heterogeneity and fragmentation of the connectivity system that hampers IoT development (Rahimi et al. 2018a). Moreover, as the 5G technology is emerging, this is considered the pioneer of IoT development. As the application of IoT is becoming widespread and diverse, the security and vulnerability of 5G IoT networks are becoming a significant concern (Haque et al. 2021). The services 5G integrated IoT provides need to be secured against cyberattacks and data loss. Even though there are no discrete security systems for IoT devices, they mostly rely on detecting vulnerabilities and taking countermeasures (Li et al. 2018).

Security is an integral part of any infrastructure. The same is true for IoT and 5G integrated architecture. Therefore, it is crucial to address the issues of cyberattack vectors, analysis, prevention, and countermeasures of 5G IoT networks. IoT devices are interconnected, combining various types of sensors, actuators, embedded software, and operating systems to run the network (Gautam et al. 2019). The application vectors have also spread across from our day-to-day hose hold toward industrial implementation. IoT devices are generally equipped with low-power capacity, relatively smaller physical architecture, and closely attached wireless and ward links. Some of these attributes of IoT are used by the cyber attackers to exploit the network by probing, inserting malicious codes, and unauthorized access to the network.

Third Generation (3G) communication and Fourth Generation (4G) communication technology are widely used in different walks of lives and around our daily environment. However, both the 3G and 4G technologies are not completely optimized for the applications of IoT. 5G technology is drastically improving and their integrations with IoT have proven to be reliable and capable. For the last 20 years, many M2M technologies have been implemented like BLE (Rahimi et al. 2018a), ZigBee (Alam et al. 2015), RPMA (Girson 2018), SigFox (Chen et al. 2017), LoRa (Palanisamy et al. 2022), etc. There are many other technologies like these, and they bring in a new set of challenges for the 5G technology (Mehbodniya et al. 2022). However, they are required to meet the necessities of IoT applications as well (Singla et al. 2021).

Motivated by the factors mentioned above, in this paper, we have comprehensively analyzed the integration factors of 5G IoT and security threats and countermeasures (Dener 2014; Teniou and Bensaber 2018; Sinha et al. 2017). We have analyzed various security threats at different levels of 5G enabled IoT devices from existing literatures and comprehensively presented the countermeasures as well. We have also outlined the research challenges associated with 5G IoT, which were not limited to reliability and communication issues. As 5G incorporates several technologies (Howe 2006; Rahimi et al. 2018b; Ahmad et al. 2018), it poses a significant impact on IoT applications. The paper's primary objective is to gain insights on the overview of the 5G network IoT devices, analyze the security issues, and develop solutions to prevent them from happening. The major contribution of this paper is listed as follows:

- This paper highlights the key features, architectural description, and extensive review of the 5G network.
- This paper provides an in-depth understanding, insights, and comprehensive description of IoT.
- This paper comprehensively discusses the integration trend, layered architectural analysis, and state-of-the-art technologies involved in 5G and IoT.
- This paper postulates various types of threats existing in 5G IoT networks.
- This paper also holistically analyzes the threat analysis and countermeasures of 5G IoT networks.
- Finally, this paper discusses the research gaps and future research directions of 5G IoT.

The paper was introduced in "Introduction" section; Second section includes the "Overview of 5G"; Section "Overview of IoT" provides a background on IoT; Section "5G Enabled IoT" contains the outline of the 5G and IoT integration; Section "5G Enabled IoT Architecture" includes the architectural description of 5G IoT integration; Section "Technologies in 5G Enabled IoT" introduces the technologies of 5G and IoT integration; threat analysis and countermeasures are comprehensively described in section "Threat Analysis in 5G IoT"; Eighth section demonstrates the "current Research for 5G and IoT"; Section "Future Research Directions and Challenges" demonstrate the current research trends and future research directions for 5G enabled IoT devices; and lastly, "Conclusion" section concludes the paper.

Overview of 5G

In telecommunications, the mobile network that has been launched is 5G which is available in different cities all over the world. After the launch of 5G, it has come to attention that it is far more significantly enhanced than 4G LTE, especially in 3 aspects. In the advanced 5G core network, a support network function virtualization and network slicing are present, which is actually, Software Defined Networking (SDN) (Bosshart et al. 2014; Lin et al. 2018; Chuang et al. 2018). The SDN in a network decouples the data, as well as the control planes, and these control planes are the ones who play a role in issuing instructions for the SDN switches in the data plane, and this control plane is present in the SDN controller.

Programmability, centralized policy management, and a global network state are a few of the advantages of SDN to the 5G system. Still, the benefits have to be properly looked at in the 5G core network. For this very reason, it is a must to maintain security within the communication channel to prevent potential attacks and safely guard the privacy and security of the data. In today's time, 5G, the state-of-the-art technology, can create new interfaces for regularly used devices and networking components. One of the essential roles of a 5G connection is to build connections between huge numbers of users so that it can provide more competent and faster communications. The design of 5G was done in a way where it can provide better coverage, bandwidth, reliability, and latency because these are what make 5G better than any other mobile network that was launched before 5G. However, even being better than other mobile networks, like any other, there are security issues, and there are several issues that look into these 5G security issues.

Ferrag et al. (2018) show 5G authentication and privacy-preserving surveys. Prasad et al. (2018) show privacy surveys, replay, bidding down, attacks on control, and user planes. But there was, this study in Basin et al. (2018) gives a formal analysis of the authentication of 5G. Jover and Marojevic 2019 showed a few unrealistic assumptions made by 5G specifications, which caused the adversarial attacks because of the vulnerability in the 5G systems until other optional security features were added. And potential security attacks or threats were shown by Teniou and Bensaber (2018) which measures are to be taken for the security of 5G. It also indicates that IPsec, a security protocol, is used primarily on LTE and to make IPsec more secured for the communication of 5G, IPsec tunneling can be done by authentication integration, integrity, and encryption.

As the mobile communication network is actually on the way to completing the 5G network cycle, it is becoming capable of supporting novel usage scenarios with stringent performance requirements. But 5G is not just stuck to seamless broadband connectivity only. It is already on the verge of advancement toward the vision of IoT. It has been prepared in order to be able to enable a wide range of machine-type applications. In today's world, wireless media is the way to have most communications that are actually open to various attackers. For this reason, efficient security operations must include and have (Sinha et al. 2017).

Overview of IoT

Nowadays, the advancement of technology, the IoT, is a significant paradigm shift in mobile service providers and the manufacturer of electronic devices. It helps to contemplate their business models and innovation context. IoT helps create an interconnection of billions of devices via the internet and has a tremendous growth rate. IoT is a network of devices consisting of sensors and actuators and helps to enable multiple applications and services by exchanging data between each other. The end-user does need compute-intensive operations, and along with it, there is a need for huge storage and real-time communication, and it is thought to be not an efficient way by the cloud service providers. And one of the essential features of all is the authentication of legitimate IoT devices (Howe 2006). And the authentication is addressed with the help of certificates given by a specific Certificate Authority (CA) but is one of such which offers a lightweight solution.

Smart city, among a large number of applications, is an integral field of IoT that is increasing the number of smart services within IoT systems (Rahimi et al. 2018b; Ahmad et al. 2018). The IoT application focuses on cities that are always composed of and controlled by computing units (Bosshart et al. 2014). Different definitions are given to smart cities like intelligent and digital cities (Lin et al. 2025). Smart cities focus on improving the service quality of the people and taking advantage of resources or decreasing the costs of public administration (Chuang et al. 2018). Smart lighting or smart traffic, and many other services are seen to grow exponentially (Ferrag et al. 2018). But with efficiency and advanced technology comes security. Security is the most crucial and the most significant feature of any smart device with an IoT architecture. Data confidentiality, authorization, trust, and client privacy are security challenges IoT systems face nowadays. So, to challenge these security problems, secured taxonomy is applied in order to handle cyberattacks and all other vulnerabilities using forensic techniques (Prasad et al. 2018).

Without efficient and strong security, IoT devices can create trouble rather than making life of people easier and more efficient because these devices end up endangering the privacy and safety of the people. There might be no trusted security, but the advancement and growth in the IoT systems and their services depend on recognizing potential security breaches and not being able to defy them. Security breaches occur due to various communication technologies being used in different layers of the wireless sensor network. The security and privacy issues were looked at with more details of the three-layer IoT architectures (Basin et al. 2018), and those defects were further investigated in Jover and Marojevic (2019).

5G Enabled IoT

Various types of research from the academic and industrial point of view focusing on 5G and IoT have been conducted (Ahmad et al. 2017; Liyanage et al. 2018; Bhushan and Sahoo 2017). Currently, advances are seen in theory, applications, and standardizations, especially in the implementations of the technologies related to 5G. The most crucial focus is to offer them a place to grow within IoT scenarios. And in the past few years, various work has been done on 5G and IoT as well (Ahmad et al. 2017). On 5G, a wireless research project was initiated by CISCO, Intel, Verizon combinedly to launch a novel set, "Neuroscience-Based Algorithms," and for the requirement of the human eye, adaptive video quality was launched where a hint was shown that it even has wireless networks which would include built-in human intelligence (Liyanage et al. 2018).

5G played a crucial role in the advancement of IoT because of 5G, billions of smart devices could create an inter-connection and interact and share data without the help of any users (Bhushan and Sahoo 2017). But recently, different application domains are making it more complicated for IoT to recognize devices that meet the application needs requirements (Bhushan and Sahoo 2017). IoT system which exists vastly uses fixed application domain like

- BLE
- ZigBee, etc.

There are other technologies like

- WiFi
- LP-WA networks
- Cellular communications, e.g., MTC using 3GPP and so.

IoT is constantly evolving quickly but with evolved proposals and new application domains. But IoT is growing and becoming more efficient to make people's lives more efficient and fast paced and trying to make more efficient inter-connection networks between smart devices. But with the growth of Industry IoT (IIoT), new challenges and obstacles are also coming in the way, like, the need for new advanced addition to the existing business models and products and solutions for the betterment (Ta-Shma et al. 2018). And there are technical challenges in Industry IoT:

- Reliabilities
- Timeless
- Connection robustness, and so on.

There are most used communication techniques within the IoT connectivity and are, 3GPP and LTE (Rathore et al. 2016), which are offered to the IoT systems also like (Zanella et al. 2014),

- Wide coverage
- Low Deployment costs
- High-security level
- Access to dedicated spectrum
- Management simplicity.

However, MTC communication cannot bear the cellular networks present because those present cellular networks are the primary key in IoT, which is one of the problems. But this isn't a problem when instead of MTC communication, the 5G network is used because the present cellular networks are making it faster in terms of data rate, and it occurs because of low latency and the better version of MTC communication with respect to current 4G (LTE) and which results in more efficient IoT applications and devices.

5G Enabled IoT Architecture

More efficient and advanced architecture is needed, which will help achieve more sustainable and efficient new technologies. More scalable architectures of IoT devices will be better than the present 5G IoT architecture (Jin et al. 2014). The advantage of using new technology during the development is that it makes the architecture

- More simple
- Convenient for scalability
- Analysis
- Modularity
- Efficiency
- Agility
- Accessibility to high-demand services
- Eight layers interconnected along with data exchange capability, two-way, this architecture has been designed, explained below (Jin et al. 2014).

L1 Physical Device Layer

The general layer of the architecture of the IoT includes

- Wireless sensors
- Actuators
- Controller

L2 Communication Layer

The two sub-layers discussed below.

Device to Device Communication Sub-Layer: 5G enhances D2D communication in this sub-layer and is an important participant that provides connectivity for devices with Machine-Type Communication (MTC) (Mohammadi et al. 2018).

Connectivity Sub-Layer: Cell phones, tablets, etc., are the devices connected to the BSs communication centers within the sub-layer. These devices proceed with data analysis and are then sent to the storage units through centers with an Intranet connection (Millr 2015; Conti et al. 2018). Those are specified with

- High Reliability
- Performance
- Agility.

L3 Fog Computing Layer

Within this layer, in order to make decisions, edge processing is applied on the data by the nodes (Kumar and Patel 2014).

L4 Data Storage Layer

Physical devices send the information of edge processing to the data storage units here in this layer (Zhao and Ge 2013). And here, large amounts of data are handled and the traffic of the future devices and applications but not without the data security, which is the key of this layer.

L5 Management Service Layer

Here in this layer, there are three sub-layers, and these are

Network Management Sub-Layer: Communication purposes occur between devices and data centers in this layer. 5G IoT or ZigBee are communication protocols where the network type is consistent with the technology present in this layer: Wireless Network Functionality Virtualization (WNFV). IoT networks are managed, and network reconfiguration is enabled because of the Wireless Software Defined Network (WSDN) (Xu et al. 2014) technology. Because of this technology, traditional network monitoring for performance enhancement is unnecessary.

Cloud Computing Sub-Layer: The data from a layer from Fog Computing Layer can be reprocessed in this sub-layer, and then the processed information is generated in the final step.

Data Analytics Sub-Layer: For generating values for decision-making in this layer, new learning algorithms of data analytics can be implemented to the last sub-layer information (Millr 2015; Conti et al. 2018). Since the information from the IoT networks is collected, it starts turning more dominant and expanding with time.

L6 Application Layer

Business related people make the most of use of this layer because they need to plan with the correct data, and it also helps in revolutionization, which are (Kumar and Patel 2014).

- Vertical markets and business need by control applications
- Vertical and mobile applications
- Business intelligence and analytics.

L7 Process and Collaboration Layer

Collaborations and communications are permitted in these layers within IoT devices and services. That occurs because the data and information cannot be utilized with a single entity since they come from the previous layers (Rahimi et al. 2018a).

L8 Security Layer

This is the protection layer for all the other previous layers, and this protection is done without impacting the different previous layers' functionality. Also, the security taxonomy for blocking and foreseeing the dangers of cyberattacks is protected here in this layer.

The Fig. 7.1 added below shows a brief overview of the 5G IoT architecture.

Technologies in 5G Enabled IoT

In the last decade, much research has been done on 5G enabled IoT (Mohammadi et al. 2018). To build the state-of-the-art IoT and 5G systems, extensive research is done by academics and industry (Millr 2015; Conti et al. 2018; Kumar and Patel 2014). 5G enabled IoT devices can significantly impact the interconnections of IoT devices. Heterogeneous networks currently are unable to satisfy the needs of the application of IoT devices (Zhao and Ge 2013). Popular IoT systems include BLE, ZigBee, WiFi, LP-WA, etc. (Hošek 2016). The current systems focus on improving our regular life, making a better-quality life, and engaging interconnections between smart homes, smart cities, agriculture, and healthcare (Mohammadi et al. 2018; Millr 2015; Conti et al. 2018). 3G and LTE networks are currently the most used connectivity technologies that offer low cost and wide coverage. However, these

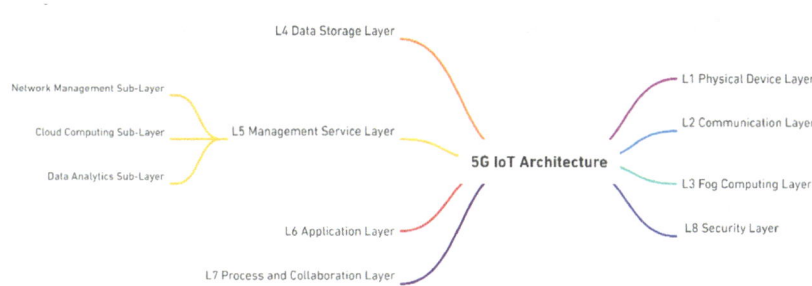

Fig. 7.1 5G IoT architecture

Table 7.1 Technologies in 5G enabled IoT

Technology	Use cases
Heterogeneous networks (HetNet)	Enables on demand transmission rate for 5G IoT (Millr 2015; Conti et al. 2018)
Direct device to device (D2D)	D2D was proposed to create communications between short ranges. power efficient, optimized power and communication load occur here and are expected to provide an efficient spectrum (Millr 2015; Conti et al. 2018)
Spectrum sharing	To enable the technology in covering the traffic load imbalance, spectrum management is the key. Massive MIMO is centerpieces of spectrum sharing (Mohammadi et al. 2018; Millr 2015)
Zigbee	It's a power optimized and cost efficient mesh network widely used in WSNs and primarily used in Industrial IoT applications (Millr 2015; Conti et al. 2018)
Other technologies	Other enabling technologies include machine-type communication (MTC), mmWave, SDN, NFV, and NB-IoT

present networks cannot manage to support MTC, which is the key to enabling the factor in IoT devices (Jin et al. 2014; Mohammadi et al. 2018).

Several 5G enabled IoT technologies have been developed in the last few years, and some are developing. Some of these are described in Table 7.1.

Threat Analysis in 5G IoT

As security in IoT is emerging as one of the significant factors in security schemes, it is essential to analyze and mitigate the attacks. In this section, categorization of various attacks and countermeasures to emphasize the contribution to this paper are done.

Eavesdropping

The attackers of eavesdropping try to intercept some of the confidential information but detecting the legitimacy of the transmitters or receivers is challenging to locate or trace since the attackers do not transmit any signals (Xu et al. 2014; Kaplan 2018).

Interception

The attackers can easily detect the authentication of the communication since they snoop within the nearby wireless environment. With this technique the attacker can capture the information about the network. The network information can include the configuration, sensory data transfer protocol, etc. Eavesdropping through interception is one of the most effective and oldest techniques to exploit the security (Hošek 2016; Liyanage et al. 2018).

Traffic Analysis

Cryptographic algorithms may encrypt the critical information in legitimate communication. In this case, attackers intercept the transmitted signal. Still, they cannot obtain significant content, but traffic analysis can come in handy for the tracking of communication patterns in order to realize other forms of attacks (Wang et al. 2019).

Contaminating

In this type of attack the attackers try to gain illicit access to the network and contaminate the channel estimation stage as well (Astely et al. 2013; Palattella et al. 2016). This sort of attack can be categorized into two types of contamination according to different channel.

Spoofing

Attackers inject fake identity information to destroy or join communications. The attacker is able to establish a fake communication channel between two or more legitimate points. In this way, unknowingly, two legitimate parties communicate with each other through a fake entity (Liyanage et al. 2018).

Here, malicious nodes can copy other nodes, claim fake identities, and generate a random number of different identities only using hardware devices (Li et al. 2021; Lin et al. 2019). Sybil types of attacks make the system generate false reports, and that can make users get spam and lose privacy (Wang et al. 2021).

Jamming

Here the target of the attackers is to block legitimate communication using noise (Haque and Bhushan 2021), and an adversary can send continuous signals by decreasing signal to noise ratio (Xu et al. 2014) through the channel only to hamper communication. It can also prompt DoS attacks at the physical layer (Kaplan 2018). There are three types of signal jamming, in general, such as pilot jamming, proactive jamming, and reactive jamming (Hošek 2016).

Pilot jamming is launched when a channel is trained (Hasan and Hossain 2013; Ge et al. 2014; Ahmad et al. 2020) and aims to create an illegitimate connection without the exact pilot sequences. An adversary can launch the attack when he knows the pilot length and sequence. Pilot jamming is also very efficient as only the signals need to be corrupted (Millr 2015; Conti et al. 2018; Haque et al. 2020).

Physical Layer Security

5G and IoT are the fundamental paradigms of today's time, and for the security of wireless communication, physical layer security is becoming a growing prospect. PLS protects the confidentiality of data by exploiting the intrinsic randomness of the communication medium (Padmavathi and Shanmugapriya 2009; Shiu et al. 2011). This technique plays an aid in improving 5G IoT security from two main aspects,

The network latency on the Internet of Vehicles (IoV) and Unmanned Aerial Vehicles (UAV) can be reduced. The vehicles can randomly join and leave the network, making the UAV highly dynamic (Steinmetzer et al. 2018). PLS will offer an efficient and quickest authentication by exploring radio frequency (RF) fingerprint otherwise, roaming in different networks will lower communication performance. Different schemes in PLS can be additional protection that cooperates with the existing security architecture to provide better protection for 5G IoT devices.

Random wireless channel use cases are done to generate keys in PLS schemes that can release the burden (Zhou et al. 2012) in 5G IoT networks; it becomes difficult or rather challenging to achieve effective key distribution and management. Reinforcing communication security can be done without encryption and decryption using information theory.

Massive MiMo

Core 5G technology received considerable attention in IoT research (Nitsche et al. 2014; Ylmaz and Arslan 2015). It helps in providing numerous communication superiority based on the beamforming technology like an array that gains channel hardening and nearly orthogonal channels (Ylmaz and Arslan 2015). In the meantime, MiMo techniques for PLS are also discussed in Zeng et al. (2010), Newsome et al. (2004).

Passive eavesdropping (Xiao et al. 2009) and also the active attacks in massive MiMo were investigated by many authors. The analysis showed that PLS against passive eavesdropping could increase pilot contamination attacks (Zhang et al. 2014), and it is fatal for MiMo communication since an active attacker can send the same pilot sequence.

NoMa

IoT with NoMa and non-orthogonal resources can improve spectral efficiency and also reduce low transmission latency and signaling cost (Mpitziopoulos et al. 2009), and it can also be used where the number of sensors is huge, like in smart farming and intelligent manufacturing (Bhushan and Sahoo 2020). Allocating two users to a single orthogonal resource block for user pairing is a technique for balancing complexity and efficiency (Clancy 2011). The capacity to superpose numerous signals into an orthogonal resource is achieved via superposed coding technology.

MmWave

A 5G technology that can improve network transmission (Mpitziopoulos et al. 2009; Yang et al. 2018) helps the devices connect to higher bandwidth communication channels (Wood et al. 2007). Characteristics like blocking effect, highly directional transmissions are new in mmWave (Wood et al. 2007). The new characteristics of mmWave channels can help in improving the efficiency of traditional PLS techniques (Wang et al. 2018; Hamamreh et al. 2018; Arsh et al. 2021). Because of the tiny wavelengths of mmWave, dozens to hundreds of antenna elements may be put in an array on a small physical platform, which significantly aids the application of MiMo and the integration of different 5G technologies (Zeng 2015; Lu et al. 2014, 2018; Araujo et al. 2016).

Trust Mechanism in WSN

This mechanism in WSN has been emerging as a significant factor when it comes to security schemes, and so, it is really a necessity to analyze how these attacks can be resisted with the help of trust schemes (Zhou et al. 2012; Goyal et al. 2021; Zhu et al. 2014; Kapetanovic et al. 2015). Recently these mechanisms have been remodeled to filter the fake nodes in a sensor network. This approach was first introduced in E-commerce to choose dependable transaction objects, and many researchers in different fields have since developed it (Zhou et al. 2012). Because the evaluation of trust is entirely based on past behaviors of participants or indirectly mixed with the reputation of other recommenders, this mechanism has the potential to be more efficient, but higher standards are required to develop an effective trust framework in WSNs because of

- Limitations of energy
- Limited Storage Space
- Wireless communication's inherent vulnerabilities.

Crowdsourcing Analysis

In 5G networks, crowdsourcing is a potent tool against hackers. The major goal of this analysis is to present the problem to a participatory community that is eager to solve the problem and then anticipate a reward (Zhou et al. 2012). This concept has been used in the IoT in a variety of ways by users and their IoT devices for a variety of reasons. (Ding et al. 2017) However, there hasn't been enough discussion of how these features can help mitigate the impact of cyberattacks in truly complex networks, which are particularly vulnerable due to the wide range of technologies at various levels of abstraction, as in the case of 5G networks (Li et al. 2018; Gautam et al. 2019; Rahimi et al. 2018a). The concept of crowdsourcing connects both the 5G and the IoT network world naturally

- Participant's interests are defined (users and providers)
- Motivating mutual cooperation to stop cyberattacks

Commercial Purposes

The 5G business model requires infrastructure sharing among service providers, and in this case it is very important to use crowdsourcing analysis between the service providers to identify attacks (Millr 2015; Conti et al. 2018). It is essential to use security mechanisms to mitigate potential attacks and ensure privacy and confidentiality. Malware Information and Sharing Platform also uses a similar system to develop countermeasures for threats (Ding et al. 2018).

Removing Physical Attacks

Data coming from mobile phones are used in crowdsourcing analysis, and it is used to provide a warning system. Software and physical attacks are very different and require more research (Ding et al. 2016). A novel way of solving problems like finding the attacker's location is to expand the security controls at the edge of the user's IoT device, and crowdsourcing analysis can be implemented to identify potential attacks (Rappaport et al. 2013).

Social Media

Social media like WhatsApp, Twitter, and Facebook are the biggest platforms for crowdsourcing (Niu et al. 2015). Social media is a gateway for crowdsourcing, and it can be done in two ways: 1. by the traditional method, which involves humans but focuses on the attacks on the systems of the whole infrastructure, or 2. the other is done by extracting relevant information and identifying attack patterns (Heath et al. 2016).

Attacks at the Architecture Level

Today's world is becoming more interconnected, and smart cities are the key. In smart cities, various IoT devices are integrated, and nodes in smart cities are vulnerable to security threats like DoS attacks and manipulation of data (Wang and Wang 2016).

L1 Physical Device Layer

Various threats and attacks can damage the sensor nodes in the architecture (Gautam et al. 2019). Some attacks in the L1 are described below.

- **Unauthorized Access to Tags**: Attackers can easily access tags in RFID due to the lack of proper authentication techniques.
- **Tag Cloning**: RFID tags can be cloned in the physical layer, and reverse engineering can extract relevant information.
- **Sleep Deprivation Attack**: Unstoppable sending of control information is done in this attack and it keeps the nodes constantly in a working state in the network.

L2 Communication Layer

- **DoS Attack**: DoS attacks engage the user to overflow the victim's system with a large amount of network traffic.
- **Sybil Attack**: This attack deceives the victim to do one task multiple times as it shows pseudonymous identities in the node.
- **Replay Attack**: During eavesdropping, a valid data packet is collected from the network and every time the user connects to the network, and the attackers collect resources from them.
- **Sinkhole Attack**: The flow of data is attracted from other nodes residing nearby by using another compromised node.

L3 Fog Computing Layer

Fake gateways and attackers replace edge devices to collect data from these edges (Zhou et al. 2012).

L4 Data Storage Layer

Data privacy, confidentiality, and integrity are concerned with any IoT data storage system.

L5 Management Layer

The attackers attack the server, database, and other services in this layer.

- **VM Manipulation**: VMs run in the host system, and the adversary controls it. This can be attacked, and the range of these attacks is from extracting information and manipulating data in the VM (Xu et al. 2014).
- **Flooding Attacks in Cloud**: This attack is done in the sub-layer of the cloud, and the attackers frequently send service requests (Xu et al. 2014).
- **Cloud Malware Injection**: Malicious services can be inserted into the cloud and used to manipulate the system, and sensitive data could retrieve sensitive data.

L6 Application Layer

Attacks in this layer are mainly to access the data of users.

- **Code Injection**: Attackers insert worms and other malicious codes to exploit the errors in the program and gain system control.
- **Buffer Overflow**: Attackers use programs to violate the data buffer or codes to overflow the entire system.
- **Permission Manipulation**: This attack mainly leads to the illegitimate administration of data and violates user privacy.

Current Research for 5G and IoT

This section discusses the current research topics and solutions that can also be introduced to future research directions.

The number of research done on mobility is minimal, and it is a fascinating subject matter in terms of mobility in physical layer attacks on both user and attacker sides. The attacker may use the mobility feature to find the best area of attack and try to avoid detection. Mobility can also be used to counter this attack. But users might also have to consider the performance if mobility is implemented and investigating the 5G IoT mobility system is a current research topic (Mohammadi et al. 2018).

mmWave and NOMA are new features currently available in 5G technology. Few studies show exploitations of these new features to achieve physical layer attacks. And schemes for these new features are yet to be found (Millr 2015).

Trust models for securing data are another field where data sensing and aggregation are focused. The wide range of data types and privacy safety increases the obstacles and brings in newer problems (Haque et al. 2021; Li et al. 2018; Gautam et al. 2019). Current threats and attacks in the WSN can be identified with trust models, and also trusted models can be used to plan the attack itself. And currently, the analysis of existing and potential threats is the objective (Mohammadi et al. 2018; Conti et al. 2018; Conti et al. 2018).

Future Research Directions and Challenges

Even though 5G satisfies the requirements for IoT security, it also opens up newer sets of challenges like architecture security of IoT and verified communications between devices. In this section, we have reviewed future research areas for 5G IoT security.

Characteristics Synthesis

The 5G framework is a synthesis of many technologies (Mohammadi et al. 2018). A combination of MIMO, mmWave, and NOMA increases the spectrum efficiency. An IoT environment with low-cost and low-power different virtual channel models can help build a powerful Access Point. It can help distinguish between multiple users which can help prevent attacks like pilot contamination attacks. Researching the field of synthesis characteristics of 5G IoT can be a boost to novel solutions.

Signal Revoking

Detection of any active attack in the network is the primary step toward any kind of countermeasure. It is expected for IoT devices to maintain secure communication even if it is under any attack. And it is very challenging to eliminate the attacks even while maintaining contact. Waveform designs (Wang et al. 2018) can be an additional functionality, and they could be used to recognize if the signals are coming from the same user. Afterward, a filter mechanism might be developed to filter eavesdroppers in the network using this technology.

Location Awareness

Location awareness can be a positive factor for removing threats and preventing threats, as 5G location services can accurately provide location data (Jaitly et al. 2017). Location awareness could help mitigate threats in the network, and there are many prospective characteristics of the 5G network to attain location awareness (Goyal et al. 2021). And to achieve more efficient communication, location information is an exciting research direction.

Technical Challenges

Many works have been made to mitigate any challenges for the 5G enabled network. But there are still many technical challenges. There are design-related issues at the architecture level that includes Scalability and network management, which is a major issue in managing the state of the information (Fuentes et al. 2013).

Interoperability and heterogeneity allow devices to connect seamlessly, and it is a major concern as it is used to collect information about smart networks or applications (Zeng 2015).

Wireless Software Defined Network

Even though WSDN solves the scalability issue in the 5G network, many cases need to be resolved in SDN. It needs to provide flexibility and separation of control and data plane, which is the most challenging part of SDNs.

Security Assurance and Privacy Analysis

Next-generation 5G enabled IoT devices, security, and privacy needs to be added to the network and device levels as they will address many different complex applications, including smart cities and intelligent networks. 5G is a diverse system, and security system is very complicated, and security assurance must be considered at the device and network levels during the design process.

Standardization Issues

As 5G is being developed, it has also enabled to provide many IoT solutions. And the calibration of IoT will make the implementation of 5G IoT even easier. Lack of consistency and standardization (Li et al. 2018; Gautam et al. 2019; Rahimi et al. 2018a) has made it a big hurdle and challenge for closing the gap between humans and environment control. IoT as a service (Haque et al. 2021) might one day be a possible result.

Conclusion

This paper focuses on various security attacks and their countermeasures, the impact of 5G enabled IoT, and possible solutions for mitigating threats in a 5G enabled network. We have reviewed 5G and IoT characteristics and physical layer threats. We also categorized various types of threats with different kinds of attacking purposes. The open issues for 5G enabled IoT were also discussed, and current and future research trends were also introduced in the last section of the paper. The development of 5G enabled IoT devices will open many more gates for the future, bringing in possible data privacy and security issues. And it is essential to acknowledge what is associated with 5G enabled IoT and its security and different solutions under the wide spectrum of the 5G network. The paper's main aim was to provide a comprehensive insight into the 5G enabled IoT and threat analysis and discuss the future research areas. We hope this paper will help further research on the future of 5G enabled devices.

References

Ahmad I, Kumar T, Liyanage M, Okwuibe J, Ylianttila M, Gurtovk A (2017) 5G security: analysis of threats and solutions. In: 2017 IEEE conference on standards for communications and networking (CSCN)

Ahmad I, Kumar T, Liyanage M, Okwuibe J, Ylianttila M, Gurtov A (2018) Overview of 5G security challenges and solutions. IEEE Commun Stand Mag 2(1):36–43

Ahmad A, Bhushan B, Sharma N, Kaushik I, Arora S (2020) Importunity & evolution of IoT for 5G. In: 2020 IEEE 9th international conference on communication systems and network technologies (CSNT). https://doi.org/10.1109/csnt48778.2020.9115768

Alam KM, Saini M, El Saddik A (2015) Toward social internet of vehicles: concept, architecture, and applications. IEEE Access 3:343–357

Araujo DC, Maksymyuk T, de Almeida AL, Maciel T, Mota JC, Jo M (2016) Massive MIMO: survey and future research topics. IET Commun 10(15):1938–1946

Arsh M, Bhushan B, Uppal M (2021) Internet of Things (IoT) toward 5G network: design requirements, integration trends, and future research directions. Adv Intell Syst Comput 887–899.https://doi.org/10.1007/978-981-15-9927-9_85

Astely D, Dahlman E, Fodor G et al (2013) LTE release 12 and beyond [accepted from open call]. IEEE Commun Mag 51(7):154160

Bahalul Haque AKM, Arifuzzaman BM, Abu Noman Siddik S, Kalam A, Sadia Shahjahan T, Saleena TS, Alam M, Rabiul Islam M, Ahmmed, F (2022) Semantic Web in Healthcare: A Systematic Literature Review of Application, Research Gap, and Future Research Avenues. Int J Clin Pract 20221–27. https://doi.org/10.1155/2022/6807484

Basin D, Dreier J, Hirschi L, Radomirović S, Sasse R, Stettler V (2018) A formal analysis of 5G authentication. In: 2018 ACM SIGSAC conference on computer and communications security. Toronto, Canada, pp 1383–1396

Bhushan B, Sahoo G (2017) A comprehensive survey of secure and energy efficient routing protocols and data collection approaches in wireless sensor networks. In: 2017 international conference on signal processing and communication (ICSPC). https://doi.org/10.1109/cspc.2017.8305856

Bhushan B, Sahoo G (2020) Requirements, protocols, and security challenges in wireless sensor networks: an industrial perspective. In: Handbook of computer networks and cyber security, pp 683–713.https://doi.org/10.1007/978-3-030-22277-2_27

Bosshart P, Daly D, Gibb G, Izzard M, McKeown N, Rexford J, Schlesinger C, Talayco D, Vahdat A, Varghese G, Walker D (2014) P4: programming protocol-independent packet processors. SIGCOMM Comput Commun Rev 44(3):87–95

Chen X, Ng DWK, Gerstacker WH, Chen H-H (2017) A survey on multiple-antenna techniques for physical layer security. IEEE Commun Surv Tutor 19(2):1027–1053

Chuang C-C, Yu Y-J, Pang A-C (2018) Flow-aware routing and forwarding for SDN scalability in wireless data centers. IEEE Trans Netw Serv Manage 15(4):1676–1691

Clancy TC (2011) Efficient OFDM denial: pilot jamming and pilot nulling. In: 2011 IEEE international conference on communications (ICC). IEEE, pp 1–5

Conti M, Dehghantanha A, Franke K, Watson S (2018) Internet of things security and forensics: challenges and opportunities. Futur Gener Comput Syst 78:544–546

de Fuentes JM, González-Manzano L, González-Tablas AI, Blasco J (2013) WEVAN—a mechanism for evidence creation and verification in VANETs. J Syst Architect 59(10):985–995

Dener M (2014) Security analysis in wireless sensor networks. Int J Distrib Sens Netw 10(10):303501

Ding Z, Fan P, Poor HV (2016) Impact of user pairing on 5G nonorthogonal multiple-access downlink transmissions. IEEE Trans Veh Technol 65(8):6010–6023

Ding ZG, Lei XF, Karagiannidis GK, Schober R, Yuan JH, Bhargava VK (2017) A survey on non-orthogonal multiple access for 5G networks: research challenges and future trends. IEEE J Sel Areas Commun 35(10):2181–2195

Ding Z-G, Xu M, Chen Y, Peng M-G, Poor HV (2018) Embracing non-orthogonal multiple access in future wireless networks. Front Inf Technol Electron Eng 19(3):322–339

Ferrag MA, Maglarasc L, Argyrioud A, Kosmanosd D, Janickec H (2018) Security for 4G and 5G cellular networks: a survey of existing authentication and privacy-preserving schemes. J Netw Comput Appl 101:55–82

Gautam S, Malik A, Singh N, Kumar S (2019) Recent advances and countermeasures against various attacks in IoT environment. In: 2019 2nd international conference on signal processing and communication (ICSPC). IEEE, pp 315–319

Ge X, Cheng H, Guizani M, Han T (2014) 5G wireless backhaul networks: challenges and research advances. IEEE Netw 28(6):611

Girson A (2018) IoT has a security problem will 5G solve it? https://www.wirelessweek.com/art icle/2017/03/iot-has-securityproblem-will-5g-solve-it. Accessed 15 Jan 2018

Goyal S, Sharma N, Kaushik I, Bhushan B, Kumar N (2021) A green 6G network era: architecture and propitious technologies. Data Anal Manag 59–75.https://doi.org/10.1007/978-981-15-833 5-3_7

Hamamreh JM, Furqan HM, Arslan H (2018) Classifications and applications of physical layer security techniques for confidentiality: a comprehensive survey. IEEE Commun Surv Tutor 21(2):1773–1828

Haque AB, Bhushan B (2021) Security attacks and countermeasures in wireless sensor networks. In: Integration of WSNs into internet of things. CRC Press, pp 17–43

Haque AKMB, Shurid S, Juha AT, Sadique MS, Asaduzzaman AS (2020) A novel design of gesture and voice controlled solar-powered smart wheel chair with obstacle detection. In: 2020 IEEE international conference on informatics, IoT, and enabling technologies (ICIoT). https://doi.org/10.1109/iciot48696.2020.9089652

Haque AKMB, Bhushan B, Dhiman G (2021) Conceptualizing smart city applications: requirements, architecture, security issues, and emerging trends. Expert Syst 1–23. https://doi.org/10.1111/exsy.12753

Hasan M, Hossain E (2013) Random access for machine-to-machine communication in LTE-advanced networks: issues and approaches. IEEE Commun Mag 51:86–93

Heath RW, Gonzalez-Prelcic N, Rangan S, Roh W, Sayeed AM (2016) An overview of signal processing techniques for millimeter wave MIMO systems. IEEE J Sel Topics Signal Process 10(3):436–453

Hošek J (2016) Enabling technologies and user perception within integrated 5G-IoT ecosystem. Brno, Czech Republic, VysokéUčeníTechnické v Brně, Nakladatelství VUTIUM

Howe J (2006) The rise of crowdsourcing. Wired Mag 14(6):1–4. Accessed 10 Feb 2022

I-Scoop, 5G and IoT in 2018 and beyond: the mobile broadband future of IoT. https://www.i-scoop.eu/internetof-things-guide/5g-iot/. Accessed 14 Jan 2018

Jaitly S, Malhotra H, Bhushan B (2017) Security vulnerabilities and countermeasures against jamming attacks in Wireless Sensor Networks: a survey. In: 2017 international conference on computer, communications and electronics (Comptelix). https://doi.org/10.1109/comptelix.2017.8004033

Jin J, Gubbi J, Marusic S, Palaniswami M (2014) An information framework for creating a smart city through internet of things. IEEE Internet Things J 1(2):112–121

Jover RP, Marojevic V (2019) Security and protocol exploit analysis of the 5G specifications. IEEE Access (99):1–1

Kapetanovic D, Zheng G, Rusek F (2015) Physical layer security for massive MIMO: an overview on passive eavesdropping and active attacks. IEEE Commun Mag 53(6):21–27

Kaplan K (2018) Will 5G wireless networks make every internet thing faster and smarter? https://qz.com/179794/will-5g-wireless-networks-make-every-internet-thing-faster-and-smarter/. Accessed 14 Jan 2018

Kumar JS, Patel DR (2014) A survey on internet of things: security and privacy issues. Int J Comput Appl 90(11)

Li S, Da Xu L, Zhao S (2018) 5G internet of things: a survey. J Ind Inf Integr 10:1–9

Li W, Wang N, Jiao L, Zeng K (2021) Physical layer spoofing attack detection in MmWave massive MIMO 5G networks. IEEE Access 9:60419–60432

Lin Y-B, Wang S-Y, Huang C-C, Wu C-M (2018) SDN approach for aggregation/disaggregation of sensor data. Sensors 18(7):2025

Lin YB, Huang TJ, Tsai SC (2019) Enhancing 5G/IoT transport security through content permutation. IEEE Access 7:94293–94299

Liyanage M, Ahmad I, Abro AB, Gurtov A, Ylianttila M (eds) (2018) A comprehensive guide to 5G security. Wiley, Hoboken, p 231

Lu L, Li GY, Swindlehurst AL, Ashikhmin A, Zhang R (2014) An overview of massive MIMO: benefits and challenges. IEEE J Sel Topics Signal Process 8(5):742–758

Lu X, Xiao L, Dai C (2018) UAV-aided 5G communications with deep reinforcement learning against jamming. arXiv preprint: 1805.06628

Mehbodniya A, Bhatia S, Mashat A, Elangovan M (2022) Proportional fairness based energy efficient routing in wireless sensor network. Comput Syst Sci Eng

Millr M (2015) The internet of things: how smart TVs, smart cars, smart homes, and smart cities are changing the world. Pearson Education

Mohammadi M, Al-Fuqaha A, Guizani M, Oh J-S (2018) Semisupervised deep reinforcement learning in support of IoT and smart city services. IEEE Internet Things J 5(2):624–635

Mpitziopoulos A, Gavalas D, Konstantopoulos C, Pantziou G (2009) A survey on jamming attacks and countermeasures in WSNS. IEEE Commun Surv Tutor 11(4)

Newsome J, Shi E, Song D, Perrig A (2004) The sybil attack in sensor networks: analysis & defenses. In: Third international symposium on information processing in sensor networks, 2004. IPSN 2004. IEEE, pp 259–268

Nitsche T, Cordeiro C, Flores AB, Knightly EW, Perahia E, Widmer JC (2014) IEEE 802.11 ad: directional 60 GHz communication for multi-gigabit-per-second wi-fi. IEEE Commun Mag 52(12):132–141

Niu Y, Li Y, Jin D, Su L, Vasilakos AV (2015) A survey of millimeter wave communications (mmWave) for 5G: opportunities and challenges. Wirel Netw 21(8):2657–2676

Padmavathi DG, Shanmugapriya M (2009) A survey of attacks, security mechanisms and challenges in wireless sensor networks. arXiv preprint arXiv:0909.0576

Palanisamy T, Alghazzawi D, Bhatia S, Malibari AA (2022) Improved energy based multi-sensor object detection in wireless sensor networks. Intell Autom Soft Comput

Palattella M, Dohler M, Grieco A et al (2016) Internet of things in the 5G era: enablers, architecture and business models. IEEE J Sel Areas Commun 34(3):2016

Prasad AR, Arumugam S, Sheeba B, Zugenmaier A (2018) 3GPP 5G security. J ICT 6(1&2):137–158

Rahimi H, Zibaeenejad A, Safavi AA (2018a) A novel IoT architecture based on 5G-IoT and next generation technologies. Presented at IEEE IEMCON conference, Vancouver, BC, Canada, Nov 2018a. https://arxiv.org/abs/1807.03065

Rahimi H, Zibaeenejad A, Rajabzadeh P, Safavi AA (2018b) On the security of the 5G-IoT architecture. In: Proceedings of the international conference on smart cities and internet of things, pp 1–8

Rappaport TS, Sun S, Mayzus R, Zhao H, Azar Y, Wang K, Wong GN, Schulz JK, Samimi M, Gutierrez F (2013) Millimeter wave mobile communications for 5G cellular: it will work! IEEE Access 1:335–349

Rathore MM, Ahmad A, Paul A, Rho S (2016) Urban planning and building smart cities based on the internet of things using big data analytics. Comput Netw 101:63–80

Shiu Y-S, Chang SY, Wu H-C, Huang SC-H, Chen H-H (2011) Physical layer security in wireless networks: a tutorial. IEEE Wirel Commun 18(2)

Singla R, Kaur N, Koundal D, Lashari SA, Bhatia S (2021) Optimized energy efficient secure routing protocol for wireless body area network. IEEE Access

Sinha P, Jha VK, Rai AK, Bhushan B (2017) Security vulnerabilities, attacks and countermeasures in wireless sensor networks at various layers of OSI reference model: a survey. In: 2017 international conference on signal processing and communication (ICSPC). https://doi.org/10.1109/cspc.2017.8305855

Steinmetzer D, Ahmad S, Anagnostopoulos N, Hollick M, Katzenbeisser S (2018) Authenticating the sector sweep to protect against beam-stealing attacks in IEEE 802.11 ad networks. In: Proceedings of the 2nd ACM workshop on millimeter wave networks and sensing systems, conference proceedings. ACM, pp 3–8

Ta-Shma P, Akbar A, Gerson-Golan G, Hadash G, Carrez F, Moessner K (2018) An ingestion and analytics architecture for IoT applied to smart city use cases. IEEE Internet Things J 5(2):765–774

Teniou A, Bensaber B (2018) Efficient and dynamic elliptic curve qu-vanstone implicit certificates distribution scheme for vehicular cloud networks. Secur Privacy 1(1):e11

Wang C, Wang H-M (2016) Physical layer security in millimeter wave cellular networks. IEEE Trans Wirel Commun 15(8):5569–5585

Wang D, Bai B, Zhao W, Han Z (2018) A survey of optimization approaches for wireless physical layer security. IEEE Commun Surv Tutor 21(2):1878–1911

Wang N, Jiao L, Zeng K (2018) Pilot contamination attack detection for NOMA in mm-wave and massive MIMO 5G communication. In: 2018 IEEE conference on communications and network security (CNS), conference proceedings. IEEE, pp 1–9

Wang N, Wang P, Alipour-Fanid A, Jiao L, Zeng K (2019) Physical-layer security of 5G wireless networks for IoT: challenges and opportunities. IEEE Internet Things J 6(5):8169 8181

Wang N, Jiao L, Wang P, Li W, Zeng K (2021) Exploiting beam features for spoofing attack detection in mmwave 60-GHz IEEE 802.11 ad networks. IEEE Trans Wirel Commun 20(5):3321–3335

Wood AD, Stankovic JA, Zhou G (2007) DEEJAM: defeating energy-efficient jamming in IEEE 802.15.4-based wireless networks. In: 4th annual IEEE communications society conference on sensor, mesh and ad hoc communications and networks, 2007. SECON'07, conference proceedings. IEEE, pp 60–69

Xiao L, Greenstein LJ, Mandayam NB, Trappe W (2009) Channel-based detection of sybil attacks in wireless networks. IEEE Trans Inf Forensics Secur 4(3):492–503

Xu LD, He W, Li S (2014) Internet of things in industries: a survey. IEEE Trans Ind Inform 10(4):2233–2243

Yang H, Shi M, Xia Y, Zhang P (2018) Security research on wireless networked control systems subject to jamming attacks. IEEE Trans Cybern 49(6):2022–2031

Ylmaz MH, Arslan H (2015) A survey: spoofing attacks in physical layer security. In: 2015 IEEE 40th conference proceedings on local computer networks conference workshops (LCN workshops). IEEE, pp 812–817

Zanella A, Bui N, Castellani A, Vangelista L, Zorzi M (2014) Internet of things for smart cities. IEEE Internet Things J 1(1):22–32

Zeng K (2015) Physical layer key generation in wireless networks: challenges and opportunities. IEEE Commun Mag 53(6):33–39

Zeng K, Govindan K, Mohapatra P (2010) Non-cryptographic authentication and identification in wireless networks [security and privacy in emerging wireless networks]. IEEE Wirel Commun 17(5)

Zhang K, Liang X, Lu R, Shen X (2014) Sybil attacks and their defenses in the internet of things. IEEE Internet Things J 1(5):372–383

Zhao K, Ge L (2013) A survey on the internet of things security. In: 2013 9th international conference on computational intelligence and security (CIS), pp 663–667

Zhou X, Maham B, Hjorungnes A (2012) Pilot contamination for active eavesdropping. IEEE Trans Wirel Commun 11(3):903–907

Zhu J, Schober R, Bhargava VK (2014) Secure transmission in multicell massive MIMO systems. IEEE Trans Wirel Commun 13(9):4766–4781

Chapter 8
Energy Efficiency and Scalability of 5G Networks for IoT in Mobile Wireless Sensor Networks

Smriti Sachan, Rohit Sharma, and Amit Sehgal

Abstract A widespread deployment of 5G technology with the Internet of Things (IoT) will be there in future years. The implementation of 5G technology perhaps becomes fortuitous for IoT as IoT has different variants of applications in the field of tracking data, and security systems. It is also applicable to applications like smart cities and smart buildings etc. Further, the introduction of the new frequency band in the present communication system gardened the interest of researchers in the area of optimization of energy in a mobile environment with dense traffic. This paper aims to represent the basics of 5G system along with IoT implementations. Also different techniques for energy efficiency are comparatively analyzed with their pros and cons for mobile wireless sensor networks.

Keywords Internet of things (IoT) · 5G technology · Energy efficiency · Mobile wireless sensor networks

Introduction

The new expansion of cellular network reaches to 5G technology. 5G is affected by several factors like high mobility access in a dense area, lifespan of battery, communication system, traffic scenarios, low latency, etc. 5G technology also enters to the industrial sector with the help of IoT technique. As the clients of mobile servers increases day by day, the apprehension for network issues also arises simultaneously. The introduction of IoT systems based on 5G technology resolves various problems related to network and the combination of these technologies works as a backbone to the growing economy of the emerging field (Alsharif et al. 2018).

S. Sachan (✉) · R. Sharma
Department of Electronics and Communication, SRM Institute of Science and Technology, NCR Campus, Ghaziabad, India
e-mail: smritisachan@gmail.com

A. Sehgal
School of Engineering and Technology, Sharda University, Greater Noida, India

© The Author(s) 2023

B. Bhushan et al. (eds.), *5G and Beyond*, Springer Tracts in Electrical and Electronics Engineering, https://doi.org/10.1007/978-981-99-3668-7_8

There are various unresolved challenges in the system such as security at the level of edge devices, large-scale deployments, low latency, communication issues and network connectivity. The IoT technology made the system more efficient and smart. It is applied for different areas like transportation system, electricity supply system, water reservation, smart homes, and the whole smart city. IoT incorporation with 5G improves the quality of service, communication system, network connectivity, etc. Also the cellular networks associated with IoT contributes toward the growth of economy of country as well as efficiency of communication system for human lives (Ericsson Mobility Report 2018; Yaqoob et al. 2019). There is an exponential increase in mobile subscribers which also enhances the data traffic and the average utilization of data per user also increases (Sah et al. 2019). The data operators have to supervise the performance of the whole network and energy efficiency has to be maintained while retaining the connectivity (Alsharif and Nordin 2017). According to today's environmental and economic conditions, the most focused area is energy efficiency for the operators (Abrol et al. 2018; Gautam et al. 2019). Researchers are working in the field of "green communication system" nowadays. Carbon neutrality is a heavily desired feature for network providers based upon energy savings. Also, due to energy savings, it is a cost effective method to make the system more reliable and sustainable in terms of the financial aspect of a network (Mowla et al. 2017). The combination of 5G communication and IoT system addresses some of the crucial issues of the network which is not only helpful to the particular mobile user but also to the overall communication system. 5G is not to substitute other running technologies in a cellular system but to provide stability and improvement in the current network so that a strong, reliable, sustainable and fast communication system can be established globally (Hasan et al. 2011).

The above given background shows that there are lots of issues that can be solved by using the combination of various techniques of 5G, IoT, and MWSN. Also there are lots of challenges in the field of energy efficiency in 5G communication system. So, this motivates us to do a review and study of various parameters which is helpful to understand the accountability of energy efficiency and IoT services in 5G system for mobile wireless sensor networks. The major contribution of this work can be summarized as follows.

- Identifying the issue based upon rigorous literature study.
- To make understand the basic concept of energy efficiency, 5G system along with IoT in MWSN.
- To summarize the concept of energy efficiency, 5G network through its frequency bands, technologies, IoT-based services, etc. and IoT techniques, its challenges and security threats, etc.

The remainder of the work is organized as follows. Section "Energy Efficiency in Mobile Wireless Sensor Networks" discusses the energy efficiency in terms of IoT in MWSN. Further, the section presents the basics of energy efficiency in MWSN through its architecture and existing schemes. The evolution of 5G networks, its frequency band, security issues, etc. are addressed in third section. Section "IoT Services Based on 5G" shows the IoT services, challenges, industry supported techniques followed by the "Conclusion" section.

Energy Efficiency in Mobile Wireless Sensor Networks

One of the most desired features of sensor networks is energy efficiency. The optimization of energy with respect to mobile nodes is a very tough task. The below-given Fig. 8.1 is of basic architecture of a mobile wireless sensor network. In this type of network, sensor nodes are mobile, and at every time instance, they change their position resulting in frequent connection make-break with the access points. This makes it really tedious to analyze these kinds of systems.

The evolution of sensor nodes has seen an increase in low-power devices with reduced power utilization thus making them suitable for power constraint systems. The sensors work together to form a network known as a wireless sensor network. They are used for data processing at high-risk zones as well for environmental factors and many more. The critical issue associated with sensors is that battery lifetime is

Fig. 8.1 Basic architecture of mobile wireless sensor network (MWSN)

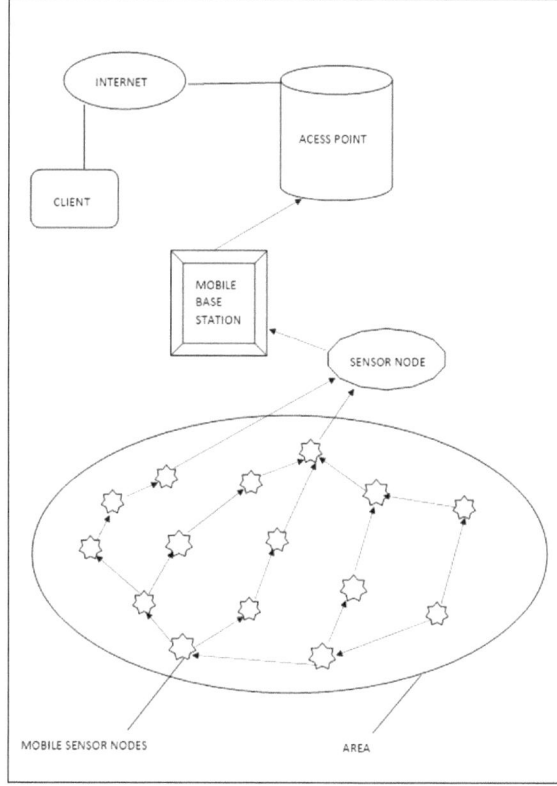

limited and uninterrupted connectivity must be maintained (Mostafaei et al. 2018). The energy efficiency is maintained by optimizing power consumption. Also, the network designing plays a vital role to utilize the resources efficiently (Liu et al. 2018; Feng et al. 2013). The percolation subject is there to learn about network issues (Wang et al. 2017). The researchers explore the probability of network connectivity while optimizing the energy consumption. Also the battery life time is enhanced by reducing the transmit power. In MWSN the major task is to continue communication with the same energy level as in mobile wireless sensor network at every instance of time even when the point of contact changes. The modern sensor network is deployed in association with coverage of the network for ad-hoc networks. Moreover, the network is divided into smaller cell with single and multiple hops (Naghibi and Barati 2020). The reinforcement technique is used to observe the natural conditions to identify the transmission and reception of the signal in the mobile wireless sensor network. It is validated with the help of simulation (Banerjeea et al. 2020). A selective method for hoping is used to maximize the lifetime of the network (Kakhandki et al. 2018). Some of the researchers work on the capacity and cost of the system, they give a power management system to enhance the capacity of the system which is cost effective (Sofi and Gupta 2018). The authors of Sofi and Gupta (2018) developed a scheduling algorithm and energy-based routing protocol for the evaluation of residual energy. It was also compared with other pre-existing techniques (Buzzi et al. 2016; Usama and Erol-Kantarci 2019). In Liu and Wu (2019), the authors presented an algorithm for retaining the connectivity within the network, and some of the researchers presented the survey on clustering algorithm (Buzzi and Poor 2016; Minoli and Occhiogrosso 2019). The paper proposed a security-based routing protocol for enhancing the network lifetime (Shafiabadi et al. 2021). Numerous other published works are available which are based upon energy efficiency, network issue, connectivity, and network lifetime in which the authors used different techniques to improve like graph theory, transmission model for node density, etc. (Usama and Erol-Kantarci 2019; Wang et al. 2019; Kaur and Kaur 2021; Tahiliani and Dizalwar 2018). The below-given Table 8.1 is to show the summary of some of the researchers work in the field of mobile wireless sensor networks in terms of energy efficiency.

Basic Concept of 5G Technology

The vital development in living style leads to the new developments in communication techniques. The growing era shows a significant change in communication traffic volume. Industry observers identified that the role of 5G technology makes the communication system more advanced and fast as compared to the existing telecom network (GSMA 2018). The devices in 5G network are connected in such a manner that it processes through 5G core network via 5G access network. 5G is an expansion of 4G network with new radio features. The 5G core network is implemented to support IoT system with improvements in network slicing and services in comparison to the 4G network. Also, the 3GPP (3rd Generation Partnership Project) base

Table 8.1 Review on existing energy efficiency schemes and challenges

Sr. No.	Reference	Researchers work	Limitations
1	Mostafaei et al. (2018)	The researchers proposed a dynamic energy-efficient routing protocol (DEER) to enhance the network lifetime	The technique can be combined with energy harvesting to improvise
2	Liu et al. (2018)	The paper presents a comparative analysis of different energy efficiency protocols	Real-time data processing, security analysis, and reliability should be considered
3	Feng et al. (2013)	The researchers proposed an SINR model for connectivity issues	Data dissemination for communication should be taken care of
4	Wang et al. (2017)	The paper focuses on enhancing the lifetime of the network along with load balancing	Security mechanism to protect from threat in WSN
5	Naghibi and Barati (2020)	The authors present algorithms for task scheduling	Some other parameters can also be included
6	Banarjee et al. (2020)	The authors present a survey on routing protocols	Some more real-time applications can be focused
7	Kakhemdki et al. (2018)	The paper presents a survey on data acquisition and mobility	The other parameters can be considered as speed and security of the network
9	Sofi and Gupta (2018)	The authors proposed a multi-tier network architecture for spectrum sharing, use of massive MIMO, and solution for hardware changes	The overall proposed system architecture is well defined for design and planning
10	Buzzi et al. (2016)	The researchers present the concept of energy efficiency for 5G technology in terms of energy harvesting, deployment of the network, traffic density and offloading parameters, etc.	The paper presents good network planning issues but not addressed any virtualization technique to solve them
11	Usama and Erol-Kantarci (2019)	The authors present a survey on energy efficiency at radio access level	The edge network problems are not discussed in the paper for energy efficiency
12	Liu and Wu (2019)	The paper presents an energy balancing scheme to reduce the power consumption	More description of power line cables can be added
14	Buzzi and Poor (2016)	The paper shows a survey on designing a 5G network	Interference and Randomness can also be included as parameters for challenges
16	Minoli and Occhiogrosso (2019)	The paper reviews the practical aspect of the 5G network	Overall coverage of topics is good

(continued)

Table 8.1 (continued)

Sr. No.	Reference	Researchers work	Limitations
17	Shafiabadi et al. (2021)	The researchers have a given a new routing protocol for energy management	The overall energy optimization is done but the latency constraint is not managed
18	Usama and Erol-Kantarci (2019)	The paper presents a survey of various issues present in 5G networks	The SINR ratio is shown by 1 bit/joule notion
19	Wang et al. (2019)	The major focus of the paper is to show the balancing of network connectivity with communication	Simulation can also be done
20	Kaur and Kaur (2021)	The authors present a survey on network lifetime, energy optimization, and energy efficiency	Some comparisons can be shown
21	Zanaj et al. (2021)	A survey of IoT associated with LPWANS and LPSANS is presented by researchers	Use of Optimal Relay Selection
22	Usama and Erol-Kantarci (2019)	Paper surveys the recent works on energy efficiency and wireless networks	Less study of self-learning mechanism
23	Tahiliani and Dizalwar (2018)	The trends of IoT technology	Practical WSN-based IoT networks to study
24	Yaohua et al. (2019)	Paper focuses on energy management issues in 5G	More techniques can be shown to optimize the EE network
25	Bakht et al. (2019)	The paper focuses on power transmission, wireless information transfer, and routing techniques	Energy harvesting can be used in more efficient manner

stations are included in 5G access network. The important aspects of 4G systems like low-power system, low latency, and energy optimization in NB-IoT are used in 5G systems. The slicing of network allows users to use network as a service and the wireless traffic is enhanced by using these services like virtual services, augmented reality, IoT techniques, entertainment with high resolution, etc. The implementers of network constructed the design in such a way that it covers all the remote areas as well as the urban one. The signals spread all over the network uniformly, in many countries the transition in cellular network is used by new generation as compared to the former technology used. Therefore, the 5G is evolved with the new era and it is faster than existing 4G networks (Goyal et al. 2021; Ahmad et al. 2020). The fundamentals of 5G networks are laid down in some of the advanced countries and but the technology used before 5G like 4G/LTE is also existing still there. The realization of the 5G with various situations is described in Table 8.2.

Table 8.2 Realization of 5G technology for different cases

Sr. No.	Parameters	Accomplished network	Precis
1	Movability	–	Basics of 5G and industry 4.0 (Rao and Prasad 2018)
2	Mobile edge computing	Demonstration in physical world	Robotic arm control on MEC (Tsokalo et al. 2019)
3	Compatibility	Demonstration in physical world	Industrial application of 5G in real world (Voigtländer et al. 2018)
4	Slicing of network	Simulation analysis	Simulation and framework of autonomous vehicles based on 5G (Chekired et al. 2019)
5	Reliability	Network infrastructure	Manufacturing system based on 5G communication (Karrenbauer et al. 2019)
6	Slicing of network	Network architecture	Network slicing for industry 4.0 in 5G system (Taleb et al. 2019)
7	Compatibility	Measurement and demonstration in physical world	Distributed control system for robotics (Voigtländer et al. 2017)
8	Network function virtualization	Framework prototyping	Demonstration of smart manufacturing through NFV (Peuster et al. 2019)
9	Mobile edge computing	System infrastructure	IoT smart manufacturing in 5G system (Cheng et al. 2018)
10	–	–	Construction management IoT (Reja and Varghese 2019)
11	mmWave	None	Localization using 5G (Lu et al. 2018)
12	Network function virtualization	Network architecture	Prototype of smart production through NVF (Schneider et al. 2019)
13	Reliability	System architecture	Palpable internet for 5G system (Tsokalo et al. 2019)
14	Slicing of network	Network infrastructure	Tele surgery description for 5G (Ahmad et al. 2017)

Evolution of 5G

5G network is the enhancement of 4G technology and has a new radio access that is also known as 5G "New Radio (NR)". The new network of 5G system has different features like controlling of user pane, virtualization of network, slicing of network, low latency, and high speed data (Priyadarshini et al. 2021a). The designing of 5G NR model is made in such a way that it should be compatible with existing LTE system. The configuration of new 5G NR system is dual frequency in nature. This is the reason it is compatible with LTE system and also with narrow band IoT. It might be happened that different elements have to be inserted in 5G system with different access, to do

Table 8.3 5G technology security threats (Dhiman and Sharma 2021)

Sr. No.	Elements of network	Security issues	Particulars that are effected		
			Medium	Seclusion	Analogy of cloud
1	Identity of client	IMSI attack	Yes	Yes	No
2	Location of client	Data access attack	Yes	Yes	No
3	Controlled centralized elements	Service denial attack	No	No	Yes
4	SDN controller	Hijacking	No	No	No
5	Core element of 5G	Signaling	Yes	No	Yes
6	Cloud resource sharing	Stealing of resources	No	No	Yes
7	Routers	Attack on configuration	No	No	No
8	Data base of users	Clients identity tampering	No	Yes	Yes
9	Switch communication	TCP attack	Yes	Yes	No
10	Decrypted medium	Key exposure	Yes	No	No

this basically two techniques are in fashion right now, i.e., non-standalone (NSA) and standalone (SA). The standalone technique contains all the core part of 5G radio access and the Non-standalone technique uses dual frequency mode to access with existing LTE packet core. The network deployment in different parts of countries may take a long time might be a decade as every area has its own situation and migration configuration will vary according to the various situations of the different areas (Dhiman and Sharma 2021). The implementation of IoT requires information related to 5G features and new radio access. The service provider continues with LTE and existing network features along with 5G NR to access the IoT-based network. The 5G network also supports 28 GHz millimeter-wave (mmWave) spectrum. Also the system is threatened by various types of threats that can damage the network in various ways. The effect of threats in a 5G network is discussed in Table 8.3 with respect to the different elements of the system (Varga et al. 2017; Arora et al. 2020).

Frequency Band of 5G System

The spectrum of 5G spreads over multiple frequency bands starting from sub-GHz to millimeter-wave. The sub-bands are categorized into three macro groups, i.e., 1, 1–6, and above 6 GHz. 5G spectrum very well uses the millimeter waves for a higher data rate. The 5G generation has more bandwidth and higher frequency rate. Earlier without modified MIMO antenna only around 10 bits per hertz was the channel bandwidth (Sachan et al. 2021). Now the adaption of new radio techniques like D2D, massive MIMO, ultra-dense networks, etc. enhances the data rate of the new mobile generation and IoT environment. The mm-wave spectrum is from 30 to 300 GHz and also called as extremely high frequency (EHF). The 5G network mostly uses 3.5 GHz, mm frequency bands ranging from 30 to 73 GHz, also some other frequency bands. To fulfill the scarcity of spectrum the service provider has to use the spectrum effectively and the utilization of smaller cell has to be keenly observed. The 5G technology is about to use beam-tracking and beam forming while the cell antenna is focused on the signal when device is tracked when in mobility (Ghanem et al. 2021). The throughput and directivity are optimized by using the beamforming technique as it uses a huge number of antennas to access the signal.

5G Innovative Technologies

- **Millimeter Waves**: These are used for shorter distance and has a range of 30 GHz to 300Ghz. These are used for smaller cells at shorter distance.
- **Massive MIMO**: This makes an efficient spectrum utilization as large number of antennas are connected together to cover huge area. It has less interference due to beamforming capability which makes it more efficient to use.
- **Heterogeneous Network (HetNet)**: It provides good capacity and coverage with different technologies.
- **Software Define Radio (SDN)**: It divides the plane as data and control to provide the high speed network. It manages the network more efficiently to do the further processing.
- **Network Functions Virtualization (NFV)**: It transfers the functions to virtual networks like servers, switches, hardware, etc., this is efficient as it full the requirement of hardware changes also. It decouples the hardware from the system which enhances the scalability of the network.

5G Network slicing is also one of the important parameters of the 5G system. It is the type of configuration that allows various networks to work on the same platform. The network is divided into various slices and each slice is according to the need of the application. Figure 8.2 shows the 5G network slicing example as per the applications.

IoT Services Based on 5G

The total revenue of IoT worldwide is expected to increase by 23% by 2025. The integration of features of both NB-IoT and LTE-M along with 5G enhances the capability of IoT structure. There are lots of commercial deployments of IoT networks at the global level. It will be a tough task for operators to meet the data rate requirement

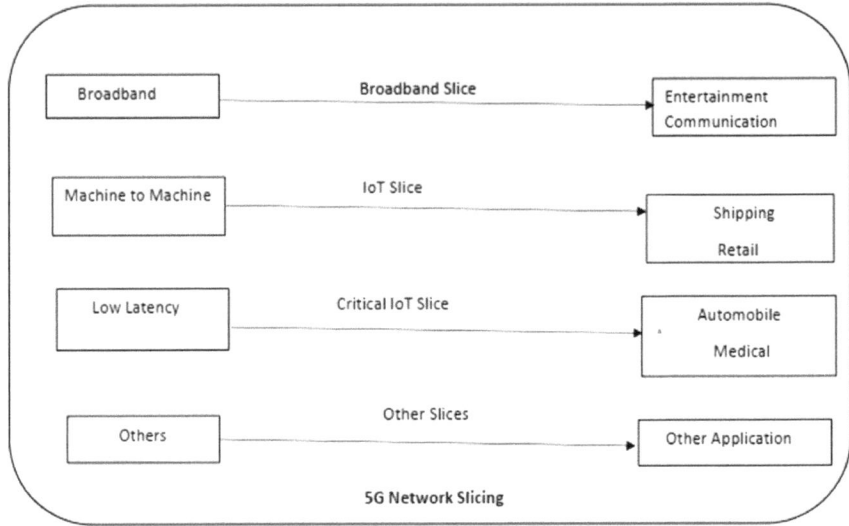

Fig. 8.2 5G network slicing

of the user as per transmission demands. The operators have to work hard on their service models to enhance the efficiency of the network. The major market (around 70%) of IoT networks will be covered by its platforms, services, and applications by 2025. Mostly the services which are provided professionally will be enhanced by 25% in the near future by using IoT networks.

Figure 8.3 shows the basic architecture of the internet of things. It shows that data is sent through the devices via gateway to the cloud. It can be processed there and clients can use that data. Each and every data is stored in the cloud and one can process it by deriving it from there only.

IoT Industry Supported by 5G Technology

The modern techniques and application of IoT are supported by 5G technology. In this section cases of industries with the adoption of modern methodology are discussed which have the specific role of 5G in it. Table 8.4 summarizes the resistance of various wireless technologies to the Internet of Things (IoT).

Industry 4.0

5G supports Industry 4.0 for most of the applications and methodologies incorporated in the industry. In the manufacturing industry, the recent trend is information

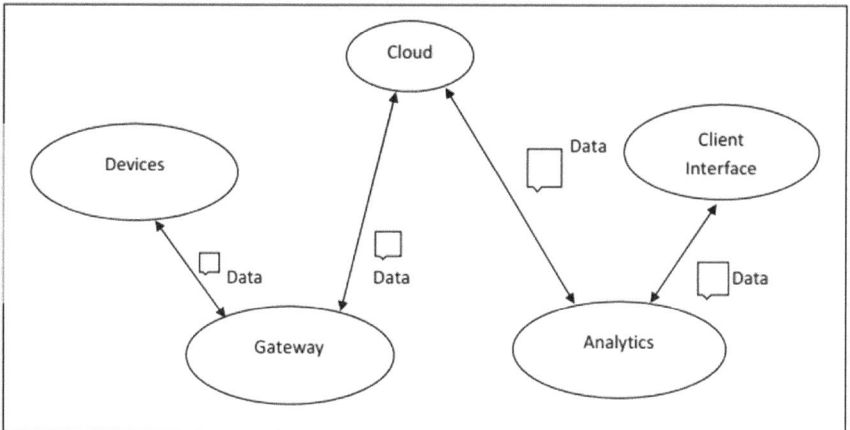

Fig. 8.3 Basic architecture of IoT network

exchange and automation which is known as industry 4.0 (Sachan et al. 2021). The major area of focus is cyber-physical system, the Internet of things, and cloud cognitive computing. It creates a smart factory with a modern structure in which cyber-physical system processes the physical network. The limitations in 3G/4G are omitted with the help of 5G technology like the huge population of devices, large energy consumption, more delay and efficacy of wireless network, etc. The researchers focus on 5G industrial applications for various features like high data rates, low latency, robustness, etc. The 5G network shows a better option as compared to the wired network. It also has good industrial applications with heterogeneous data resources. It has a variety of features that emphasis on energy efficiency, mobility in network, virtualization and mesh network, etc. The researchers further listed the communication network features, processing, infrastructure issues, and reference architecture of Industry 4.0. The distributed robotics capabilities are shown with the help of 5G technology by the authors. The mobile robots are used to detect critical issues in real-time 5G scenarios. The communication is set up between mobile robot and cloud server to process the critical loads. The complexity of NB-IoT performance is also calculated with the help of cloud server in harsh environment, also it has been observed that LTE cannot full fill the requirement of industry (Sachan et al. 2021). There is various technique that enable LTE system to work under 10 ms of delay (Priyadarshini et al. 2021b), but at other end in 5G technology the process can be done under 5 ms of delay with more than 50% of traffic load (Priyadarshini et al. 2021c; Singh et al. 2021) which shows the capabilities of the 5G technology in comparison to the existing ones.

Table 8.4 Persistency of wireless technology for IoT (Sachan et al. 2021; Priyadarshini et al. 2021b)

Sr. No.	Technology used	Key features	Accessibility in indoor	Accessibility in outdoor
1	Zigbee	• Applications related to home and industry • Less power consumption and better battery optimization • Low data rates	Accessibility (30–300 feet)	No Accessibility
2	Wi-Fi	• Different bands • Indoor IoT usage (mostly adopted) • Better functionality	Accessibility (300 feet)	To some extent
3	Bluetooth	• Application in the industrial and medical field • Real-time location detection usage (for limited range) • Less bandwidth	Accessibility (30 feet)	No accessibility
4	Sigfox	• Narrowband • Less power • Less bandwidth	To some extent	Accessibility (30 miles in rural and 1–6 miles in urban)
5	LTE-M	• Less power modem used • Cellular architecture deployment • Used for tracking mobile objects • Use 4G-LTE bands below 1 GHz	Accessibility	Accessibility (10–20 miles)
6	NB-IoT	• Licensed spectrum • Less power • Less cost • Less penetration in buildings	Accessibility	Accessibility (20 miles)
7	5G	• Various bands • Not widely deployed • Cellular network architecture • Cost effective • Licensed spectrum	Accessibility	Accessibility (10–15 miles)

Palpable Internet

The palpable Internet performs all the operations of physical interaction while incurring manipulations. The concept of the palpable Internet is to operate with health care system, smart grids, infrastructure, education, etc. (Sahu et al. 2021). It enables the technique to control virtual and real platform by humans. It can be controlled

Table 8.5 Challenges and solutions supported by IoT (Sharma et al. 2021a)

Sr. No.	Urban challenges	Pertinency of 5G	Reliability/latency	Solutions supported by IoT
1	Storm and flood control	Exists	Good/average	Sensors are placed to detect early fault and warning is placed
2	Nature monitoring	Exists	Average/average	Environmental parameters are monitored by placing temperature sensors, humidity sensors, pressure sensor, etc.
3	Monitoring of pollution	Exists	Good/average	Sensors are placed to identify toxic elements generated from factories, plants, vehicular pollution, etc.
4	Mobility management	Exists	Good/poor	System sensors are placed to monitor traffic flow, logistics, good transportation, etc.
5	Power supporting utilities	Exists	Good/average	Infrastructure with smart grid solutions
6	Security of the network	Exists	Good/average	Sensors and drones are placed to detect gunshots, biohazards, crowd monitoring, and plated reading for license
7	Real estate management	Exists	Average/poor	Sensors and drones are placed to check building functionality, water resources, electric meters, smart parking, etc.
8	Quality of life	Exists	Average/average	Sensors are deployed to detect real-time health problems (air quality), multi-modal infrastructure, transportation systems, etc.

by human reaction of listening, visualizing, and manual interaction. The need for palpable Internet is to maintain connectivity in critical situation to achieve user's requirements. The basics of 5G features and services are elaborated to deal with palpable internet (Sharma et al. 2021a). The 5G system architecture is based on software-based design on the basis of distributed cloud structure. Table 8.5 shows the different types of issues and their solutions that are presented in the IoT system.

Security of IoT in 5G System

To deal with security concerns is the most critical task in the area of IoT. In today's era the security concern is the most important aspect of day-to-day life, so many IoT devices are placed for residential as well as industrial security. The deployment of IoT devices is widespread to achieve the solutions for threats and security issues of

Table 8.6 Security menaces in IoT with their alleviation possibility (Sharma et al. 2021b)

Sr. No.	Type of menaces	Layer	Alleviation
1	Information access common channel application of client Multiple client access	Application layer	Trackable Verification, authentication Filtering of anti-virus Design, planning, and process
2	Spyware Social planning Enervation	Data layer	Detection of spyware Spreading threat awareness Monitoring of congestion
3	Bug Service denial	Network layer	Detection and encryption Firewall usage
4	Service denial	Physical	Use of spread spectrum technique

the network (Ha et al. 2021). As compared to the conventional network IoT structure has different security issues, also the IoT devices are low-power device with low storage capacity that's why all types of solutions cannot be used for end to end security issues (Arsh et al. 2021). The network has a heterogeneous system for the security threats and relates it with cloud servers to process the IoT services (Sharma et al. 2021b). Table 8.6 summarizes the security issues in IoT with their alleviation possibility for different layers.

Conclusion

The chapter presents a holistic approach to show the review on energy efficiency with their challenges in mobile wireless sensor networks. The basics of 5G network are very well analyzed in the paper in terms of their evolution, technologies, security issues, and frequency bands. The role of IoT networks w.r.t to 5G technology is also discussed and shows that it has a very high impact on overall 5G system associated with IoT and mobile wireless sensor network (MWSN).

References

Abrol A, Jha RK, Jain S, Kumar P (2018) Joint power allocation and relay selection strategy for 5G network: a step towards green communication. Telecommun Syst 68:201–215

Ahmad I, Kumar T, Liyanage M, Okwuibe J, Ylianttila M, Gurtov A (2017) 5G security: analysis of threats and solutions. In: Proceedings of the IEEE conference on standards for communications and networking (CSCN), Helsinki, Finland

Ahmad A, Bhushan B, Sharma N, Kaushik I, Arora S (2020) Importunity and evolution of IoT for 5G. In: 2020 IEEE 9th international conference on communication systems and network technologies (CSNT), pp 102–107. https://doi.org/10.1109/CSNT48778.2020.9115768

Alsharif MH, Nordin R (2017) Evolution towards fifth generation (5G) wireless networks: current trends and challenges in the deployment of millimetre wave, massive MIMO, and small cells. Telecommun Syst 64:617–637

Alsharif MH, Nordin R, Abdullah NF, Kelechi AH (2018) How to make key 5G wireless technologies environmental friendly: a review. Trans Emerg Telecommun Technol 29:e3254

Arora S, Sharma N, Bhushan B, Kaushik I, Ahmad A (2020) Evolution of 5G wireless network in IoT. In: 2020 IEEE 9th international conference on communication systems and network technologies (CSNT), pp 108–113. https://doi.org/10.1109/CSNT48778.2020.9115773

Arsh M, Bhushan B, Uppal M (2021) Internet of things (IoT) toward 5G network: design requirements, integration trends, and future research directions. In: Hassanien AE, Bhattacharyya S, Chakrabati S, Bhattacharya A, Dutta S (eds) Emerging technologies in data mining and information security. Advances in intelligent systems and computing, vol 1286. Springer, Singapore. https://doi.org/10.1007/978-981-15-9927-9_85

Bakht K, Jameel F, Ali Z, Khan WU, Khan I, Sardar Sidhu GA, Lee JW (2019) Power allocation and user assignment scheme for beyond 5G heterogeneous networks. Wirel Commun Mobile Comput Hindawi

Banerjeea PS, Mandala SN, Deb D, Maiti B (2020) RL-sleep: temperature adaptive sleep scheduling using reinforcement learning for sustainable connectivity in wireless sensor networks. Sustain Comput Inform Syst 26:100380. Elsevier

Buzzi S, Poor V (2016) A survey of energy efficient techniques for 5G Networks and challenges ahead. arXiv:1604.00786v1 [cs.IT]

Buzzi S, Chih-Lin I, Klein TE, Poor HV, Yang C, Zappone A (2016) A survey of energy-efficient techniques for 5G networks and challenges ahead. IEEE J Sel Areas Commun 34(4):697–709

Chekired DA, Togou MA, Khoukhi L, Ksentini A (2019) 5G-slicing-enabled scalable SDN core network: toward an ultra-low latency of autonomous driving service. IEEE J Sel Areas Commun 37:1769–1782

Cheng J, Chen W, Tao F, Lin CL (2018) Industrial IoT in 5G environment towards smart manufacturing. J Ind Inf Integr 10:10–19

Dhiman G, Sharma R (2021) SHANN: an IoT and machine-learning-assisted edge cross-layered routing protocol using spotted hyena optimizer. Complex Intell Syst. https://doi.org/10.1007/s40747-021-00578-5

Ericsson Mobility Report (2018) Available online

Feng D, Jiang C, Lim G, Cimini LJ, Feng G, Li GY (2013) A survey of energy-efficient wireless communications. IEEE Commun Surv Tutor 15:167–178

Gautam S, Malik A, Singh N, Kumar S (2019) Recent advances and countermeasures against various attacks in IoT environment. In: 2019 2nd international conference on signal processing and communication (ICSPC). https://doi.org/10.1109/icspc46172.2019.8976527

Ghanem S, Kanungo P, Panda G et al (2021) Lane detection under artificial colored light in tunnels and on highways: an IoT-based framework for smart city infrastructure. Complex Intell Syst. https://doi.org/10.1007/s40747-021-00381-2

Goyal S, Sharma N, Kaushik I, Bhushan B, Kumar N (2021) A Green 6G network era: architecture and propitious technologies. In: Khanna A, Gupta D, Pólkowski Z, Bhattacharyya S, Castillo O (eds) Data analytics and management. Lecture notes on data engineering and communications technologies, vol 54. Springer, Singapore. https://doi.org/10.1007/978-981-15-8335-3_7

GSMA (2018) Road to 5G: introduction and migration. https://www.gsma.com/futurenetworks/wp-content/uploads/2018/04/Road-to-5G-Introduction-and-MigrationFINAL.pdf

Ha DH, Nguyen PT, Costache R et al (2021) Quadratic discriminant analysis based ensemble machine learning models for groundwater potential modeling and mapping. Water Resour Manag. https://doi.org/10.1007/s11269-021-02957-6

Hasan Z, Boostanimehr H, Bhargava VK (2011) Green cellular networks: a survey, some research issues and challenges. IEEE Commun Surv Tutor 13:524–540

Kakhandki AL, Hublikar S, Kumar P (2018) Energy efficient selective hop selection optimization to maximize lifetime of wireless sensor network. Alexandria Eng J 57:711–718. Elsevier

Karrenbauer M, Ludwig S, Buhr H, Klessig H, Bernardy A, Wu H, Pallasch C, Fellan A, Hoffmann N, Seelmann V et al (2019) Future industrial networking: from use cases to wireless technologies to a flexible system architecture. At-Autom 67:526–544 (In Germany)

Kaur L, Kaur R (2021) A survey on energy efficient routing technologies in WSNS focusing IoT Applications and enhancing for computing paradigm. In: Global transitions proceedings, vol 2, issue 2, pp 520–529

Liu X, Wu J (2019) A method for energy balance and data transmission optimal routing in wireless sensor networks. Sensors 19(13):3017

Liu Z, Tsuda T, Watanabe H, Ryuo S, Iwasawa N (2018) Data driven cyber-physical system for landslide detection. Mob Netw Appl

Lu Y, Richter P, Lohan ES (2018) Opportunities and challenges in the industrial internet of things based on 5G positioning. In: Proceedings of the 2018 8th international conference on localization and GNSS (ICL-GNSS), Guimaraes, Portugal, pp 1–6

Minoli D, Occhiogrosso B (2019) Practical aspects for the integration of 5G networks and IoT applications in smart cities environments. In: Integration of 5G networks and internet of things for future smart city, vol 2019

Mostafaei H, Chowdhury MU, Obaidat MS (2018) Border surveillance with WSN systems in a distributed manner. IEEE Syst J 12:3703–3712

Mowla MM, Ahmad I, Habibi D, Phung QV (2017) A green communication model for 5G systems. IEEE Trans Green Commun Netw 1:264–280

Naghibi M, Barati H (2020) EGRPM: energy efficient geographic routing protocol based on mobilesink in wireless sensor networks. Sustain Comput Inform Syst 25:100377. Elsevier

Peuster M, Schneider S, Behnke D, Müller M, Bök P, Karl H (2019) Prototyping and demonstrating 5G verticals: the smart manufacturing case. In: Proceedings of the 2019 IEEE conference on network softwarization (NetSoft), Paris, France, pp 236–238

Priyadarshini I, Mohanty P, Kumar R et al (2021b) A study on the sentiments and psychology of twitter users during COVID-19 lockdown period. Multimed Tools Appl. https://doi.org/10.1007/s11042-021-11004-w

Priyadarshini I, Kumar R, Sharma R, Singh PK, Satapathy SC (2021a) Identifying cyber insecurities in trustworthy space and energy sector for smart grids. Comput Electr Eng 93:107204

Priyadarshini I, Kumar R, Tuan LM et al (2021c) A new enhanced cyber security framework for medical cyber physical systems. SICS Softw Intens Cyber Phys Syst. https://doi.org/10.1007/s00450-021-00427-3

Rao SK, Prasad R (2018) Impact of 5G technologies on Industry 4.0. Wirel Pers Commun 100:145–159

Reja V, Varghese K (2019) Impact of 5G technology on IoT applications in construction project management. In: ISARC. Proc. Int. Symp. Autom. Robot. Constr., vol 36, pp 209–217

Sachan S, Sharma R, Sehgal A (2021) SINR based energy optimization schemes for 5G vehicular sensor networks. Wirel Pers Commun. https://doi.org/10.1007/s11277-021-08561-6

Sachan S, Sharma R, Sehgal A (2021) Energy efficient scheme for better connectivity in sustainable mobile wireless sensor networks. Sustain Comput Inform Syst 30:100504

Sah DK, Kumar DP, Shivalingagowda C, Jayasree P (2019) 5G applications and architectures. In: 5G enabled secure wireless networks. Springer, New York, NY, USA, pp 45–68

Sahu L, Sharma R, Sahu I, Das M, Sahu B, Kumar R (2021) Efficient detection of Parkinson's disease using deep learning techniques over medical data. Expert Syst e12787. https://doi.org/10.1111/exsy.12787

Schneider S, Peuster M, Behnke D, Müller M, Bök P, Karl H (2019) Putting 5G into production: realizing a smart manufacturing vertical scenario. In: Proceedings of the 2019 European conference on networks and communications (EuCNC), Valencia, Spain, pp 305–309

Shafiabadi MH, Gafi AK, Manshandy DD, Nouri N (2021) New method to improve energy savings in wireless sensor network. Annals RSCB 25(1):4321–4328. ISSN: 1583–6258

Sharma R, Kumar R, Sharma DK et al (2021a) Water pollution examination through quality analysis of different rivers: a case study in India. Environ Dev Sustain. https://doi.org/10.1007/s10668-021-01777-3

Sharma R, Gupta D, Polkowski Z, Peng S-L (2021b) Introduction to the special section on big data analytics and deep learning approaches for 5G and 6G communication networks (VSI-5g6g). Comput Electr Eng 95:107507. ISSN: 0045 7906. https://doi.org/10.1016/j.compeleceng.2021.107507

Singh R, Sharma R, Akram SV, Gehlot A, Buddhi D, Malik PK, Arya R (2021) Highway 4.0: digitalization of highways for vulnerable road safety development with intelligent IoT sensors and machine learning. Saf Sci 143:105407. ISSN 0925–7535

Sofi IB, Gupta A (2018) A survey on energy efficient 5G green network with a planned multi-tier architecture. J Netw Comput Appl 118:1–28

Tahiliani V, Dizalwar M (2018) Green IoT systems: an energy efficient perspective. In: International conference on contemporary computing (IC3). https://doi.org/10.1109/IC3.2018.8530550

Taleb T, Afolabi I, Bagaa M (2019) Orchestrating 5G network slices to support industrial internet and to shape next-generation smart factories. IEEE Netw 33:146–154

Tsokalo IA, Wu H, Nguyen GT, Salah H, Fitzek HPF (2019) Mobile edge cloud for robot control services in industry automation. In: Proceedings of the 2019 16th IEEE annual consumer communications networking conference (CCNC), Las Vegas, NV, USA, pp 1–2

Usama M, Erol-Kantarci M (2019) A survey on recent trends and open issues in energy efficiency of 5G. Sensors 19(14):3126

Varga P, Plosz S, Soos G, Hegedus C (2017) Security threats and issues in automation IoT. In: Proceedings of the IEEE 13th international workshop on factory communication systems (WFCS), Trondheim, Norway

Voigtländer F, Ramadan A, Eichinger J, Lenz C, Pensky D, Knoll A (2017) 5G for robotics: ultra-low latency control of distributed robotic systems. In: Proceedings of the 2017 international symposium on computer science and intelligent controls (ISCSIC), Dudapest, Hungary, pp 69–72

Voigtländer F, Ali AR, Eichinger J, Grotepaß J, Ganesan K, Canseco FD, Pensky DH, Knoll A (2018) 5G for the factory of the future: wireless communication in an industrial environment. arXiv 2018, arXiv:1904.01476

Wang L, Han T, Li Q, Yan J, Liu X, Deng D (2017) Cell-less communications in 5G vehicular networks based on vehicle-installed access points. IEEE Wirel Commun Lett 24:64–71

Wang L, Yan J, Han T, Deng D (2019) On connectivity and energy efficiency for sleeping-scheduled based wireless sensor networks. Sensors (Basel) 19(9):2126

Yaohua S, Mugen P, Shiwen M (2019) Deep reinforcement learning-based mode selection and resource management for green fog radio access networks. IEEE Internet Things J 6:1960–1971

Yaqoob I, Hashem IAT, Ahmed A, Kazmi SA, Hong CS (2019) Internet of things forensics: recent advances, taxonomy, requirements, and open challenges. Future Gener Comput Syst 92:265–275

Zanaj E, Caso G, De Nardis L, Mohammadpour A, Alay Ö, Di Benedetto M-G (2021) Energy efficiency in short and wide-area IoT technologies—a survey. Technologies 9(1):22

Chapter 9
Security Services for Wireless 5G Internet of Things (IoT) Systems

Ayasha Malik, Veena Parihar, Bharat Bhushan, Rajasekhar Chaganti, Surbhi Bhatia, and Parma Nand Astya

Abstract The Internet of Things (IoT) is an emerging field that has evolved in recent past years and tends to have a major effect on our lives in the coming future. The development of communication techniques is very rapid and tends to achieve many innovative results. With the invention of 5th Generation mobile networks, i.e., 5G, it is becoming an exciting and challenging topic of interest in the field of wireless communication. 5G networks have the ability to connect millions or billions of nodes without affecting the quality of throughput and latency. The 5G technology can develop a truly digital society in which every device may be connected through the Internet. IoT is an emerging technology in which everything can be connected and communicated via the Internet, the term everything may include computing devices, humans, software, platforms, and solutions. The development of this technology leads to the advent of a number of solutions that are helpful for humankind, for example, smart retailing creation of smart cities, smart farming, intelligent transport systems, smart eco-systems, etc. While IoT is a revolutionary technology in the progression of the Internet, it still has some significant challenges for implementation like ensuring security, performance issues, quality of support and saving of energy, etc. Furthermore, the paper elaborates on the motivation of combining two

A. Malik
Delhi Technical Campus (DTC), GGSIPU, Greater Noida, India

V. Parihar (✉)
KIET Group of Institutions Delhi-NCR, Ghaziabad, India
e-mail: veena2parihar@gmail.com

B. Bhushan · P. N. Astya
Department of Computer Science and Engineering, School of Engineering and Technology (SET), Sharda University, Greater Noida, India
e-mail: parma.nand@sharda.ac.in

R. Chaganti
University of Texas, San Antonio, USA

S. Bhatia
College of Computer Science and Information Technology, King Faisal University, Hofuf, Saudi Arabia
e-mail: sbhatia@kfu.edu.sa

© The Author(s) 2023
B. Bhushan et al. (eds.), *5G and Beyond*, Springer Tracts in Electrical and Electronics Engineering, https://doi.org/10.1007/978-981-99-3668-7_9

technology together named IoT and 5G for better communication. Additionally, the paper illustrates the basic architecture of IoT enabling 5G and discussed various solutions to provide communication. Moreover, the paper also discussed the various challenges and research gaps of 5G-IoT technology.

Keywords 5G · Internet of Things · Security · Communication · Devices · Data · Network · Layer · Challenges · Performance · Application

Introduction

The history of mobile communication systems' developments aims at meeting the requirements of humanity. With the passing of time, data rates of mobile communication have been improved and achieved great results compared to previous ones. Generations of mobile communications have progressed through five stages, i.e., from 1 Generation (1G) to the 5 Generation (5G) which is the current generation (Zhang et al. 2019). The generations between 1 to 3G have sequentially evolved on the basis of service qualities and speed factors. In the early stages of the year 2000, the concept of 4G was invented and it was the first communication generation that was totally built on the top of the IP packet switching technique (Liyanage et al. 2018). As the implementation of 4G technology improved over the years, the advantages of previous times started to convert into drawbacks. Currently, the 4G network's speed has turned out to be very low with respect to high latency (Maier and Reisslein 2019). A present time, kinds of solutions are needed which may connect at the data rates up to Gbps. In this sequence, the invention of the next-generation network, i.e., 5G in the duration of the initial 2021s, affected the whole digital world. Particularly, with the advent of the 5G concept, a new area of research has been generated, called as Internet of Things (IoT) (Bonfim et al. 2019). IoT technology can be defined as an integrated solution involving innovative technologies, allowing to connect together the people, platforms, software, services, devices, etc. via the Internet (Kitanov et al. 2021). According to a study by Cisco, by the year 2035, 1000 billion or more devices will be linked together through the Internet. Such devices, forming the network of IoT will be having all the advanced modules of IoT for allowing and implementing the Device to Device (D2D) communications with one another. The applications of IoT may be implemented in almost all the dimensions of humanity such as smart grids, smart farming (Han et al. 2019), self-driven cars (Zhang 2019), healthcare industries (Sarraf 2019), smart homes, supply chain industries, and many more (Architecture 2019). Moreover, some of the main features of 5G technology enabled with IoT are shown in Fig. 9.1.

The history of the development of all network generations presents an important insight that every single generation is evolved for correcting and modifying the drawbacks of earlier generations and for incorporating some new concepts things that the past generations could not apply (Malik et al. 2019). In the early stages year 2020, the concept of IoT has been proposed parallelly with the advent of 5G communication. The prediction according to a study state that by the year 2020 almost 40 billion nodes

Fig. 9.1 Features of 5G-IoT

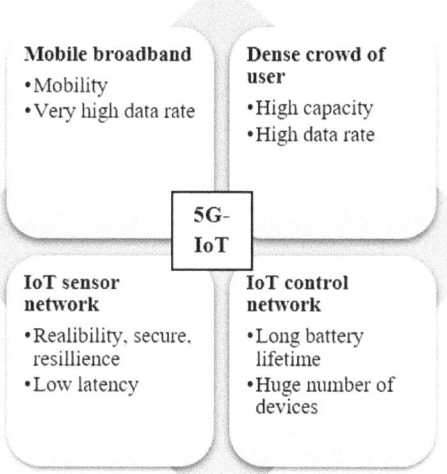

are connected as part of an IoT network (Gautam et al. 2019). A critical challenge of implementing efficient IoT is the requirement of transferring huge data volumes among devices, which derives the need of improving the already existing infrastructures. The IoT technology has transformed the ubiquitous computing by considering various sensors-based applications (Shahabuddin et al. 2018). A huge amount of activity has been noted in IoT-based products and also it is expected to grow in the coming years at a high rate, i.e., billion devices and almost 15 devices on an average per person by the year 2024 (Shafique et al. 2020). In the past research works, most of the issues of protocol-level or device level have been resolved and presently the recent trend involves working on the issues of integration of sensor-based systems and D2D communications. IoT is becoming a central part of 5G wireless communication systems as IoT is the integration of various computing and non-computing, it will form a major part of this 5G network (Chettri and Bera 2020). IoT like D2D communication with the integration of data analytics may drastically transform the framework of many industries. With the rise of cloud computing technology inclusive of fog computing and the generation of intelligent devices, there are much more chances for future innovations in the field of IoT (Wang et al. 2018a). These innovations are motivations for the researchers to study and analyze current research, develop new frameworks, and incorporate them into the new applications. There are also various challenges associated with implementing IoT with the integration of 5G technology (Wang et al. 2018b).

Machine to Machine (M2M) communication plays an important role in the emerging paradigm of IoT. The integration of IoT and 5G technology may be extended to design and develop robots, drones, actuators for distributed coordination, and also the low latency execution tasks (Huang et al. 2018). Though the research studies also show the various significant challenges of IoT with 5G networks, for example, performance enhancement, quality of service support, security and privacy concerns, etc. (Malik and Steganography, 2020). Various solutions related to protocols, routing algorithms, architecture, or spectrum have been suggested for solving the significant problems (Esfahani et al. 2019). Moreover, the key motivation of the study is as follows:

- The work familiarizes with the inspiration and ideas of implementing 5G technology in cooperation with IoT technology for providing better communication and connection between smart devices and sensors.
- The work illustrates the vision of the IoT in 5G, the work also described the basic architecture of IoT in collaboration with 5G technology.
- The work deliberates the inclusive survey on the various recent communication technologies for IoT in 5G, where various technologies are explained such as SigFox, LoRa, and ZigBee.
- The work discussed the various challenges or vulnerable issues as well as numerous research gaps and future research directions to enable this technology in other technologies to provide better communication.

The rest of the paper is organized as follows. Section "Inspiration and ideas for 5G based on IoT" elaborates on the details and motivation to enable the IoT with 5G technology for better communication and connection in smart cities. Moreover, section "Structure of IoT in co-operation with 5G" discussed the architecture of IoT in cooperation with 5G, where all the five layers are discussed in detail. Here a table has also explained the use of IoT sensors in different smart areas. Additionally, section "Current resolutions for communication for IoT in 5G" explains the recent resolutions for communication and connection for IoT-enabled 5G technology where BLE, SigFox, IEEE 802.15.4, LoRa, Wi-Fi, ZigBee, and NB-IoT are explained along with a differencing table. Furthermore, section "Challenges and Security Vulnerabilities" highlights some challenges and vulnerable concerns of 5G enabled IoT technology. Moreover, section "Future Research Directions" elaborates on the various research gaps in numerous technologies and future research directions in 5G-IoT. Finally, section "Conclusion" devotes itself to the conclusion of the whole study.

Inspiration and Ideas for 5G Based on IoT

By looking into different challenges associated with 5G integrated with IoT, there is a deep motivation of providing a thorough study on 5G wireless communication that enables IoT (Henry et al. 2020). Since a huge number of researchers and communication industries are involved in researching the field of 5G-IoT, thus it gives us a kind of motivation and encouragement for providing the latest research perspectives on IoT. The network and communication technologies are the base of investigation for providing new insightful direction on 5G-IoT (Yarali 2020). At present, security is the most important concern in IoT as it is prone to cybercrimes. Thus, IoT has huge opportunities and possibilities for research and it also covers all technologies of 5G relevant to IoT. 5G-IoT is an architecture containing five layers. For enabling effective communication among devices and resource sharing, a generalized network framework is to be developed in 5G-IoT. This kind of generalized network may be able to decrease the cost and the complexity (Bikos and Sklavos 2018). Moreover, the evaluation of 5G from 1G is shown in Fig. 9.2 where IoT is enabled from 4G onwards.

In the current era of technology and advancements, the Internet is a vital component as communication between any devices or machines can only be established through it without any interruption from humans. 5G IoT is an immense technology that is implemented over critical communications and complex network technologies for example mm-wave technology, 5G New Radio (NR), Multiple Inputs Multiple Outputs (MIMO), etc. (Slamnik-Kriještorac et al. 2020). 5G technology runs at a very fast speed in comparison to the existing technologies and also comparatively huge number of devices can be connected within a network and reliable communication can be established (3GPP 2018).

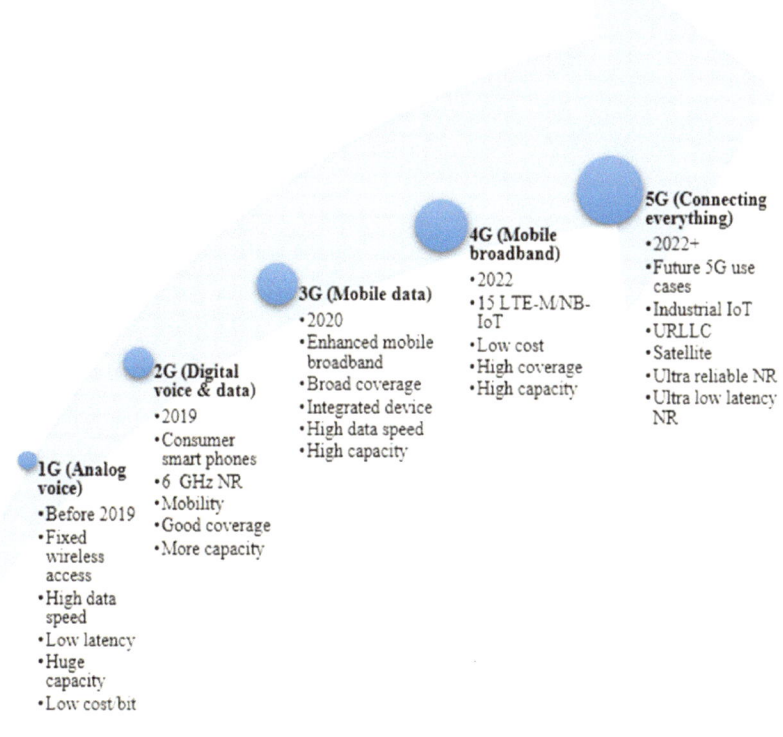

Fig. 9.2 Evaluation of 5G-IoT

Structure of IoT in Cooperation with 5G

IoT integrated with 5G technology is primarily implemented on of five-layered archi-tecture. It includes the tasks of data collection, data processing, data analysis, and information sharing among various devices and communication links. In this type of architecture, the smart sensors remain connected through the gateway to the low-powered networks like Low-Power Wide Area Network (LPWAN), LoRa, NB-IoT, or SigFox that are used to establish long distanced communications (Le et al. 2021). The main task of this efficient gateway is to collect data from all the devices connected and transmit this information to base stations through the 5G communication links. By the use of 5G NR technologies with MM wave communication and effective numerology selection, the communication links of 5G technology can be designed. Furthermore, the 5G cellular base station processes the IoT signals that involve the MIMO antenna with the added capacity of spatial multiplexing and beam creation (Garro et al. 2020). The MM wave communication technology enables us to transfer the radio signals on higher frequencies, i.e., more than 6 GHz. There are numerous applications that can be implemented through 5G NR technologies (Kumar et al. 2019). Furthermore, the basic architecture of IoT in cooperation with 5G is shown in Fig. 9.3.

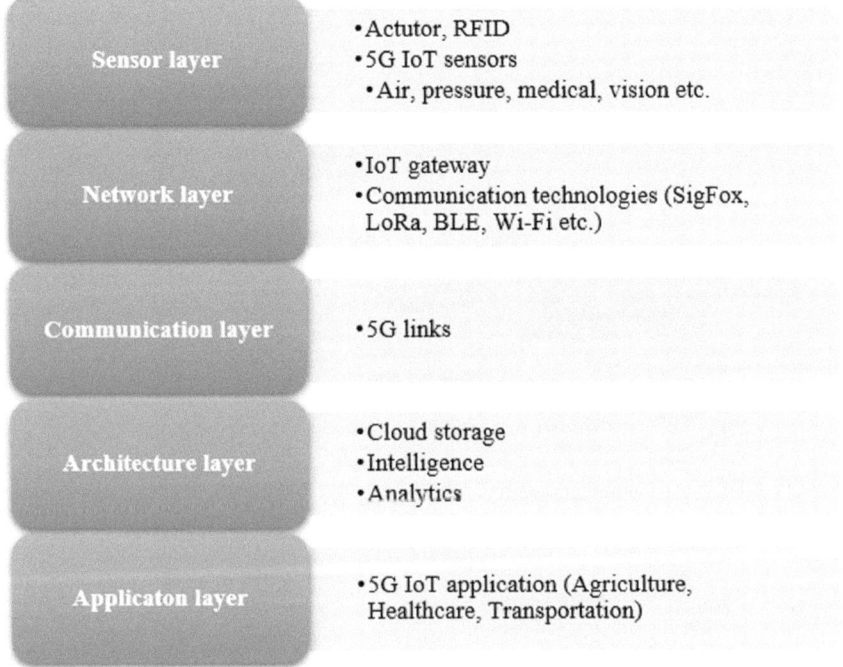

Fig. 9.3 Architecture of 5G-IoT

Sensor Layer

Sensors and other data collection devices create the first layer of the architecture. This sensor layer can be considered as a physical layer that includes devices, sensors, and actuators, and connects to the next layer, i.e., network layer. The present era is a technological era and utilization of the technology is everywhere. With the progression of electronic devices, semiconductor manufacturing industries, and automation solutions, the utilization of smart sensors is growing. the combination of sensors and integrated computing resources is called smart sensors. The smart sensors are able to communicate in a two-way manner between the network layer and sensors and analyze data to make useful decisions (https://www.5gradar.com/features/5g-sec urity-5g-networks-contain-security-flaws-from-day-one). In an IoT architecture, the first layer accomplishes Machine Type Communication (MTC) and connects with the upper layer, i.e., network layer. The new smart sensors are more advantageous than the old conventional sensors like Smart protocols for establishing communication between devices and sensors, reduced communication through cables, easy to set up and maintain, flexible connectivity, reduced cost and less power solution, highly reliable, and effective performance (Anamalamudi et al. 2018). Some of the famous sensors based on 5G-IoT, which are used in different scenarios of IoT are shown in Table 9.1.

Network Layer

The second layer of IoT architecture is the network layer which comprises LPWAN like Sigfox, LoRa alliance, NB-IoT, ZigBee, etc. In 5G communication technology, the main task of the network layer is to deliver long-range and low-power connectivity for applications of IoT. A various number of connections may be established for achieving huge and complex connectivity by utilizing LPWAN. LPWAN is the technology that is mostly used for IoT applications as it provides a wide range of coverage areas, increased energy efficacy, less power consumption, and higher data transfer rates (Adame et al. 2019).

Communication Layer

The communication layer may be considered the support system of the architecture of IoT because its main task is of transferring all the information over all the layers. The Radio Access Technology (RAT) is used in the communication layer for 5G-IoT applications. 5G NR technology is an effort of 3GPP for developing a new standard for the wireless communication technologies of the next generations (Dutta and Hammad 2020). 5G NR technology is standardized as per the releases 15 and 16

Table 9.1 IoT sensors used in various applications

Application-based on IoT (Le et al. 2021; Garro et al. 2020; Kumar et al. 2019; https://www.5gradar.com/features/5g-security-5g-networks-contain-security-flaws-from-day-one; Anamalamudi et al. 2018; Adame et al. 2019; Dutta and Hammad 2020; Salem et al. 2022; Fang et al. 2018; Malik and Bhushan 2022; Frustaci et al. 2018)	Sensor-based on 5G-IoT
E-healthcare	X-Ray, E-wearable, Temperature, Pressure, SpO2, Ultrasound, ECG, EEG, BP, Pulse rate, etc
Public security	Location sensor radars, chemical operations, temperature observation, light monitoring, gyroscope
Smart industries	Proximity sensor, air quality sensor, fiber optic sensor, smoke sensor
Environment	Humidity, temperature, light, energy, harmful rays, chemical, pollution
Smart homes	Temperature, electronics device operators, LPG monitoring, robotics
Smart transportation	Traffic sensor, proximity, smart parking, position sensor, prior accident sensor, automatic driving sensor
E-study	3D presentation, robotic devices, sensor-based labs

of 3GPP. 5G NR technology is a subpart of RAT that has a composition of 5G NR and Long-Term Evolution (LTE) technology. There are two operational categories of 5G NR technology named 6 GHz and millimeter wave (mm-wave) range. Various intricate technologies such as NR support IoT with the inclusion of massive MIMO technology, waveforms, and frame structure. Radio access offers the complexities and opportunities, both, in the structure of Radio Access Network (RAN) specifically in IoT platforms like smart farming, smart home, critical services, smart factories, health care applications, and many more. 5G NR technology not only eases the market prospects for small base stations and small cells like picocells or femtocells but also facilitates the smart sensors for various kinds of IoT applications (Salem et al. 2022).

Architecture Layer

It is one of the layers of the IoT framework in which the architectures are included such as Big Data Analytics (BDA), cloud computing technology, etc. The cloud computing technology is mostly considered for 5G-IoT, as it is one of the trending technologies of IoT, and is mainly related to the Information Technology (IT) solutions. The system programming may also be embedded in it. The architecture of cloud technology is deployed with devices like a smartphone, Personal Computers (PC), host machines, and laptops. The integration of cloud technology with IoT architecture provides ubiquitous services which can be distributed to the clients with increased efficacy and less service management. Thus, IoT is a technology that is implemented with Big Data (BD), and data management is done by cloud computing technology. It is a kind of interface where all the services such as storage, data servers, authentication and authorization processes, the user interface for registration and login, are made available through the cloud (Fang et al. 2018). The cloud computing technology is categorized into three models which are as follows.

- **Infrastructure as a Service (I-a-a-S)**—This service is also called hardware as a service. This kind of service removes the need of installing any kind of hardware at our ends like server, storage, or computing resources. This service virtually provides all the services related to the infrastructure that is used to be present at data centers traditionally like network hardware, storage, maintenance, privacy, backup and recovery services, security services, and many more.
- **Platform as a Service (P-a-a-S)**—This kind of service provides hardware and software services virtually that are required for developing an application. The service may involve interfaces, development environments, and databases for maintaining data. Embedded systems that involve programming interfaces can also be implemented through this service. P-a-a-s frees the developers from the responsibility of installing and managing software or hardware services and enable them to focus on the application development only. The service provider maintains the services and resources.
- **Software as Service (S-a-a-S)**—S-a-a-S is a kind of software distribution service that works on the accomplishment of clients' demands. In this configuration, there is no need to install any software physically on the system but services related to the software can be provided to the clients according to their requirements. Also, the update or addition of new services in the software is installed automatically without any intervention from clients (Malik and Bhushan 2022).

Application Layer

The application layer works as an interface for integrating the network devices with network configurations. It provides integration of all the devices and information to the network or Internet. 5G MTC offers a large variety of services and applications. In future advanced technologies of wireless networks and communication, machines, and devices will be able to communicate without any intervention from humans (Frustaci et al. 2018). Nowadays, there is a variety of applications that require high latency and speed, high data rates, and connectivity of multiple devices.

Current Resolutions for Communication for IoT in 5G

With the advancement of semiconductor industries, electronics devices, and automation solutions, there is a rise in the development of communication solutions for 5G-IoT. These solutions are more reliable, smarter, have high data rates, are robust, and are energy efficient. So as an outcome, various communication technologies having low-power configurations are proposed for the 5G IoT, for example, LoRa, SigFox. The research studies state that low-power technologies have some unique characteristics like large coverage area, low-power consumption, higher data rates, energy efficient, etc., that are 5G-IoT. The following sections contain a discussion about the currently developed communication solutions for 5G-IoT (Wang et al. 2019).

Bluetooth Low Energy (BLE)

Bluetooth communication is generally implemented on laptops, PCs, cars, keyboards, wireless mouses, and earphones. This particular protocol is implemented on a Personal Area Network (PAN) and is able to establish communication between the devices over a short range of distances, i.e., almost 10 m. On the other hand, a modified protocol is required to be integrated into IoT as there is a need of decreasing power consumption to be able the execution with very short-sized batteries. Special Interest Group (SIG) of Bluetooth technology has inherited BLE technology while launching Bluetooth 4.0 version for identifying the market scenario (Chen et al. 2018). The BLE standard is designed with the ability to transfer small data packets periodically unlike the traditional Bluetooth technology. The key variation between the traditional Bluetooth technology and the new BLE technology is visible in the Physical (PHY) layer. The standard Bluetooth technology offers 79 channels containing 1 MHz bandwidth whereas the new BLE standard offers 40 channels from which 37 are data channels and the remaining 3 are advertising channels with 2 MHz bandwidth. In both the standards, the Radio Frequency (RF) is classified into two categories, i.e., data RF channels and advertising RF channels. In the first category, data channels are used for transferring data among the Bluetooth linked devices whereas the second category of a channel is utilized for connection, streaming, and device acquisition. Both kinds of channels are activated on an unregistered Industrial, Scientific, and Medical (ISM) frequency band of 2.4 GHz. In traditional Bluetooth standards, the process of switching varies from Gaussian Frequency Shift Keying (GFSK) to the Phase Shift Keying (PSK), i.e., 4-PSK and 8-PSK, whereas, in BLE standard, the GFSK optimization is done. The GFSK generates a less value of Peak-to-Average Power Ratio (PAPR), which infers lower power consumption due to the power amplification (Salimibeni et al. 2020).

SigFox

SigFox is a new technology, invented in the year 2010 with the purpose of connecting low-powered components or devices like smartwatches, meters of electricity, regulators, etc., which are required to be operated continuously at a very low rate of data. SigFox utilizes the RF band of ISM. It operates on a frequency of 868 MHz in Europe and at 902 MHz frequency in the US with 100 MHz channel bandwidth. SigFox signals can be transmitted easily through dense objects. These are called ultra-narrow band signals that provide low energy consumption. It is also called LPWAN technology due to the low power and energy requirements implemented in a single-hop star topology. SigFox technology is utilized for covering huge areas and reaching underground devices. The cells of SigFox provide a coverage range of about 30–50 km in less crowded areas and about 20–40 km in crowded areas like urban areas. Thus, conclusively, SigFox is developed for providing a Wide Area Network (WAN) with low consumption of power. At present, around 72 countries have been covered by the SigFox IoT system having a population of almost 1.4 billion (Ikpehai et al. 2019).

There are various applications as described in this section that is built on the top of SigFox communication. Mazhar et al. (2021) have developed an independent SigFox sensor node that is capable of collecting data from the sensors and passing data to the cloud platform for implementing smart agricultural applications. For enhancing the system, the sensors were designed in such a way, so as to consume solar energy. The experimental analysis states that the system enables transmitting data every five minutes even in cloudy conditions. Mroue et al. (2018) work on the analysis of the SigFox performance under various scales and density situations of IoT sensors. By the analysis, it is shown that approximately, a maximum of 100 sensors are able to transmit data at the same time moment. The outcomes show that, with the increasing number of sensors above 100, the performance of the network may be reduced. This particular study also presents solutions for the performance improvement of high-density sensors in SigFox. Lavric et al. (2019) developed an independent Sigfox-sensor-node which is able to transmit the sensor data to the cloud server directly. The solar cell is used for providing power backup and it is capable of transferring data in every 5 min in the cloudy weather at a data rate of around

5000 lx. This much of high data transmission rate has not been recorded till now for a completely autonomous setup. For the actual implementation, two sensor nodes were placed at the vineyard for collecting atmospheric parameters.

IEEE 802.15.4

IEEE 802.15.4 standard involves the specifications for Medium Access Control (MAC) and PHY layer. The PHY operates in various ISM groups which permit it to the region of operation. The worldwide standard band capacity is 2.4 GHz but there are other bands of data rates also exist such as 915 MHz in North America. IEEE 802.15.4 standard was intended to be developed for PAN. It was primarily used for the applications of the organizations such as ecological, agricultural, and engineering. If compared to the IEEE 802.11 standard, IEEE 802.15.4 is not prominent for higher rates of data, and also it does not emphasize on linking of the devices. Thus, this provides lower data rates for wireless communication for portable, fixed, or less battery-moving devices. Zigbee can be an example that effectively used IEEE 802.15.4 standard. The major advantage of Zigbee as compared to others is that it is proficient in working with multi-hop structures and can also perform well under network failures (Musaddiq et al. 2021). There are seven different categories of working modes proposed in the IEEE 802.15.4 standard. The primary methods for lower energy consumption from the IoT point of view are Offset Quadrature Phase Shift Keying with Direct Sequence Spread Spectrum (OQPSK-DSSS), Differential Quadrature Phase Shift keying variation with Chirp Spread Spectrum (DQPSK-CSS), and Gaussian Frequency Shift Keying (GFSK) with non-virtual distribution (Aboubakar et al. 2020).

LoRa

The LoRa is an emergent and one of the most important LPWAN communiqué technology. It has the capability to provide connectivity to the energy-constrained devices that are distributed over a wide area and that too at lower costs. The LoRa technology makes use of the LPWAN modulation process and unlicensed bands of frequency such as 433 and 868 MHz in the Europe region, 915 MHz in North America and Australia region, and 923 MHz for the Asia region. LoRa provides transmissions over a wide range with low consumption of power. The LoRa technology is operated on PHY and the protocols like Long Range Wide Area Network (LoRaWAN) operate over the network layer. It may achieve the data rates between 0.3 and 27 kbps (around) dependent on the distribution factor. Though, as per the studies, implementing a flexible LoRa network with a cost-effective feature is still a significant challenge (Leonardi et al. 2019).

Ma et al. (2021) proposed and developed an open-access LoRa framework for IoT. This particular development process involves the design framework and implementation of hardware for the LoRa gateway which uses open-source codes of LoRa available on GitHub. The LoRa server can be improved by the utilization of the messages system to create interaction among different modules for scalability and flexibility enhancement. The outcomes of this experiment show that there is a significant improvement in the performance of the LoRa network when compared to the traditional LoRa. Lee et al. (2018) analyzed the LoRa mesh network for examining its ability to be applied in urban areas. For this analysis work, 19 LoRa mesh nodes were installed in a range of 800×600 m in the campus area of a university and a gateway is also installed which performed the data collection at the time interval of 1 min. The results of the experiment disclosed that the average packet delivery ratio achieved from the mesh LoRa framework is almost 93.19%, while the star topology of LoRa was able to achieve only 67.9% delivery ratio under the similar circumstances. The LoRa is considered to be the most effective out of all the LPWAN technologies. It is capable of establishing robust communications in IoT applications over large distances with very low-power consumption. This technology is also promising from the industrial perspective of IoT. Though the LoRa framework also has a limitation in that it does not provide support for data flows in real time. For overcoming this limitation, Senewe et al. (Senewe and Suryanegara 2020) designed a new strategy related to media access, called Real-Time LoRa (RT-LoRa), which aims to provide real-time support for IoT applications based on LoRa. The experimental outputs state that RT-LoRa is capable of supporting real-time data flows.

Wi-Fi

Wi-Fi technology is a very well-acknowledged and well-used technology of wireless communication based on the IEEE 802.11 standard. It is generally used for accessing the Internet within the range of 100 m. Its operational frequency band is 2.4–5 GHz. Since Wi-Fi is appropriate for communication within a short range that's why it is the right solution for establishing communication in IoT networks. Jiang et al. (2021) proposed a solution for smart homes IoT applications based on queue management where access points are linked together through Wi-Fi. The main goal of this work was to propose an admission control mechanism at the access point of Wi-Fi for reducing the time of response. The results of the experiment proved that the new system based on Wi-Fi is extra reliable and stable than the earlier smart home IoT applications. Qi et al. (2020) invented a new solution based on WIOTAP to propose an energy-saving communication for Wi-Fi-based IoT systems. This solution works on an intelligent access point of Wi-Fi. It presented a mechanism for reducing contention of downlink channel access and delay in queuing process of stations, called downlink packet scheduling mechanism. The outputs of the experiment stated that the current system is an improved one from the old one. Thus, the consumption of energy was improved by 38% whereas delay was enhanced by 41%.

The most crucial challenge of IoT is tracking and locating in real-time scenarios. Systems or applications based on Global Positioning System (GPS) are very well recognized for the outside surroundings and it is not appropriate for interior scenarios. Pokhrel et al. (2020) projected a new Wi-Fi signal-based IoT solution to locate and track interior environments. The work uses a type of message which is built upon the 802.11–REVmc2 standard of Wi-Fi. To enhance the accuracy and capability of the positioning system, the time of roundtrip and signal strength are analyzed. The results of the experimental setup have presented that the current system improvised the performance and attained the positional accuracy average of 1.43 m for 0.19 s update time for interior scenarios.

ZigBee

ZigBee is a technology of wireless communication that is based on the IEEE 802.15.4 standard and is operational in the ISM RF bands. It was designed for providing low-power and low-cost wireless communication for IoT infrastructure. ZigBee technology is advantageous over other communication techniques related to IoT networks due to its reduced costs, simplicity of implementation, and flexibility feature. The ZigBee is able to transmit data over a distance of 100 m at a data rate of 250 kbps, depending upon environmental and power attributes. ZigBee technology is typically applicable in the scenarios of lower data rate networks, long-term battery life, and short-range communication like smart homes, healthcare devices, manufacturing equipment control, etc. Franco et al. (Franco de Almeida and Leonel Mendes 2018) developed a MIMO-based ZigBee receiver for controlling jamming attacks in IoT networks. This work also proposed a learning method to reduce unfamiliar interference. The results of the proposed experiment confirmed that the developed system may provide the jamming mitigation capacity of 26.7 dB on an average in comparison to the previous version of ZigBee receiver.

Yu et al. (2019) proposed a time-stamp-based security framework for handling replay attacks for ZigBee. This proposed solution significantly improvises the consumption of energy. Also, for enhancement of feasibility, this framework makes use of powered devices for providing energy to the devices that are power constrained along with the present time-stamp. The framework is designed in such a way that it is appropriate for all the ZigBee systems. The experimental setup significantly enhances the handling of replay attacks in IoT networks based on ZigBee. Karie et al. (2021) proposed a design of smart sensors by combining two different communication modules, i.e., LoRa and ZigBee for measuring the factors like humidity and temperature and humidity for the IoT applications. By using transceiver modules of LoRa or ZigBee, the sensor data is transferred to the central receiver. The real-world design and experimental statistics represent the advantages of high range and low-power communication frameworks for the applications of IoT.

NarrowBand IoT

NarrowBand IoT (NB-IoT) is a new radio technology based on LPWAN which is invented by the 3GPP group for supporting a huge number of connections, wide coverage area, lower cost, and low-power consumption in 5G IoT (Chen et al. 2019). It is an emerging and evolving communication technique for 5G IoT (Migabo et al. 2018). The main focus of NB-IoT is indoor area coverage, less expense, extended battery life, and massive connections (Loulou et al. 2020; Ghazali et al. 2021). The utilized bandwidth is narrow, i.e., 200 kHz. For downlink communication, the OFDM modulation technique is used and SC-FDMA is applied for the uplink communication. Thanh et al. (2019) proposed an open-source prototype of an NB-IoT network for 5G IoT applications. This experiment represents a process of utilizing the already existing module of commercial NB-IoT for transmitting the sensor data through open-source NB-IoT. Cao et al. (2020) performed the performance evaluation and modified the NB-IoT protocol for improvement in 5G IoT. The main motive of this work is to evaluate the "delay metric" by using the stochastic network and to improvise the NB-IoT protocol by improving the k means clustering algorithm to categorize the devices and perform a priority-driven scheduling strategy. The outcomes of the experiment showed that the proposed scheduling method for uplink traffic has improved the performance over the already existing scheduling schema. Goyal et al. (2021) proposed a methodology for designing the PHY of the NB-IoT device. The main goal of this work is to present the features, uplink, and downlink scheduling of physical channels present at the base station and devices of the user end for helping users to know the specifications of 3GPP without much reading. Hence, a summarized view of the above-discussed current solution for communication in 5G-IoT is shown in Table 9.2.

Challenges and Security Vulnerabilities

This particular study highlights the various innovative contributions of 5G IoT in a variety of fields for serving humanity. Low power technologies are the main support system for implementing IoT applications commercially. These applications may be related to various domains like environmental conditions, smart cities, homes, buildings, and smart farming (Nurlan et al. 2022). The communication technologies, increasingly applied in IoT applications have various characteristics like low-power consumption, wide-area coverage, higher data rates, wide frequency bands, etc. which make these technologies suitable to be applied in IoT architecture. Examples of these technologies include BLE, ZigBee, IEEE 802.15.4, SigFox, LoRa, and many more. The goal of these communication services is to establish connectivity in 5G IoT applications. Although these technologies are useful, there are a variety of challenges related to implementation as it connects a billion IoT devices. The

Table 9.2 Difference between current solutions for communication in 5G-IoT

Technology (Salimibeni et al. 2020; Ikpehai, et al. 2019; Mazhar et al. 2021; Mroue et al. 2018; Lavric et al. 2019; Musaddiq et al. 2021; Aboubakar et al. 2020; Leonardi et al. 2019; Ma et al. 2021; Lee and Ke 2018; Senewe and Suryanegara 2020; Jiang, et al. 2021; Qi et al. 2020; Pokhrel et al. 2020; Franco de Almeida and Leonel Mendes 2018; Yu et al. July 2019; Karie et al. 2021; Chen et al. 2019; Migabo et al. 2018; Loulou, et al. 2020; Ghazali et al. 2021; Thanh et al. 2019; Cao, et al. 2020; Goyal et al. 2021)	BLE	SigFox	IEEE 802.15.4	LoRa	Wi-Fi	ZigBee	NB-IoT
Range	100 m	Many Kms	100 m	2–5 km	Many Kms	< 1 km	1–10 km
Bandwidth	1–10 Mb/s	250/500 kHz	2.4 GHz, 2 MHz	100 Hz	20/40 MHz	2 MHz	200 kHz
Standardization	IEEE 802.15.1 alliance	Collaboration of ETSI	LR-WPAN	LoRa alliance	IEEE 802.11 alliance	ZigBee alliance	3GPP
Cost	Low	Medium	Low	Medium	Low	Medium	Low
Frequency band	2.4 GHz	<1 GHz	2.4 GHz	<1 GHz	<1, 2.4, 5 GHz	902–928 MHz, 2.4 GHz	700.800, 900 MHz
Maximum data rate	1 Mb/s	100 b/s	0.252 Mb/s	18 b/s-37.5 kb/s	1–54 Mb/s	250 kb/s	200 kb/s
Power	Low	Low	Low	Low	Medium	Low	Low
Spectrum strategy	Wideband	Ultra narrow band	Wideband	Wideband	Wideband	Wideband	Wideband

(continued)

Table 9.2 (continued)

Technology (Salimibeni et al. 2020; Ikpehai, et al. 2019; Mazhar et al. 2021; Mroue et al. 2018; Lavric et al. 2019; Musaddiq et al. 2021; Aboubakar et al. 2020; Leonardi et al. 2019; Ma et al. 2021; Lee and Ke 2018; Senewe and Suryanegara 2020; Jiang, et al. 2021; Qi et al. 2020; Pokhrel et al. 2020; Franco de Almeida and Leonel Mendes 2018; Yu et al. July 2019; Karie et al. 2021; Chen et al. 2019; Migabo et al. 2018; Loulou, et al. 2020; Ghazali et al. 2021; Thanh et al. 2019; Cao, et al. 2020; Goyal et al. 2021)	BLE	SigFox	IEEE 802.15.4	LoRa	Wi-Fi	ZigBee	NB-IoT
Modulation	GFSK	DBPSK	O-QPSK, CCK/DSSS	LoRa	256-QAM	BPSK, QPSK	QPSK
Sensitivity	−95 dBm	−126 dBm	−97 dBm	−149 dBm	−95 dBm	−85 dBm	−141 dBm
1-Hop latency	3 ms	2 s	1.5/1./20 ms	500 ms	NA	140 ms	1 Mb/s

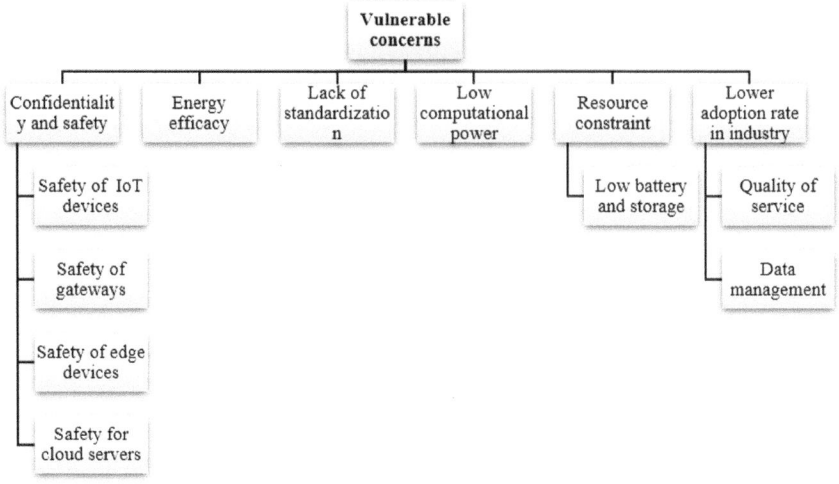

Fig. 9.4 Issues of 5G-IoT that need to be addressed

two key issues are related to energy efficiency and security (Ahmad et al. 2020). Additionally, Fig. 9.4 shows the various vulnerable concerns in 5G-IoT technology.

Confidentiality and Safety

The main aim of IoT is to establish connectivity between everything. The invention of IoT infrastructure has formed an open world in which everything is linked via the Internet. But there are always cons associated with pros. As everything is connected to the Internet, the nodes or devices are very much prone to security threats and attacks. Thus, security and privacy issues are the most crucial factors for promoting the development of IoT infrastructure to be implemented practically (Arora et al. 2020). The security attacks may be injected into multiple layers of IoT architecture.

- **Safety for IoT devices**: The devices included in IoT infrastructure are of low computing capacity and huge in numbers and are not appropriate for framing a robust and secure system. Therefore, the main focus of attackers is to exploit the weaknesses of the IoT devices.
- **Safety for gateway devices**: The gateway is a communication interface between the PHY devices and the higher layers. That's why it is called the central part of IoT infrastructure. The attacks like Denial of Service (DoS) or spoofing of data generally target the gateway device of IoT.
- **Safety for edge devices**: The technology of edge computing is a core part of the newly proposed solutions for reducing the response time of services in real-time IoT. Therefore, securing the edge servers from attacks is a key challenge.

- **Safety for cloud servers**: Cloud technology is a probable solution for storing and analyzing the vast volume of data generated from IoT devices. Thus, ensuring the security feature of cloud-based servers is also the main challenge (Malik et al. 2018).

Energy Efficacy

Though the applications of IoT are considered to be energy efficient, the energy consumption of their own is so much. It is assumed from the study results that as the 5G IoT applications are becoming widespread, the billions of IoT-connected devices will be operational and transmitting data endlessly at every moment. So as a consequence, there will be a massive amount of energy consumption and it will also increase with every passing moment. Therefore, implementing such solutions that are feasible and energy efficient is a major challenge (Atakora and Chenji 2018).

Future Research Directions

The need for 5G communication technology at present time is to deliver standards for establishing communication among a huge number of devices over a wide area for fulfilling the requirements of industrial as well as social applications related to IoT. For the successful implementation of such technology, it is vital to identify the technical and practical challenges along with ensuring the QoS (Generation Partnership Project (3GPP 2018). In particular, the section tries to represent some significant challenges related to 5G IoT and some ideas for future research, which are shown in Fig. 9.5.

Structure of Network Based on Big Data

The present structure of wireless communication networks is to facilitate data transmission and establish communication among the devices of the network. For accessing the key benefits of BD in the field of 5G IoT, the new solutions and frameworks considering the BD have to be proposed. This new framework is capable of accommodating a huge volume of data and integrating that BD into the network very efficiently. This novel solution focuses on ignoring the unexploited data and processing the desired information at a suitable location (Ijaz et al. 2018). Another feature of research related to big-data analytics is personalized networking. This new approach includes Service-Function Chain (SFC) or network partitioning which supports various services of BD by developing a service concerned networking on the top of the physical network. Further customization of network partitioning can

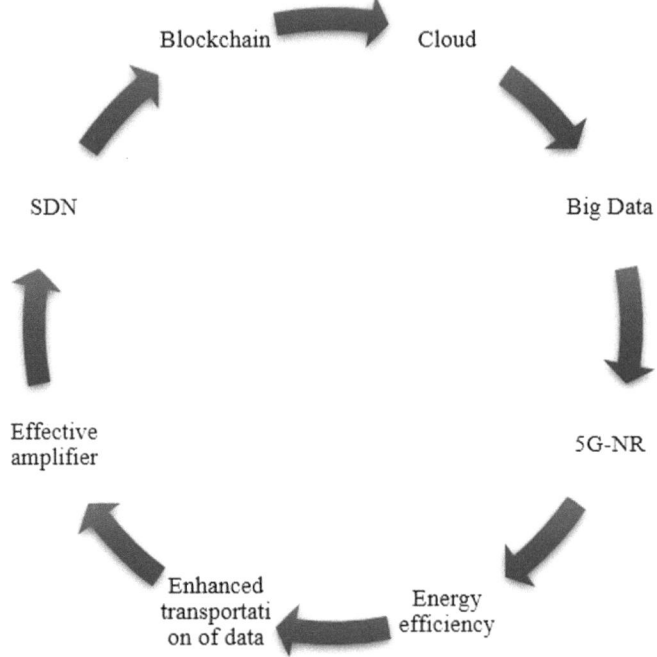

Fig. 9.5 Research gaps of 5G-IoT

be done according to the requirements of services. To make the most utilization of network resources, multiple partitioning or SFC should be used. In 5G technology, SFC should have the capability of identifying variations in the network status and service needs (Sharma et al. 2021).

5G New Radio (NR): A New Wave-Form Design Consideration

The selection of wave-form is the most crucial part of the design of 5G NR technology. The primary choice for designing LTE was OFDM but it is not appropriate for 5G wave-form due to its higher Inter-Channel Interference (ICI), increased Inter Symbol Interference (ISI), and higher PAPR. So, all these OFDM wave-form limitations can be taken as challenges for future research in 5G. The primary design characteristic of the new wave-form must be less latency (<1 ms) for enabling new services. The feature of low latency is useful for IoT applications and very less latency is utilized for enhanced Mobile Broadband (eMBB) and crucial communications such as automatic driving. Another aspect of designing a new wave is applying a cyclic prefix. It can be used in both modes, normal as well as extended. To design the wave-form, selection of numerology is considered and it uses dissimilar numeric values. By considering all mentioned aspects of the 5G wave-form, different kinds of the wave-form such as Filter Bank Multi-Carrier (FBMC) or Generalized Frequency Division Multiplexing (GFDM) can be generated (Khapre et al. 2020; Alfian et al. 2018).

Energy Efficacy

As per the thorough review studies, the key consideration to design and develop 5G wireless networks is energy consumption. As the 5G technology evolved, a billion devices are likely to be connected within a single network having various base stations in comparison to LTE networks. Thus, to accommodate such a huge number of devices, there is a need of proposing an energy-efficient solution. The first assumption for overcoming this problem may be to set up a small-cell base station. Such kind of base station is able to enhance the capacity in the areas of higher density. It is capable to improvise coverage area, battery life, and data rates by reducing the consumption of power. Energy efficacy may be attained through the framework as follows.

- Deployment and energy efficacy trade-off: It is important in achieving reduced consumption of energy and less cost in the proposed network.
- Spectrum and energy efficacy trade-off: It is utilized for balancing the consumption of energy.
- Bandwidth and power trade-off: For a target rate of transmission, it is utilized for balancing the bandwidth consumption.
- Delay power trade-off: Use of it is for balancing average service delay with respect to power consumption (Nie et al. 2020; Latif et al. 2019).

Interchange Between Communication, Gathering, and Computing

5G wireless communication is a kind of diverse communication. The tasks of collecting data and computing resources should be performed intelligently for supporting BD in a heterogeneous network of 5G-IoT. Thus, a balancing factor is important for communication, catching process, and resource computing. For reducing the cost of storage, the results of computation must be saved on a temporary basis. For providing optimal resources, the balance between the heterogeneous resources is essential. The 5G IoT technology has evolved with the vast volume of data resources. This data is gathered from various resources that generate data distribution in a non-uniform manner. Thus, for storing, retrieving, and analyzing this huge volume of data, the proposed solution is cooperative edge catching. For the processing of data, edge computing technology is needed (Escolar et al. 2021; Yang et al. 2020).

Design of Synchronized Multi-Band, and High-Power Effective Amplifier

The importance of a multi-band power amplifier is for reducing the cost and size of the base station in 5G-IoT. It is capable of supporting multi-band frequency signals concurrently which enables all the functions to execute at the same time (Failed 2020). The most capable amplifiers are concurrent and parallel single-band power amplifiers. In 5G NR, the MIMO and mmwave are used for communication at the base station. The linear RF power amplification is important for the consumption of energy at the base station. The application of a power amplifier is also helpful to reduce heat dissipation at the base station. Reduction in energy consumption of radio stations also tends to decrease the environmental input of RAT (Tikhvinskiy et al. 2020).

Conclusion

The most important vision of 5G-IoT technology is to establish communication among a various number of devices under a similar network. There is a lot of application based on 5G wireless communication such as smart farming smart homes, smart cars, smart city, and smart medical devices that led to the revolution in IoT. The implementation of such applications requires the establishment of huge connectivity at a higher speed under 5G wireless technology. On the basis of the analysis of essential components of 5G-IoT, this review study represents a brief discussion about power-constrained communication technologies. The IoT will be considered the new future for the world where everything will be linked together by the Internet like devices, software, people and systems, etc. The invention of 5G-IoT technology has developed a variety of applications like intelligent farming, smart cities, smart homes, smart healthcare devices, green energy systems, etc. This study has presented a complete overview of all the communication techniques of 5G-IoT that are characterized by huge coverage, increased energy efficacy, and low consumption of power. By concluding the current research work, the aspects related to energy consumption and security will be an interesting area of future research and will obtain attention from industry as well as a research perspective. Research in the field of 5G-IoT may be proven to be a social service in the development of a better world.

References

Aboubakar M, Roux P, Kellil M, Bouabdallah A (2020) An efficient and adaptive configuration of IEEE 802.15.4 MAC for communication delay optimization. In: 2020 11th international conference on network of the future (NoF), pp 1–7. https://doi.org/10.1109/NoF50125.2020.9249218

Adame T, Bel A, Bellalta B (2019) Increasing LPWAN scalability by means of concurrent multiband IoT technologies: an industry 4.0 use case. IEEE Access 7:46990–47010. https://doi.org/10.1109/ACCESS.2019.2909408

Ahmad A, Bhushan B, Sharma N, Kaushik I, Arora S (2020) Importunity & evolution of IoT for 5G. In: 2020 IEEE 9th international conference on communication systems and network technologies (CSNT). https://doi.org/10.1109/csnt48778.2020.9115768

Alfian G, Syafrudin M, Ijaz M, Syaekhoni M, Fitriyani N, Rhee J (2018) A personalized HEALTH-CARE monitoring system for diabetic patients by UTILIZING Ble-based sensors and real-time data processing. Sensors 18(7):2183. https://doi.org/10.3390/s18072183

Anamalamudi S, Sangi AR, Alkatheiri M, Muhaya FT, Liu C (2018) 5G-Wlan security. A comprehensive guide to 5G security, pp 143–163. https://doi.org/10.1002/9781119293071.ch7

Ahad A, Tahir M, Yau K-LA (2019) 5G-based smart healthcare network challenges and future research directions. IEEE Access 7:100747–100762. https://doi.org/10.1109/ACCESS.2019.2930628

Arora S, Sharma N, Bhushan B, Kaushik I, Ahmad A (2020) Evolution of 5G wireless network in IoT. In: 2020 IEEE 9th international conference on communication systems and network technologies (CSNT). https://doi.org/10.1109/csnt48778.2020.9115773

Atakora M, Chenji H (2018) A multicast technique for fixed and mobile optical wireless backhaul in 5G networks. IEEE Access 6:27491–27506. https://doi.org/10.1109/access.2018.2832980

Bikos AN, Sklavos N (2018) Architecture design of an area efficient high speed crypto processor for 4G LTE. IEEE Trans Dependable Secure Comput 15(5):729–741. https://doi.org/10.1109/TDSC.2016.2620437

Bonfim MS, Dias KL, Fernandes SF (2019) Integrated nfv/sdn architectures. ACM Comput Surv 51(6):1–39. https://doi.org/10.1145/3172866

Cao J et al (2020) A survey on security aspects for 3GPP 5G networks. IEEE Commun Surv Tutor 22(1):170–195. Firstquarter. https://doi.org/10.1109/COMST.2019.2951818

Chen J, Chen Y, Jiang Y (2018) Energy-efficient scheduling for multiple latency-sensitive bluetooth low energy nodes. IEEE Sens J 18(2):849–859. https://doi.org/10.1109/JSEN.2017.2759327

Chen S, Yang C, Li J, Yu FR (2019) Full lifecycle infrastructure management system for smart cities: a narrow band IoT-based platform. IEEE Internet of Things J 6(5):8818–8825. https://doi.org/10.1109/JIOT.2019.2923810

Chettri L, Bera R (2020) A comprehensive survey on Internet of Things (IoT) toward 5G wireless systems. IEEE Internet Things J 7(1):16–32. https://doi.org/10.1109/JIOT.2019.2948888

Dutta A, Hammad E (2020) 5G security challenges and opportunities: a system approach. In: 2020 IEEE 3rd 5G world forum (5GWF) Bangalore, India, pp 109–114. https://doi.org/10.1109/5GWF49715.2020.9221122

Escolar AM, Alcaraz-Calero JM, Salva-Garcia P, Bernabe JB, Wang Q (2021) Adaptive network slicing in multi-tenant 5G IoT networks. IEEE Access 9:14048–14069. https://doi.org/10.1109/ACCESS.2021.3051940

Esfahani A et al (2019) A lightweight authentication mechanism for M2M communications in industrial IoT environment. IEEE Internet Things J 6(1):288–296. https://doi.org/10.1109/JIOT.2017.2737630

Fang D, Qian Y, Hu RQ (2018) Security for 5G mobile wireless networks. IEEE Access 6:4850–4874. https://doi.org/10.1109/ACCESS.2017.2779146

Franco de Almeida IB, Leonel Mendes L (2018) Linear GFDM: a low out-of-band emission configuration for 5G air interface. In: 2018 IEEE 5G world forum (5GWF). https://doi.org/10.1109/5gwf.2018.8516993

Frustaci M, Pace P, Aloi G, Fortino G (2018) Evaluating critical security issues of the IoT world: present and future challenges. IEEE Internet of Things J 5(4):2483–2495. https://doi.org/10.1109/JIOT.2017.2767291

Gupta N, Sharma S, Juneja PK, Garg U (2020) SDNFV 5G-IoT: a framework for the next generation 5G enabled IoT. In: 2020 international conference on advances in computing, communication & materials (ICACCM), pp 289–294. https://doi.org/10.1109/ICACCM50413.2020.9213047

Garro E et al (2020) 5G mixed mode: NR multicast-broadcast services. IEEE Trans Broadcast 66(2):390–403. https://doi.org/10.1109/TBC.2020.2977538

Gautam S, Malik A, Singh N, Kumar S (2019) Recent advances and countermeasures against various attacks in IoT environment. In: 2019 2nd international conference on signal processing and communication (ICSPC). https://doi.org/10.1109/icspc46172.2019.8976527

3rd Generation Partnership Project (3GPP) (2018) Technical specification group services and system aspects; network architecture, release 8. In: 3rd generation partnership project (3GPP), Technical Report 3GPP TS 23.002. https://doi.org/10.1109/COMST.2019.2916180

3rd Generation Partnership Project (3GPP) Technical specification group services and system aspects (SA3); TS 33.501: security architecture and procedures for 5G system, release 15, 3rd generation partnership project (3GPP), Technical Report 33.501

Ghazali MHM, Teoh K, Rahiman W (2021) A systematic review of real-time deployments of UAV-based lora communication network. IEEE Access 9:124817–124830. https://doi.org/10.1109/ACCESS.2021.3110872

Goyal S, Sharma N, Kaushik I, Bhushan B, Kumar N (2021) A green 6g network era: architecture and propitious technologies. Data Anal Manage 59–75. https://doi.org/10.1007/978-981-15-8335-3_7

Han B, Wong S, Mannweiler C, Crippa MR, Schotten HD (2019) Context-awareness enhances 5g multi-access edge computing reliability. IEEE Access 7:21290–21299. https://doi.org/10.1109/access.2019.2898316

Henry S, Alsohaily A, Sousa ES (2020) 5G is real: evaluating the compliance of the 3GPP 5G new radio system with the ITU IMT-2020 requirements. IEEE Access 8:42828–42840. https://doi.org/10.1109/ACCESS.2020.2977406

https://www.5gradar.com/features/5g-security-5g-networks-contain-security-flaws-from-day-one

Huang J, Xing CC, Shin SY, Hou F, Hsu C-H (2018) Optimizing M2M communications and quality of services in the IoT for sustainable smart cities. IEEE Trans Sustain Comput 3(1):4–15. https://doi.org/10.1109/TSUSC.2017.2702589

Ijaz M, Alfian G, Syafrudin M, Rhee J (2018) Hybrid prediction model for type 2 diabetes and Hypertension USING DBSCAN-BASED outlier DETECTION, SYNTHETIC minority over sampling technique (SMOTE), and random forest. Appl Sci 8(8):1325. https://doi.org/10.3390/app8081325

Ikpehai A et al (2019) Low-power wide area network technologies for Internet-of-Things: a comparative review. IEEE Internet of Things J 6(2):2225–2240. https://doi.org/10.1109/JIOT.2018.2883728

Jiang X et al (2021) Hybrid low-power wide-area mesh network for IoT applications. IEEE Internet of Things J 8(2):901–915. https://doi.org/10.1109/JIOT.2020.3009228

Karie NM, Sahri NM, Yang W, Valli C, Kebande VR (2021) A review of security standards and frameworks for IoT-based smart environments. IEEE Access 9:121975–121995. https://doi.org/10.1109/ACCESS.2021.3109886

Khapre SP, Chopra S, Khan A, Sharma P, Shankar A (2020) Optimized routing method for wireless sensor networks based on improved ant colony algorithm. In: 2020 10th international conference on cloud computing, data science & engineering (Confluence), Noida, India, pp 455–458. https://doi.org/10.1109/Confluence47617.2020.9058312

Kitanov S, Popovski B, Janevski T (2021) Quality evaluation of cloud and Fog computing services in 5g networks. In: Research anthology on developing and optimizing 5G networks and the impact on society, pp 240–275. https://doi.org/10.4018/978-1-7998-7708-0.ch012

Khan R, Kumar P, Nalin D, Liyanage M (2019) A survey on security and privacy of 5G technologies: potential solutions. Recent Adv Future Dir IEEE Commun Surv Tutor. https://doi.org/10.1109/COMST.2019.2933899

Latif G, Shankar A, Alghazo JM, Kalyanasundaram V, Boopathi CS, Arfan Jaffar M (2019) I-cares: advancing health diagnosis and medication through IoT. Wirel Netw 26(4):2375–2389. https://doi.org/10.1007/s11276-019-02165-6

Lavric A, Petrariu AI, Popa V (2019) SigFox communication protocol: the new era of IoT? International conference on sensing and instrumentation in IoT era (ISSI) 2019:1–4. https://doi.org/10.1109/ISSI47111.2019.9043727

Le TK, Salim U, Kaltenberger F (2021) An overview of physical layer design for ultra-reliable low-latency communications in 3GPP releases 15, 16, and 17. IEEE Access 9:433–444. https://doi.org/10.1109/ACCESS.2020.3046773

Lee H, Ke K (2018) Monitoring of large-area IoT sensors using a LoRa wireless mesh network system: design and evaluation. IEEE Trans Instrum Measure 67(9):2177–2187. https://doi.org/10.1109/TIM.2018.2814082

Leonardi L, Battaglia F, Lo Bello L (2019) RT-LoRa: a medium access strategy to support real-time flows over LoRa-based networks for industrial IoT applications. IEEE Internet of Things J 6(6):10812–10823. https://doi.org/10.1109/JIOT.2019.2942776

Liyanage M, Ahmad I, Abro AB, Gurtov A, Ylianttila M (2018) A comprehensive guide to 5G security. Wiley Publishing. https://doi.org/10.1002/9781119293071

Loulou A et al (2020) Multiplierless filtered-OFDM transmitter for narrowband IoT devices. IEEE Internet of Things J 7(2):846–862. https://doi.org/10.1109/JIOT.2019.2945186

Ma H, Cai G, Fang Y, Chen P, Han G (2021) Design and performance analysis of a new STBC-MIMO LoRa system. IEEE Trans Commun 69(9):5744–5757. https://doi.org/10.1109/TCOMM.2021.3087122

Maier G, Reisslein M (2019) Transport SDN at the dawn of the 5G ERA. Opt Switch Netw 33:34–40. https://doi.org/10.1016/j.osn.2019.02.001

Malik H, Pervaiz H, Mahtab Alam M, Le Moullec Y, Kuusik A, Ali Imran M (2018) Radio resource management scheme in NB-IoT systems. IEEE Access 6:15051–15064. https://doi.org/10.1109/access.2018.2812299

Malik A, Bhushan B (2022) Challenges, standards, and solutions for secure and intelligent 5G internet of things (IoT) scenarios. In: Smart sustainable approaches for optimizing performance of wireless networks: real-time applications. Wiley, pp 139–165 https://doi.org/10.1002/9781119682554.ch7

Malik A (2020) Steganography: step towards security and privacy of confidential data in insecure medium by using LSB and cover media (December 12, 2020). SSRN Electron J https://doi.org/10.2139/ssrn.3747579

Malik A, Gautam S, Abidin S, Bhushan B (2019) Blockchain technology-future of IoT: including structure, limitations and various possible attacks. In: 2nd international conference on intelligent computing, instrumentation and control technologies (ICICICT), Kannur, India, pp 1100–1104. https://doi.org/10.1109/ICICICT46008.2019.8993144

Mazhar N, Salleh R, Zeeshan M, Hameed MM (2021) Role of device identification and manufacturer usage description in IoT security: a survey. IEEE Access 9:41757–41786. https://doi.org/10.1109/ACCESS.2021.3065123

Migabo E, Djouani K, Kurien A (2018) A modelling approach for the narrowband IoT (NB-IoT) physical (PHY) layer performance. In: IECON 2018—44th annual conference of the IEEE industrial electronics society, pp 5207–5214. https://doi.org/10.1109/IECON.2018.8591281

Mroue H, Nasser A, Hamrioui S, Parrein B, Motta-Cruz E, Rouyer G (2018) (2018), MAC layer-based evaluation of IoT technologies: LoRa, SigFox and NB-IoT. In: IEEE Middle East and North Africa communications conference (MENACOMM), pp 1–5. https://doi.org/10.1109/MENACOMM.2018.8371016

Musaddiq A, Rahim T, Kim D-S (2021) Enhancing IEEE 802.15.4 access mechanism with machine learning. In: 2021 Twelfth international conference on ubiquitous and future networks (ICUFN), pp 210–212. https://doi.org/10.1109/ICUFN49451.2021.9528725

Nie X, Fan T, Wang B, Li Z, Shankar A, Manickam A (2020) Big data analytics and IoT in OPERATION safety management in under water management. Comput Commun 154:188–196. https://doi.org/10.1016/j.comcom.2020.02.052

Nurlan Z, Zhukabayeva T, Othman M, Adamova A, Zhakiyev N (2022) Wireless sensor network as a mesh: vision and challenges. IEEE Access 10:46–67. https://doi.org/10.1109/ACCESS.2021.3137341

Pokhrel SR, Vu HL, Cricenti AL (2020) Adaptive admission control for IoT applications in home WiFi networks. IEEE Trans Mob Comput 19(12):2731–2742. https://doi.org/10.1109/TMC.2019.2935719

Qi N, Miridakis NI, Xiao M, Tsiftsis TA, Yao R, Jin S (2020) Traffic-aware two-stage queueing communication networks: queue analysis and energy saving. IEEE Trans Commun 68(8):4919–4932. https://doi.org/10.1109/TCOMM.2020.2988278

Salem RMM, Saraya MS, Ali-Eldin AMT (2022) An industrial cloud-based IoT system for real-time monitoring and controlling of wastewater. IEEE Access 10:6528–6540. https://doi.org/10.1109/ACCESS.2022.3141977

Salimibeni M, Hajiakhondi-Meybodi Z, Malekzadeh P, Atashi M, Plataniotis KN, Mohammadi A (2020) IoT-TD: IoT dataset for multiple model BLE-based indoor localization/tracking. In: 2020 28th European signal processing conference (EUSIPCO), pp 1697–1701. https://doi.org/10.23919/Eusipco47968.2020.9287547

Sarraf S (2019) 5G emerging technology and affected industries: quick survey. Am Sci Res J Eng Technol Sci (ASRJETS) 55(1):75–82. https://www.researchgate.net/publication/334282546_5G_Emerging_Technology_and_Affected_Industries_Quick_Survey

Senewe KF, Suryanegara M (2020) Innovative design of Internet of Things LoRa to determine radio refractivity in real-time. In: 2020 3rd international conference on computer and informatics engineering (IC2IE), pp 399–403. https://doi.org/10.1109/IC2IE50715.2020.9274626

Shafique K, Khawaja BA, Sabir F, Qazi S, Mustaqim M (2020) Internet of Things (IoT) for next-generation smart systems: a review of current challenges, future trends and prospects for emerging 5G-IoT scenarios. IEEE Access 8:23022–23040. https://doi.org/10.1109/ACCESS.2020.2970118

Shahabuddin S, Rahaman S, Rehman F, Ahmad I, Khan Z (2018) Evolution of cellular systems. A comprehensive guide to 5G security, pp 1–29. https://doi.org/10.1002/9781119293071.ch1

Sharma P, Shankar A, Cheng X (2021) Reduced paper Model predictive control Based Fbmc/oqam signal for NB-IoT paradigm. Int J Mach Learn Cybern. https://doi.org/10.1007/s13042-020-01263-8

Slamnik-Kriještorac N, Kremo H, Ruffini M, Marquez-Barja JM (2020) Sharing distributed and heterogeneous resources toward end-to-end 5G networks: a comprehensive survey and a taxonomy. IEEE Commun Surv Tutor 22(3):1592–1628. Third quarter. https://doi.org/10.1109/COMST.2020.3003818

Thanh TQ, Covaci S, Magedanz T (2019) VISECO: an annotated security management framework for 5G. Mob Secur Programm Netw 251–269. https://doi.org/10.1007/978-3-030-03101-5_21

Tikhvinskiy V, Deviatkin E, Aitmagambetov A, Kulakaeva A (2020) Provision of IoT services for co-located 4G/5G networks utilisation with dynamic frequency sharing. In: 2020 international conference on engineering management of communication technology (EMCTECH), pp 1–4. https://doi.org/10.1109/EMCTECH49634.2020.9261551

Wang N, Wang P, Alipour-Fanid A, Jiao L, Zeng K (2019) Physical-layer security of 5G wireless networks for IoT: challenges and opportunities. IEEE Internet Things J 6(5):8169–8181. https://doi.org/10.1109/JIOT.2019.2927379

Wang W, Xu P, Yang LT (2018a) Secure data collection, storage and access in cloud-assisted IoT. In: IEEE cloud computing, vol 5, no 4, pp 77–88. https://doi.org/10.1109/MCC.2018.111122026

Wang D, Chen D, Song B, Guizani N, Yu X, Du X (2018b) From IoT to 5G I-IoT: the next generation IoT-based intelligent algorithms and 5G technologies. IEEE Commun Mag 56(10):114–120. https://doi.org/10.1109/MCOM.2018.1701310

Yang M, Lim S, Oh SM, Shin J (2020) An uplink transmission scheme for TSN service in 5G industrial IoT. In: 2020 international conference on information and communication technology convergence (ICTC), pp 902–904. https://doi.org/10.1109/ICTC49870.2020.9289303

Yarali A (2020) Issues and challenges of 4G and 5G for PS. Public Safety Networks from LTE to 5G. Wiley, pp 189–194. https://doi.org/10.1002/9781119580157.ch11

Yu C et al (2019) Full-angle digital predistortion of 5G millimeter-wave massive MIMO transmitters. IEEE Trans Microw Theory Tech 67(7):2847–2860. https://doi.org/10.1109/TMTT.2019.291 8450

Zeqiri R, Idrizi F, Halimi H (2019) Comparison of algorithms and technologies 2G, 3G, 4G and 5G. In: 3rd international symposium on multidisciplinary studies and innovative technologies (ISMSIT), Ankara, Turkey, pp 1–4. https://doi.org/10.1109/ISMSIT.2019.8932896

Zhang S (2019) An overview of network slicing for 5g. IEEE Wirel Commun 26(3):111–117. https://doi.org/10.1109/mwc.2019.1800234

Zhang P, Yang X, Chen J, Huang Y (2019) A survey of testing for 5G: solutions, opportunities, and challenges. China Commun 16(1):69–85. https://doi.org/10.12676/j.cc.2019.01.007

Chapter 10
Securing the IoT-Based Wireless Sensor Networks in 5G and Beyond

N. Ambika

Abstract The previous contribution uses the k-means procedure to create clusters. It converts into a chain route when the threshold content goes beyond the energy of the devices in the system. The information transmitter fuel includes the power of the machine circuitry and the magnitude of facts communication and blowout. The vibrancy helps in communication circuitry. The knowledge packages ship to the destination. The architecture has two stages. The groups form during the clustering stage. The Optimal CBR method uses the k-means procedure to construct groups. It selects the cluster head based on the Euclidean length and device fuel. The verge posted by the group head to the individual set associates is the characteristic weight above which the machine transmits the data to the head. When two-thirds of the devices are lifeless, the instruments use the greedy procedure to construct a chain-like multiple-hop methodology to reach the base station. A beacon transmission is sent by the base station to the active devices in the chaining stage (when the energy of the nodes is lower). The base station creates the path using multiple-hop chain routing and the greedy technique. The devices send the notification to the base station using the chain track. The proposed work increases security by 9.67% when transmitting data and by 11.38% (device getting compromised).

Keywords KNN algorithm · Security · IoT · 5G communication · Hashing · Sensor network · Public key

Introduction

Sensors (Ambika 2020, 2021) are tiny devices deployed to accumulate information from an object of interest. The number of sensor hubs in a detector organization is higher than the number of devices in an impromptu organization. Sensing element hubs are inclined to disappointment. The geography of a sensor network changes much of the time. Sensor hubs fundamentally utilize a transmission correspondence

N. Ambika (✉)
Department of Computer Science and Applications, St. Francis College, Bangalore, India
e-mail: Ambika.nagaraj76@gmail.com

B. Bhushan et al. (eds.), *5G and Beyond*, Springer Tracts in Electrical and Electronics Engineering, https://doi.org/10.1007/978-981-99-3668-7_10

worldview, while most impromptu organizations depend on the money-to-point interchanges. Sensor hubs are restricted in power, computational limits, and memory. Detector hubs might not have worldwide recognizable proof (ID) given the measure of upward and countless tiny devices.

Smart sensors with actuators make Internet of Things (Ambika 2019; Dian et al. 2020). The idea of the Internet is to associate PC gadgets changing to a bunch of associates. It encompasses things of human residing space, like home apparatuses, machines, transportation, business capacity, products, etc. The quantity in the living space is more than the quantity of the total populace. Research continues to make these things speak with one another using the Internet. The correspondence among these things alludes to the Internet of Things.

The recommendation (Jothikumar et al. 2021) uses a k-means algorithm. It converts into a chain route when the threshold content goes beyond the energy of the devices in the system. The information transmitter fuel includes the power of the machine circuitry and the magnitude of facts communication and blowout. The vibrancy helps in communication circuitry. The knowledge packages ship to the destination. The architecture has two stages. The groups form during the clustering stage. The Optimal CBR method uses the k-means procedure to construct groups. It selects the cluster head based on the Euclidean length and device fuel. The verge posted by the group head to the individual set associates is the characteristic weight above which the machine transmits the data to the head. When two-thirds of the devices are lifeless, the instruments use the greedy procedure to construct a chain-like multiple-hop methodology to reach the base station. A beacon transmission is sent by the base station to the active devices in the chaining stage (when the energy of the nodes is lower). The base station creates the path using multiple-hop chain routing and the greedy technique. The devices (Nagaraj 2021) send the notification to the base station using the chain track.

The suggestion employs a hashing methodology. The base station broadcasts to the public for every session. The devices generate the hash codes for the sensed data and create the outcome using the public key. This methodology secures the devices and data during transmission.

The contribution of the work:

- The devices send the notification to the base station using the chain track. The suggestion employs a hashing methodology.
- The base station broadcasts to the public for every session.
- The devices develop the hash codes for the sensed information and generate the outcome using the public key.
- The proposed work increases security by 9.67% when transmitting data and 11.38% details when the device is compromised.

The work is divided into sections. Literature survey is summarized in segment 2. IoT-based wireless network is detailed in division 3. Different kinds of attacks in IoT are explained in section "Different Types of Attacks". Importance of 5G is briefed in segment 5. Background is discussed in section "Background". The proposed work is detailed in division 7. Analysis of work is detailed in section "Analysis of the proposal". The work concludes in section "Conclusion".

Literature Survey

The following sections briefs the contribution made by various authors. The recommendation (Jothikumar et al. 2021) uses a k-means algorithm. It converts into a chain route when the threshold content goes beyond the energy of the devices in the system. The information transmitter fuel includes the power of the machine circuitry and the magnitude of facts communication and blowout. The vibrancy helps in communication circuitry. The knowledge packages ship to the destination. The architecture has two stages. The groups form during the clustering stage. The Optimal CBR method uses the k-means procedure to construct groups. It selects the cluster head based on the Euclidean length and device fuel. The verge posted by the group head to the individual set associates is the characteristic weight above which the machine transmits the data to the head. The instruments use the greedy procedure to construct a chain-like multiple-hop methodology to reach the base station When two-thirds of the devices are lifeless. A beacon transmission is sent by the base station to the active devices in the chaining stage (when the energy of the nodes is lower). The base station creates the path using multiple-hop chain routing and the greedy technique. The devices send the notification to the base station using the chain track.

It is M2M traffic mode (Fu et al. 2018). It further develops traffic adjusting strategies. This model is reasonable in speaking to the current promising mass of gadgets. 3GPP has made a record 3GPP TR 37.868, which gives a way to deal with demonstrating M2M traffic in the LTE organization. The existing traffic models depict a fixed arbitrary process. It has a limited time stretch. M2M gadgets produce traffic. This approach offers two traffic models and double crosses intervals. The first model portrays the ordinary condition of the organization, where each M2M gadget for 60 s communicates one message. The subsequent model shows the condition of the expanded network load. This heap prompts the mass enactment of M2M gadgets.

WSN (Fu et al. 2018) can work at 900 MHz/2.4 GHz to help the 5 GHz, recurrence groups. The concentrator contains associate WSN pointing to the 5G versatile interchanges world. The methodology develops the framework execution. It uses the UAV as a transfer station. SINK UAV BS improves on the framework model. The concentrator communicates a sign to the BS with one UAV as the hand-off. This framework model can more readily uncover the connection between the place of the UAV-based hand-off and the framework energy consumption. The limited transmission distance between the Base station and the concentrator can limit the sending force of the concentrator. The ideal flight path is not set in stone by AI.

The organization (Lynggaard and Skouby 2015) contains an assortment of homes furnished with IoT. It deals with administration like lighting, warming, security, and theater setups for its users. These IoT gadgets interconnect with the home organization, which associates with the web cloud services. The network interconnects the brilliant homes and interfaces using cloud administrations that consume the enormous information produced by the home IoT. The IoT gadgets create a tremendous measure of data to be handled by the city CoT administrations. It contains an assortment of associated sensor hub bunches where each gathering ends in a sensor end gadget. These end gadgets speak with a switching hub which thus courses correspondence through the network.

The gridlock situation (Sachan et al. 2021) is an examination of two D2D correspondence modes. The work distinguishes the hubs with lesser responsibilities from the previous information saved in the control unit of the base station. The controlling unit has every one of the subtleties of commitment in a specific gadget. It very well may be distinguished what hubs are with a lesser burden. It infers that such notes are moving next to zero data. The hubs with low loads can be worked at lower communication power levels all at once and additionally works on the SINR because of diminished by and large impedance in the framework and further develops the battery duration of the portable hubs. The encompassing hubs have a lesser burden at a specific time. There will be proficient correspondence while decreasing the impedance.

The work (Sekaran et al. 2021) is an integrated spectrum selection and spectrum access using a greedy and AI-based framework to allow the forthcoming and subsequent demands on 5G and beyond to be presented. A fractional Knapsack Greedy-based strategy is introduced, and Lagrange Hyperplane-based approach is utilized to realize the AI-based strategies for spectrum selection and spectrum allocation for IoT-enabled sensor networks. This framework is called Fractional Knapsack and Lagrange Hyperplane Spectrum Access (FK-LHSA). The First Fractional Knapsack Multi-band spectrum selection (FKMSS) model is designed along with an energy consumption model to optimize channel or spectrum throughput. A Lagrange Hyperplane (LH) spectrum access model minimizes spectrum access delay and improves access accuracy. The simulation results show that the proposed FKM and LH model can effectively reduce the spectrum access delay (along with the improvement of throughput and spectrum access accuracy).

The proposal (Shin and Kwon 2020) cures security weaknesses in light of the framework engineering in WSNs for 5G-coordinated IoT. The proposed conspire parts into five stages. The framework arrangement stage incorporates the statement of the framework boundaries and entryway and sensor hub enrollment before sending. The client enrollment stage starts when a client sends a solicitation message for enlistment to the confirmation server over a secure channel. The user needs to get to the WSN responsible for the entryway the accompanying advances perform with the client, verification server, and passage over a public channel. With the assistance of the verification server, the client and passage commonly validate one another and lay out a typical meeting key for future correspondence. The client can acquire the tangible information progressively from the WSN that matches entrance honors. The key and biometric update stage permits a client to refresh the secret key and biometrics without connection with the validation server. The messages communicate over a channel in the entrance honor update stage.

It is an energy harvest Markovian battery model (Mahmoud et al. 2017) of 100 states, which addresses the fuel of the optional client, and infers the throughput while considering detecting a power error reap CRWSN. The energy reaps supplementary client can send bundles on the channel Whenever the pipeline is passive. The auxiliary client neglects to communicate information when the channel is inactive because erroneous detection prompts throughput corruption. The throughput improves the fruitful transmission boundaries. The conditions address a restricted battery limit. It is a non-complete M/M/1 model of the energy gather CR-WSN model because of the inconsistent worth of the primary state. The progress of the states relies upon the entrance likelihood. It views that there is general information at the optional client to send. The range states change in light of the traffic of the essential client. The range state stays unaltered by involving a client with probability or travels to sit with the likelihood in the past schedule opening.

IoT-Based Wireless Networks

The IoT is reconciliation and correspondence between clever devices. IoT's incomparability contributes to new advancements and applications. Such detectors and actuators collaborate with different handsets, microcontroller gadgets, and conventions for the correspondence of control and sensor information. Such constant modules communicate detected information to the unified storehouses. In contrast with traditional wired or remote systems administration frameworks, the highlights of IoT using remote advances are unique as the number of specialized gadgets is very high. Figure 10.2 is the representation of the same.

Different Types of Attacks

The following are the different security concerns:
(Fig. 10.1).

Perception Layer

- *Eavesdropping* (Khattak et al. 2019)—Assailants can sniff the traffic produced by IoT information stream to accumulate client's data by setting up comparable IoT gadgets.
- *Malicious Data Injection* (Alromih et al. 2018)—Bogus sensor information infusion is a type of assault where the sensor information utilized in IoT applications is produced or altered for malevolent purposes.

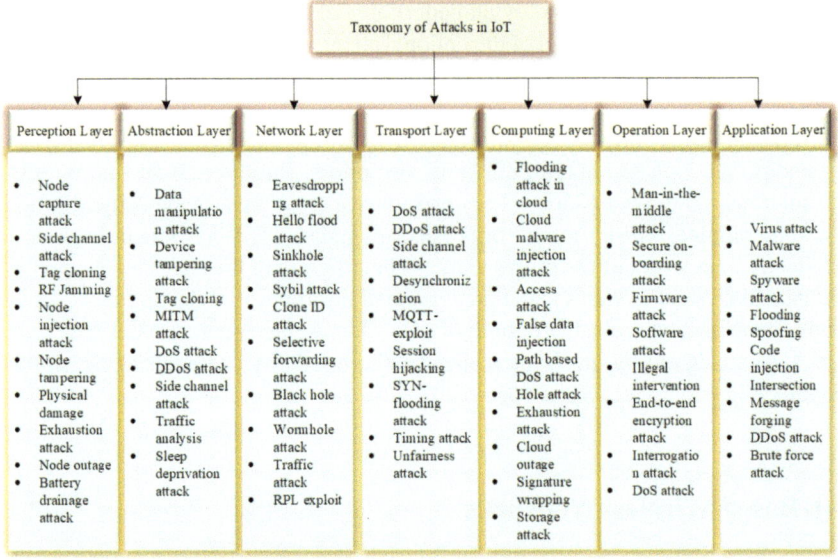

Fig. 10.1 Taxonomy of threats in IoT (Krishna et al. 2021)

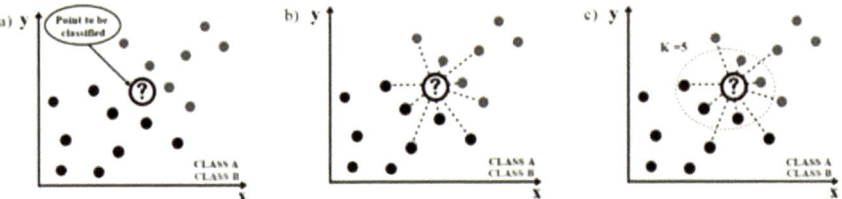

Fig. 10.2 K-nearest Neighbor algorithm (Pacheco et al. 2021)

- *Sybil Attack* (Mishra et al. 2018)—The noxious hubs in this can have numerous personalities of a veritable hub by either imitating it or with a phony character through duplication.
- *Disclosure of Critical Information* (Zhang Et Al. 2017)—Sensors utilized in IoT devices can reveal delicate data, for example, passwords, secret keys, charge card certifications, etc. These subtleties disregard client security or fabricate an information base for future assaults.
- *Side-Channel Attacks* (Kumar et al. 2017)—The aggressor assembles data and plays out the figuring out cycle to gather the encryption accreditations of an IoT gadget while the encryption interaction is in progress. This data is not gathered from plaintext or ciphertext during the encryption cycle. Side-channel goes after the utilization of information to gain the key the gadget utilizes.
- *Malicious Data Injection* (Alromih et al. 2018)—Assailants exploit defects in correspondence conventions to embed information into the organization. The

gateway will mess with the data expected to control the gadget on the off chance. The infusion assault might bring about code execution or framework control from a remote place.

- *Node cloning* (Khattak et al. 2019)—For unapproved purposes, the gadgets can be effectively fashioned and recreated. It is called the cloning of hubs.
- *Exhaustion attack* (Aarika et al. 2020)—Depletion is a spot assault. It is associated with deactivation attacks. It decreases the size of the organization and eliminates hubs for all time from the organization.

Abstraction Layer

- *Illegal access* (Alramadhan and Sha 2017)—The unlawful access and vindictive difference in information might emerge when handling delicate information.
- *Man-in-the-Middle*—A framework (Navas et al. 2018) tunes in on rush hour gridlock between a savvy gadget and an entryway. All traffic steers utilizing the assailant's PC using the ARP harming procedure
- *Spoofing*—To start a caricaturing assault (Mohammadnia and Slimane 2020), an aggressor can imitate a node. A transmission could record utilizing a convenient per user.
- *Threat to communication protocols* (Failed 2017)—OSI layered convention engineering and the actual layer encryption aren't supported. It requires extra security techniques in the upper layers.
- *Tag cloning*—The attack (Dimitriou 2005) can mimic.
- *Denial-of-Service (DoS)*—It is a kind of assault (Liang et al. 2016) where a gadget or application is malevolently denied typical activity.
- *DDoS*—Any IoT gadget, organization, or programming system could be closed somewhere around a disseminated forswearing of administration (DoS) assault (Zhang and Green 2015), delivering the assistance out of reach to its shoppers.
- *Traffic analysis*—Invaders (Hafeez et al. 2019) distinguish the base station, close by hubs, or bunch heads to uphold forswearing of administration assault or bundle listening in.
- *Sleep deprivation*—The forswearing of a rest assault (Brun et al. 2018) on a battery-fueled gadget will bring about energy consumption.

Network Layer

- Hello flood (Srinivas and Manivannan 2020)—The hubs in the organization decipher a welcome message coming from the inside and imprint it as a correspondence course.
- Sinkhole (Pundir et al. 2020)—By utilizing this methodology, an aggressor compromises an organization's focal hub and supersedes to deliver it inaccessible.

- Blackhole—Assuming that the noxious hub encounters a Blackhole assault (Sahay et al. 2018), it will drop all bundles experienced.
- Traffic Analysis—The assailant investigates the traffic and saves a duplicate for later use in this assault.
- Wormhole—This organization's assault (Pongle and Chavan 2015) would catch traffic in one area and divert it to another.
- Selective forwarding—An aggressor dispatches an assault (Hariri et al. 2019) by entering an organization and dropping parcels.
- RPL exploit (Airehrour et al. 2019)—The angry hubs can try to divert ways when information is moved.
- Transport Layer
- Desynchronization—Desynchronizing the transmissions between two hubs permits an aggressor to break real connections between them
- Session hijacking (Humaira et al. 2020)—The aggressor takes the meeting ID and professes to be the genuine client to assume control over a client's Internet-based meeting

Computing Layer

- Malicious Attack (Ahmed et al. 2018)—As laborers in the organization download vindictive programming programs from the Internet, there is a decent opportunity for the machine to get hacked. The malware would spread across the organization, putting the entire organization under its impact.
- SQL injection (Uwagbole et al. 2017)—SQL infusion is a web security blemish that permits an assailant to interfere with a web application's data set inquiries. It allows an aggressor to get to the data that they wouldn't ordinarily have the option to recover.
- Illegal Access (Deebak et al. 2019)—If IoT contraptions don't have an expected design, the whole organization is harmed. The organizations utilizing cloud-based registering to need unlimited authority over their organizations, which requires designing and safeguarding their cloud arrangements on security controls given by their cloud specialist co-ops.
- Storage Attack—The programmers will dial back the movement of the gadget as they utilize the distributed storage assets.
- Access Attack—An unapproved individual or foe accesses the IoT network here of assault.
- Software modification (Wurster et al. 2005)—An IoT gadget can be undermined by altering its product or firmware by utilizing physical or remote admittance to make unapproved moves.

Operation Layer

- Illegal Intervention—Even though cloud specialist co-ops are engaged in determinedly developing APIs and points of interaction, this blast has stretched out security perils connected with them.
- Unauthorized Access (Failed 2013)—Whenever varied clients can adjust the plans of various sections of the IoT systems, synchronous execution of arrangement changes and synchronous adjusting of course of action records prompts unpredictable system status.

Application Layer

- Malicious code—Pernicious codes or focused on malware can undoubtedly take advantage of the weaknesses of IoT widgets through the Internet, which permits programmers to think twice about gadgets.
- Software Modification—The assailant will want to reinvent IoT gadgets remotely. This action could bring about the IoT network being hacked.
- Data tampering (Huang et al. 2021)—During an assault of this kind, the data on the end gadget is distorted by an assailant.
- Cross-site script (Failed 2006)—It is a procedure aggressors use to embed vindictive code into a trusted site.
- Identity Thefts—IoT frameworks manage a lot of individual and touchy data. This information can be taken.
- Virus attack—The target of these assaults is to break the classification of the framework. The gamble of these assaults is fundamentally higher for cell phones, sinks, or entryways in IoT organizations.
- Spyware attack (Wazid et al. 2013)—Introduced on IoT gadgets without assent, spyware is an established program that gathers data.
- Code Injection (Ray and Ligatti 2012)—Assailants ordinarily utilize the most straightforward method for breaking into a device or organization. Assuming the gadget is imperiled to resentful contents and confusion, it is the primary mark for an aggressor.
- Intersection (Berthold and Langos 2002)—Whenever a framework's trustworthiness is compromised, there is a high gamble of well-being and security dangers.
- Brute force attack (Knudsen and Robshaw 2011)—A power assault includes deliberately trying and speculating each conceivable passphrase or secret key blend to get sufficiently close to the framework.

Importance of 5G

The Internet of Things (Lee et al. 2017) is a unique paradigm that gives users access to wireless communication networks and artificial intelligence technology and is thought to be relevant to a wide range of disciplines and applications. The development of fifth-generation cellular network technology opens up the possibility of deploying vast sensors in the IoT and processing massive data, testing communications, and data mining capabilities.

The confluence of the Internet, intelligence, and objects is the 5G IIoT paradigm (Mavromoustakis 2016). Traditional IoT is a paradigm that integrates large network connection entities and encompasses the Internet and things. An intelligent individual combines intellect and objects. It creates high-functioning agents or gadgets to fulfill complex applications such as object identification.

The advancement of fifth era (5G) networks is more promptly available as a significant driver of the development of IoT applications. New applications and plans of action later on IoT require new execution standards such as enormous availability, security, dependability, the inclusion of remote correspondence, super low idleness, throughput, super solid, et al. for an immense number of IoT gadgets. The developing Long-Term Evolution (LTE) and 5G innovations are supposed to give new availability connection points to the future IoT applications (To meet these prerequisites).

Background

K-Nearest Neighbor (Sun and Huang 2010) is one of the complex Machine Learning calculations of the Supervised Learning procedure. It accepts the similitude between the new case/information and accessible cases into the classification. It stores every one of the accessible information and orders another information point in light of the comparability. It tends to group into a good suite class by utilizing the KNN algorithm. It can be utilized for Regression as well concerning Classification. It is a non-parametric calculation and implies it makes no presumption on the information. It stores the dataset and plays out an activity on the dataset. KNN calculation at the preparation stage keeps the dataset. It orders that information into a classification after getting new information. Figure 10.2 portrays the same. The steps are as follows:

Step-1: Select the number K of the neighbors.

Step-2: Calculate the Euclidean distance of K number of neighbors.

Step-3: Take the K closest neighbors according to the determined Euclidean distance.

Step-4: Among these k neighbors, count the quantity of the elements in every classification.

Step-5: Assign the new information focusing on that classification for which the quantity of the neighbor is most extreme.

Step-6: Our model is prepared.

Mathematical Explanation for Euclidean Distance

The distance between two points we should subtract the dimensions of each coordinate by each other, sum them all, apply power of two then square root it. Let the points be A and B. let the coordinates of A be (a_1, a_2) and B is (b_1, b_2)

$$d(A, B) = \sqrt[2]{(b_2 - a_2)^2 + (b_1 - a_1)^2} \qquad (10.1)$$

Proposed Work

The previous contribution (Jothikumar et al. 2021) uses k-means procedure to create clusters. It converts into chain route when the threshold content goes beyond the energy of the devices in the system. The information transmitter fuel includes the power of the machine circuitry and the magnitude of facts communication and blowout. The vibrancy helps in communication circuitry. The knowledge packages ship to the destination. The architecture has two stages. The groups form during the clustering stage. The Optimal CBR method uses the k-means procedure to construct groups. It selects the cluster head based on the Euclidean length and devices fuel. The verge posted by the group head to the individual set associates is the characteristic weight above which the machine transmits the data to the head. When two-thirds of the devices are lifeless, the instruments use the greedy procedure to construct a chain-like multiple-hop methodology to reach the base station. A beacon transmission is sent by the base station to the active devices in the chaining stage (when the energy of the nodes is lower). The base station creates the path using multiple-hop chain routing and the greedy technique. The devices send the notification to the base station using the chain track.

The contribution is the improvement of the previous suggestion. The dataset is generated in the trial state. The sink node generates a public key and dispatches it to the other devices in the network. The devices create the hash code using sensed data. The code is used with the public key to generate the final outcome. The methodology secures the data from the hackers. The base station uses KNN algorithm to segregate the data into groups. The method detects the security breach at an early stage.

Assumptions

- The nodes are assumed to be static by nature. They are deployed to track an object of interest. The same is communicated to the devices before deployment.
- The IoT device is designated base station.

Table 10.1 Generation of hash code

Step 1—Input the sensed data (16 bits)
Step 2—Insert '0' bits in even positions (32 bits)
Step 3—Apply right circular shift
Step 4—Divide the bits into two halves
Step 5—Insert the odd position bits of second halve into the even position of first halve
Step 6—Apply left circular shift
Step 7—Apply random symbol value to the hash code (only base station has the rights to generate the symbol)

- The nodes are embedded with a set of algorithms and credentials before deployment.
- The nodes use multi-hop methodology to transmit messages to the base station (IoT device) or the predestined location.
- The cluster heads communicate with the store nodes after authenticating themselves.
- The base station broadcast the public key to its network.

Creating Trial Data Sets

- The nodes after deployment are into trial state, where the trial readings are gathered from the nodes. This creates the trial dataset. This dataset is stored in the base station for reference.
- It generates hash code and the same is used along with public key to generate the final outcome.
- It uses KNN algorithm to classify the data sets into subsets (Table 10.1).

Transmitting the Messages

- The nodes sense the environment and generate the hash code. The public key is used to generate the final outcome.
- Any new value is recognized at an early stage.

Analysis of the Proposal

The previous architecture (Jothikumar et al. 2021) has two stages. The groups form during the clustering stage. The Optimal CBR method uses the k-means procedure to construct groups. It selects the cluster head based on the Euclidean length and devices fuel. The verge posted by the group head to the individual set associates is the characteristic weight above which the machine transmits the data to the head. When two-thirds of the devices are lifeless, the instruments use the greedy procedure to construct a chain-like multiple-hop methodology to reach the base station. A beacon transmission is sent by the base station to the active devices in the chaining stage (when the energy of the nodes is lower). The base station creates the path using multiple-hop chain routing and the greedy technique. The devices send the notification to the base station using the chain track.

A hashing computation (Pieprzyk 1993) is a cryptographic hash work. A numerical calculation maps information (of erratic size) to a hash of proper size. A hash work calculation intends to be a one-way work, infeasible to modify. Nonetheless, as of late, a few hashing calculations have been compromised.

A public key (Ambika and Raju 2010) encodes a message with the authenticity of a computerized signature. It joins a relating private key. It is known exclusively to its proprietor. Public keys are accessible from a declaration authority, which issues advanced testaments that demonstrate the proprietor's character and contain the proprietor's public key. Public keys utilize irregular calculations. It matches the shared key with a related private key. A public key is given to any individual with whom a singular need to convey, through a private key has a place with the singular it was made for and isn't shared. The public key is commonly put away on a public key foundation server and scrambles information safely before being sent on the web.

The suggestion employs a hashing methodology. The base station broadcasts public for every session. The devices generate the hash codes for the sensed data and generate the outcome using the public key. This methodology secures the data in the devices and data during transmission. The work is simulated using Python. Table 10.2 portrays the simulation parameters used in the proposal. In simulation, we have considered temperature as the parameter.

Table 10.2 Simulation parameters

Parameters used	Description
Number of temperature variants used	15 * 10 groups are created
Number of groups formed	15 groups
Length of hash code	32 bits
Length of input (sensed data)	16 bits
Length of public key	16 bits
Total length of the output	256 bits
Simulation time	60 ms

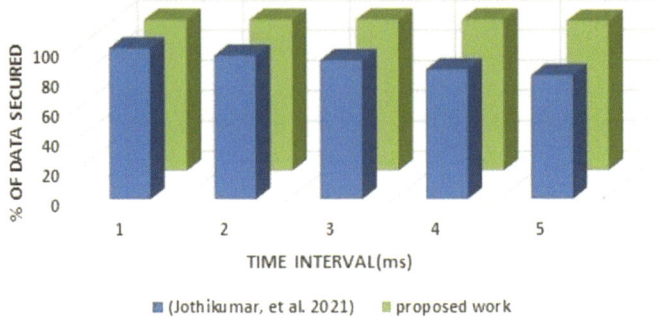

Fig. 10.3 Data security

Security

The IoT is where the Internet meets the actual world. The new aspect of protection ought to be explored as the going after danger moves from controlling data to controlling incitation. The worldview makes many worries over the securing information, benefits, and, surprisingly, the whole IoT framework. The attributes like secrecy, uprightness, verification, approval, accessibility, and protection should be guaranteed for the IoT framework to ensure security in IoT. The confidential data are necessary to be secured. Hence different kinds of security measures (Varshney et al. 2019; Sharma et al. 1286) are to be adopted. The proposed work increases security by 9.67% compared with previous work (Jothikumar et al. 2021). The same is represented in Fig. 10.3.

The nodes will get compromised, if the devices are not able to defend themselves. Sensors are cheap devices. Hence protection is a must. The proposal generates hash codes, followed by the generation of outcome based on the public key. If the adversary captures the nodes, it will not be able to figure out anything out of it. The data in the devices are 11.38% secure compared to Jothikumar et al. (2021). Figure 10.4 represents the same.

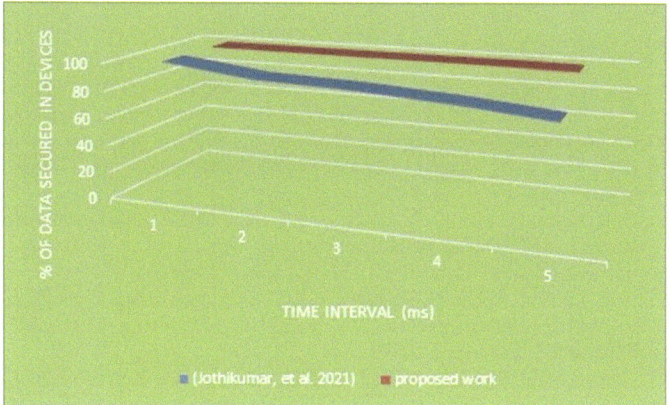

Fig. 10.4 Data secure in nodes

Conclusion

Smart sensors and actuators work together and send data to IoT devices. These devices communicate over a common platform. The instruments use 5G Internet facility to communicate with other devices or same/different caliber. The recommendation uses k-means algorithm. It converts into chain route when the threshold content goes beyond the energy of the devices in the system. The information transmitter fuel includes the power of the machine circuitry and the magnitude of facts communication and blowout. The vibrancy helps in communication circuitry. The knowledge packages ship to the destination. The architecture has two stages. The groups form during the clustering stage. The Optimal CBR method uses the k-means procedure to construct groups. It selects the cluster head based on the Euclidean length and devices fuel. The verge posted by the group head to the individual set associates is the characteristic weight above which the machine transmits the data to the head. When two-thirds of the devices are lifeless, the instruments use the greedy procedure to construct a chain-like multiple-hop methodology to reach the base station. A beacon transmission is sent by the base station to the active devices in the chaining stage (when the energy of the nodes is lower). The base station creates the path using multiple-hop chain routing and the greedy technique. The devices send the notification to the base station using the chain track. The suggestion employs a hashing methodology. The base station broadcasts public for every session. The devices generate the hash codes for the sensed data and generate the outcome using the public key. This methodology secures the data in the devices and data during transmission. The proposed work increases security by 9.67% when transmitting data and 11.38% data when the device is compromised.

References

Aarika K, Bouhlal M, Abdelouahid RA, Elfilali S, Benlahmar E (2020) Perception layer security in the internet of things. In: The 17th international conference on mobile systems and pervasive computing (MobiSPC), the 15th international conference on future networks and communications (FNC);The 10th international conference on sustainable energy information technology, Leuven, Belgium

Andy S, Rahardjo B, Hanindhito B (2017) Attack scenarios and security analysis of MQTT communication protocol in IoT system. In: 4th international conference on electrical engineering, computer science and informatics (EECSI), Yogyakarta, Indonesia

Ahmed A, Latif R, Latif S, Abbas H, Khan FA (2018) Malicious insiders attack in IoT based multi-cloud e-healthcare environment: a systematic literature review. Multimed Tools Appl 77(17):21947–21965

Airehrour D, Gutierrez JA, Ray SK (2019) SecTrust-RPL: a secure trust-aware RPL routing protocol for Internet of Things. Futur Gener Comput Syst 93:860–876

Alramadhan M, Sha K (2017) An overview of access control mechanisms for internet of things. In: 26th international conference on computer communication and networks (ICCCN), Vancouver, BC, Canada

Alromih A, Al-Rodhaan M, Tian Y (2018) A randomized watermarking technique for detecting malicious data injection attacks in heterogeneous wireless sensor networks for internet of things applications. Sensors 18(12):4346

Ambika N (2020) SYSLOC: hybrid key generation in sensor network. In: Handbook of wireless sensor networks: issues and challenges in current scenario's. advances in intelligent systems and computing, vol 1132. Springer, Cham, pp 325–347

Ambika N (2019) Energy-perceptive authentication in virtual private networks using GPS data. In: Security, privacy and trust in the IoT environment. Springer, Cham, pp 25–38

Ambika N (2021) Wearable sensors for smart societies: a survey. In: Green technological innovation for sustainable smart societies. Springer, Cham, pp 21–37

Ambika N, Raju GT (2010) Figment authentication scheme in wireless sensor network. In: Security technology, disaster recovery and business continuity, Jeju Island, Korea, Springer, pp 220–223

Berthold O, Langos H (2002) Dummy traffic against long term intersection attacks. In: International workshop on privacy enhancing technologies, Cambridge, UK

Brun O, Yin Y, Gelenbe E, Kadioglu YM, Augusto-Gonzalez J, Ramos M (2018) Deep learning with dense random neural networks for detecting attacks against IoT-connected home environments. In: International ISCIS security workshop, London, UK

Bulashenko A, Piltyay S, Polishchuk A, Bulashenko O (2020) New traffic model of M2M technology in 5G wireless sensor networks. In: 2nd international conference on advanced trends in information theory (ATIT), Kyiv, Ukraine

Deebak BD, Al-Turjman F, Aloqaily M, Alfandi O (2019) An authentic-based privacy preservation protocol for smart e-healthcare systems in IoT. IEEE Access 7:135632–135649

Dian FJ, Vahidnia R, Rahmati A (2020) Wearables and the Internet of Things (IoT), applications, opportunities, and challenges: a Survey. IEEE Access 8:69200–69211

Dimitriou T (2005) A lightweight RFID protocol to protect against traceability and cloning attacks. In: First international conference on security and privacy for emerging areas in communications networks (SECURECOMM'05), Athens, Greece

Fu S, Zhao L, Su Z, Jian X (2018) UAV based relay for wireless sensor networks in 5G systems. Sensors 18(8):2413

Gardašević G, Katzis K, Bajić D, Berbakov L (2020) Emerging wireless sensor networks and Internet of Things technologies—foundations of smart healthcare. Sensors 20(13):3619

Hafeez I, Antikainen M, Tarkoma S (2019) Protecting IoT-environments against traffic analysis attacks with traffic morphing. In: IEEE international conference on pervasive computing and communications workshops (PerCom Workshops), Kyoto, Japan

Hariri A, Giannelos N, Arief B (2019) Selective forwarding attack on iot home security kits. In: Computer security, Cham. Springer, pp 360–373

Hassan WH (2019) Current research on Internet of Things (IoT) security: a a survey. Comput Netw 148:283–294

Huang DW, Liu W, Bi J (2021) Data tampering attacks diagnosis in dynamic wireless sensor networks. Comput Commun 172:84–92

Humaira F, Luva IMSSA, Rahman MB (2020) A secure framework for IoT smart home by resolving session Hijacking. Global J Comp Sci Technol 20(2):1–13

Jothikumar C, Ramana K, Chakravarthy VD, Singh S, Ra IH (2021) An efficient routing approach to maximize the lifetime of IoT-based wireless sensor networks in 5G and beyond. Mobile Inf Syst 1–11

Kirda E, Kruegel C, Vigna G, Jovanovic N (2006) Noxes: a client-side solution for mitigating cross-site scripting attacks. In: ACM symposium on applied computing, Dijon France

Khattak HA, Shah MA, Khan S, Ali I, Imran M (2019) Perception layer security in Internet of Things. Futur Gener Comput Syst 100:144–164

Khattak HA, Shah MA, Khan S, Ali I, Imran M (2019) Perception layer security in Internet of Things. Future Gener Comput Syst 100:144–164

Knudsen LR, Robshaw MJ (2011) Brute force attacks. In: The block cipher companion. Springer, Berlin, Heidelberg, pp 95–108

Krishna R, Priyadarshini A, Jha A, Appasani B, Srinivasulu A, Bizon N (2021) State-of-the-art review on IoT threats and attacks: taxonomy, challenges and solutions. Sustainability 13:9463

Kumar S, Sahoo S, Mahapatra A, Swain AK, Mahapatra KK (2017) Security enhancements to system on chip devices for IoT perception layer. In: IEEE international symposium on nanoelectronic and information systems (iNIS), Bhopal, India

Lee SK, Bae M, Kim H (2017) Future of IoT networks: a survey. Appl Sci 7(10):1072

Liang L, Zheng K, Sheng Q, Huang X (2016) A denial of service attack method for an iot system. In: 8th international conference on information technology in medicine and education (ITME), Fuzhou, China

Lynggaard P, Skouby KE (2015) Deploying 5G-technologies in smart city and smart home wireless sensor networks with interferences. Wirel Pers Commun 81(4):1399–1413

Mahmoud HH, ElAttar HM, Saafan A, ElBadawy H (2017) Optimal operational parameters for 5G energy harvesting cognitive wireless sensor networks. IETE Tech Rev 34(sup1):62–72

Mavromoustakis CX, Mastorakis G, Mongay JM (2016) Internet of Things (IoT) in 5G mobile technologies, vol 8, Springer, Switzerland

Mishra AK, Tripathy AK, Puthal D, Yang LT (2018) Analytical model for sybil attack phases in internet of things. IEEE Internet Things J 6(1):379–387

Mohammadnia H, Slimane SB (2020) IoT-NETZ: Practical spoofing attack mitigation approach in SDWN network. In: Seventh international conference on software defined systems (SDS), Paris, France

Muslukhov I, Boshmaf Y, Kuo C, Lester J, Beznosov K (2013) Know your enemy: the risk of unauthorized access in smartphones by insiders. In: 15th international conference on Human-computer interaction with mobile devices and services, Munich Germany

Nagaraj A (2021) Introduction to sensors in IoT and cloud computing applications. Bentham Science Publishers, UAE

Navas RE, Bouder HL, Cuppens N, Cuppens F, Papadopoulos GZ (2018) Do not trust your neighbors! A small IoT platform illustrating a man-in-the-middle attack. In: International conference on ad-hoc networks and wireless, Saint-Malo, France

Pacheco A, Junior J, Ruiz-Armenteros A, Henriques R (2021) Assessment of k-nearest neighbor and random forest classifiers for mapping forest fire areas in central Portugal using landsat-8, sentinel-2, and terra imagery. Remote Sens 13:1345

Pieprzyk JSB (1993) Design of hashing algorithms. Springer, Berlin, Heidelberg

Pongle P, Chavan G (2015) Real time intrusion and wormhole attack detection in internet of things. Int J Comput Appl 121(9):0975–8887

Pundir S, Wazid M, Singh DP, Das AK, Rodrigues JJPC, Park Y (2020) Designing efficient sinkhole attack detection mechanism in edge-based IoT deployment. Sensors 20(5):1300

Ray D, Ligatti J (2012) Defining code-injection attacks. Acm Sigplan Notices 47(1):179–190

Sachan S, Sharma R, Sehgal A (2021) SINR based energy optimization schemes for 5G vehicular sensor networks. Wirel Pers Commun 1–21 (2021)

Sahay R, Geethakumari G, Mitra B, Thejas V (2018) Exponential smoothing based approach for detection of blackhole attacks in IoT. In: IEEE international conference on advanced networks and telecommunications systems (ANTS), Indore, India

Sekaran R, Goddumarri SN, Kallam S, Ramachandran M, Patan R, Gupta D (2021) 5G integrated spectrum selection and spectrum access using AI-based frame work for IoT based sensor networks. Comput Netw 186:107649

Sharma M, Bhushan B, Khamparia A (2021) Securing Internet of Things: attacks, countermeasures and open challenges. In: Emerging technologies in data mining and information security. Advances in intelligent systems and computing, vol 1286. Singapore, Springer, pp 873–885

Shin S, Kwon T (2020) A privacy-preserving authentication, authorization, and key agreement scheme for wireless sensor networks in 5G-integrated Internet of Things. IEEE Access 8:67555–67571

Srinivas TAS, Manivannan SS (2020) Prevention of hello flood attack in IoT using combination of deep learning with improved rider optimization algorithm. Comput Commun 163:162–175

Sun S, Huang R (2010) An adaptive k-nearest neighbor algorithm. In: Seventh international conference on fuzzy systems and knowledge discovery, Yantai, China

Tilwari V, Dimyati K, Hindia M, Mohmed Noor Izam T, Amiri I (2020) EMBLR: a high-performance optimal routing approach for D2D communications in large-scale IoT 5G network. Symmetry 12:438

Uwagbole SO, Buchanan WJ, Fan L (2017) Applied machine learning predictive analytics to SQL injection attack detection and prevention. In: IFIP/IEEE symposium on integrated network and service management (IM), Lisbon, Portugal

Varshney T, Sharma N, Kaushik I, Bhushan B (2019) Architectural model of security threats & their countermeasures in IoT. In: International conference on computing, communication, and intelligent systems (ICCCIS), Greater Noida, India

Wazid M, Katal A, Goudar RH, Singh DP, Tyagi A, Sharma R, Bhakuni P (2013) A framework for detection and prevention of novel keylogger spyware attacks. In: 7th international conference on intelligent systems and control (ISCO), Coimbatore, India

Wurster G, Van Oorschot PC, Somayaji A (2005) A generic attack on checksumming-based software tamper resistance. In: IEEE symposium on security and privacy (S&P'05), Oakland, CA, USA

Zhang C, Green R (2015) Communication security in internet of thing: preventive measure and avoid DDoS attack over IoT network. In: 18th symposium on communications & networking, VA, USA

Zhang G, Kou L, Zhang L, Liu C, Da Q, Sun J (2017) A new digital watermarking method for data integrity protection in the perception layer of IoT. Secur Commun Netw 1–13

Chapter 11
5G and Internet of Things—Integration Trends, Opportunities, and Future Research Avenues

A. K. M. Bahalul Haque, Md. Oahiduzzaman Mondol Zihad, and Md. Rifat Hasan

Abstract The Fifth Generation Communication System (5G) has revolutionized data (voice, text, and hybrid) transmission and communication. Advanced communication protocol and sophisticated technology open up the opportunity to integrate 5G with other state-of-the-art technologies. Similarly, the Internet of Things combines sensors, actuators, and other devices that network together to collect contextual and environmental data for application-specific purposes. Nowadays, the applications of IoT need a fast data transfer to ensure smooth service. 5G has the potential to achieve this function for IoT. However, the energy-efficient architecture and easy-to-manage 5G-enabled IoT are still developing. Hence, the potential vulnerability issues of 5G-enabled IoT architecture need to be studied. In this paper, firstly, we have comprehensively discussed the fundamental architecture and characteristics of the 5G ecosystem. Later, the paper comprehensively outlined the characteristics and layered architecture of the internet of things. Then, this chapter also explores the requirements of 5G-enabled IoT, Blockchain-based 5G IoT, and 5G with artificial intelligence. Followed by this discussion, the chapter investigates the opportunities of 5G IoT in different domains. Finally, this paper investigates and analyzes the research gaps, challenges, and probable solutions comprehensively in a tabular format.

Keywords 5G · IoT · Opportunities · Challenges · Architecture

A. K. M. Bahalul Haque (✉)
LUT School of Engineering Science, LUT University, Lappeenranta, Finland
e-mail: Bahalul.haque@lut.fi

Md. Oahiduzzaman Mondol Zihad
Department of Electrical and Computer Engineering, North South University, Dhaka, Bangladesh
e-mail: mondol.zihad@northsouth.edu

Md. Rifat Hasan
Department of Computer Science & Engineering, Fareast International University, Dhaka, Bangladesh
e-mail: rifat.cse@fiu.edu.bd

© The Author(s) 2023
B. Bhushan et al. (eds.), *5G and Beyond*, Springer Tracts in Electrical and Electronics Engineering, https://doi.org/10.1007/978-981-99-3668-7_11

Introduction

New technologies, such as Big data, Blockchain, Communication systems, the Internet of Things, Virtual Reality, etc., have revolutionized the world (Li et al. 2018a). The Internet of Things (IoT) facilitates information collection using various devices, exchanges data with the help of communication protocols, and transmits data from real-life situations for different IoT applications. By 2030, IoT modules will have connected more than 80 billion people. It creates a massive volume of data generated from different system nodes. 5G network enables IoT to transfer this data with higher speed and data quality irrespective of the data structure differences.

The IoT systems for future applications will require enormous capabilities regarding the data receiving and processing mechanism. 5G can provide the platform to achieve these capabilities. Implementing a 5G network in IoT infrastructure will pose a new challenge and open many research windows. At the same time, it will provide unrivaled potential in modern times. 5G network aims to improve drawbacks of prior communication networks including 2G/3G/4G. It will also add new capabilities to the mentioned network system forming an architecture for heterogenic use. 5G IoT applications support a variety of QoS criteria, wide spectrum of wireless connections, and large quantities of data traffic. There are various benefits and drawbacks due to the gap in convergent technologies. The installation value as well as the growth of accomplishments for these drawbacks restrains this advancement. 5G is configured to improve the speed and security of our connections. So, 5G will enable modern enterprises and applications due to security, high speed, adaptation of new technologies and applications with a low-cost deployment.

5G is envisioned as a future cellular network technology with a 99.99999 percent dependability and a 100Mbps data transfer rate. It can ensure a consistent high-speed experience for users projecting an estimated one millisecond of round-trip delay. The three primary sorts of applications that make up 5G are Ultra-Reliable Low Latency Communication (URLLC), Massive Machine Type Communication (mMTC), and Enhanced Mobile Broadband (eMBB). Haptic communication and automation integration are made possible using fixed access bands of sub-GHZ mm-Wave embedded with tiny Base Stations (BTS) and mm-Wave local nodes. Furthermore, 5G cellular networks can interface with UAVs in various ways. Blockchain technology, according to recent research, increases the security, privacy, and transparency spectrums of 5G (llah et al. 2019). In compliance with blockchain and AI, 5G can change the world of communication (Haque and Rahman 2020).

The Internet of Objects (IoT) is based on the concept of intercommunication amongst heterogeneous things that use a variety of communication standards, stations, sensors, nodes, data centers, and artificial intelligence (AI) capable devices. As a result, the next generation 5G network will be connected to billions of devices, resulting in a super IoT infrastructure.

5G integrated IoT is an idea to develop the communication and transmission process of data in the IoT environment (Arora et al. 2020). Implementation of 5G in IoT architecture will change tour lifestyle bringing a vast amount of IoT devices in one place within a second. It will provide the applications that we believe to be backdated due to response time as the fastest ones. The 5th generation wireless system is a driver for modern-day IoT applications. Shortly, the 5G will become inevitable for the advanced devices used in IoT systems. Even there is a chance of needing advanced 5G for more advanced applications of IoT like satellite research or worldwide wireless data transmission (Popovski et al. 2018). Nevertheless, along with these, there come some issues regarding the architecture. It is still a new concept and consistently evolving. Getting a low latency in a wide range is not an easy task. 5G itself is developing and has not yet been implemented worldwide. Hence, there remains a vast domain of security and structural challenges.

This chapter explains the opportunities and challenges of the 5G integrated IoT ecosystem. The domain can be extended to blockchain and artificial intelligence perspectives. The outline of the contribution of this work can be briefed as follows:

- The fundamental concept, architecture, and characteristics of the 5G ecosystem and its evolution are comprehensively discussed.
- Discussed an overview of IoT, along with its characteristics.
- The three-layered architecture for IoT systems has been speculated.
- Requirements to build a 5G integrated IoT ecosystem are outlined.
- Presents a holistic description of 5G-enabled IoT application scope in different domains.
- Finally, delves deep into analyzing the research gaps, challenges, and tentative solutions of 5G IoT.

The rest of the paper is structured as follows: section "Overview of 5G" provides an overview of 5G and its evolution and general architectural requirements. Section "An Insight into IoT" details the IoT characteristics and the layered architecture. Section "Requirements for 5G Integrated IoT Architecture" illustrates the basics of the 5G integrated IoT ecosystem in different domains. Section "portunities of 5G Integrated IoT" illustrates the opportunities of 5G IoT in various domains. Finally, section "Challenges of 5G Integrated IoT" speculates the challenges of the 5G IoT ecosystem for future research, followed by the conclusion in section "Conclusion".

Overview of 5G

5G enables the next generation of mobile networks to make a quantum leap forward in wireless communication. Day by day, the demands of many applications are increasing. The rapidly changing wireless technology is constantly trying to keep up with this evolution. Applications that are dependent on big data and other networks of multiple things like quick money transfer, detecting critical diseases, inventory management etc. use the characteristics of 5G to enhance the system increasing efficacy (Ficzere et al. 2021). 4G and prior generations cannot support the latest apps that require high data processing ability with high quality of service (QoS) and quality of experience (QoE). The transmission rates are likewise in Gbps, and the typical transmission rate is 100+ (Mbps). The tests show that 5G can give a transmission rate up to 20 Gbps, which is 100 times faster than 4G. So, introducing 5G in the network system can change the communication process entirely.

Evolution of 5G

In the 1950s, the first rudimentary portable communications were introduced in the United States. The first (1G) portable was introduced after three decades. The second generation of radio communication (2G) was spurred by digital technologies that used one's SMS efficiency with the invention of the microprocessor. After a few years, the GPRS combined with 2G network provides the benefits of sharing voice calls, MMS, pictures, etc. more smoothly. For improved systems, the third era (3G) was adopted in the sound and notifying equipment like live TV and fast internet in the twenty-first century. The Fourth Generation (4G) was developed in response to the excessively high momentum of online connection in 4G-enabled gadgets. Modern-day communication exceeds the barrier of mobile phones including iceboxes, computers, automobiles, and other modern gadgets to the architecture. A fast site and data transfer feature is necessary to understand ongoing device engagement and access additional devices. This fact also inspires the innovation of fifth era (5G) of communication. We can simply put it, 5G is a 5th generation mobile network. Here, we will discuss three types of 5G in general (Waring 2018).

Low Band 5G

Low-band 5G uses frequencies less than 2 GHz. These are the radio and television frequencies that have been around the longest. They can cover vast distances, but there are no particularly large channels. The Low Band 5G indicates the lowest possible data rate. As a result, 5G with limited bandwidth is sluggish. These channels vary from 5 to 15 MHz for different cellular networks like AT&T, T-Mobile, and Verizon. It is considered to be the worst case of 5G, which is somewhat faster than 4G.

Medium Band 5G

Mid-band 5G uses 2.5 GHz and 3.5–3.7 GHz frequencies in most countries. It is faster than Low-band 5G using less than 6 GHz frequency. It can cover the majority of frequencies used by the mobile and networks connected to Wi-Fi, as well as several frequencies slightly above them. Because these networks can cover a radius of several miles from towers built-in not more than half a mile across, they are the functioning networks with the highest 5G traffic in most nations.

High Band 5G

Compared to the 5G mentioned above, short-distance tower-enabled millimeter wave spectrum brings high-band 5G. The airwaves have mostly been in the 20–100 GHz band till now. These frequencies have never been used for consumer purposes previously. They're only used across short distances. Therefore, running the tests on several platforms suggests roughly 800 feet away in a dense urban environment from the towers. However, there are many unoccupied signals, allowing for extremely high speeds of hopefully up to 800 MHz at a time.

Characteristics and Requirements of 5G Ecosystem

5G can increase its speed up to 20 times than 4G. It is expected to offer 20 GB per second speed whereas 4G is only promised 1 GB per second. Varying to the network infrastructure and service operator, the speed can vary. According to Qualcomm, 5G has shown a speed of 4.5 GB per second in its tests and an average of 1.4 GB per second (Qualcomm 2020). This is at least 20 times improved speed than the fastest 4G network. Enabling 5G to reach that speed will change the form of HD streaming, making the 'download' button a 'play' button. For the high latency, delay time will be significantly reduced and browsing will faster than ever before. Some of the significant characteristics of the 5G network are:

- Very low latency (around 1 ms).
- Speed up to 100 Gbps ($10–100\times$ than 4G and 4.5G).
- Availability of 99.99% over the world.
- Cover 100% of the places.
- Reduction of energy up to 90%.
- Increase battery life for IoT devices with low power up to 10 year.

To have a 5G network with these characteristics in association with IoT, there should be some specific capabilities provided to the architecture (Kozma et al. 2019). Such as:

- Efficient resource management for IoT and bulk operations.
- Prioritize the quality of service and standard control.
- Network slicing and exposure.
- Energy efficacy.
- Application in cyber-physical domain.
- Positioning availability.

High dependability is a fundamental differentiator as compared to non-licensed radio spectrum designs or traditional, evolutionary-engineered heterogeneous networks (Bose et al. 2011). So, it is especially critical for 5G.

5G Architecture

The network layer, controller layer, management, and service layer are the four levels of the 5G architecture paradigm. The 5G protocol stack has two sublayers: Radio Link Control (RLC) and Pocket Data Convergence Protocol (PDCP). Instead of base stations (BS), 5G's network architecture uses adaptive, virtual, and flexible radio access network (RAN) points and a sophisticated dispersed design. To establish various data access points, these virtual RANs incorporate additional interfaces, components, and compositions (Ngo 2021). A generic architecture for 5G ecosystem must have the following function:

- Radio Access Network (RAN): 5G uses RAN to connect many technologies providing FDD frequency.
- Data Network: It provides operator services third-party services for internet access.
- Access and Mobility Management: This function ensures integrity protection, authorizes access, manages mobility, links among devices, connect ability, etc.
- Network Slice Selection: This function decides instances for user equipment and information for the assistance function.
- Server Authentication Function: It does the work of authentication for trusted and untrusted 3GPP access.
- Control Policy: This function initiates policy frameworks to control network behavior.
- Network Exposure: It exposes the network application and manages external and internal communication securing information.

There are many more functions of 5G architecture varying from application to application. However, the basic scenario of every function tends to achieve more speed, low latency, proper management, and security (Haque and Bhushan 2021a).

An Insight into IoT

IoT or the Internet of Things is a networked digital system of various electronic devices like sensors, activators, receivers, nodes that compute data, etc. By eliminating human involvement, IoT devices have transformed the data collecting and processing system. From top to bottom, IoT devices enhance the development of concepts like smart home, smart vehicle, smart agriculture (Pranto et al. 2021), smart health care, communication, cybersecurity and many more systems (Haque et al. 2021a). They have been used to conduct, monitor, and produce reactions based on the information gathered. People have been thinking of connecting devices to the Internet for a long time. The Internet of Things, on the other hand, enhances and extends network technology based on existing internet technology, allowing computing and smart objects to connect and communicate with one another. The IoT can be broadly defined as any object that communicates, produces, and interchanges data with other objects via the Internet to perform orientation tracing, tracking, intelligent recognition, and management. This process is conducted by various sensors or peripherals such as GPS, thermal sensors, RFID, etc. (Yang et al. 2011).

Characteristics of IoT

There are many functional and non-functional IoT needs for creating the infrastructure. We will discuss some of the most valuable characteristics of IoT here.

Availability

To provide customers with facilities wherever and whenever they need them, IoT availability must be implemented at the hardware and software levels. The capacity of IoT systems to give functionality to anybody in any location is referred to as software availability (Mistry et al. 2020a). The nature of computers that are always compatible with IoT features and protocols is referred to as hardware availability. To allow IoT capabilities, protocols like IPv6, 6LoWPAN, RPL, CoAP, and others need to be implemented inside the restricted devices of the single board resource. One technique for achieving high IoT service availability is to ensure the availability of critical hardware and facilities (Bahalul Haque 2019).

Mobility

Although most utilities are designed to be delivered via Smartphone devices, IoT implementation is hampered by accessibility. A key IoT premise is to keep customers connected to their preferred resources when moving. When mobile devices are relocated from one gateway to another, service interruptions may occur. Caching and tunneling for service continuity allow apps to access IoT data even if the internet is down for a short time. The vast number of smart devices available in IoT systems is usually included in any solid framework for mobility control.

Scalability

Scalability in the Internet of Things refers to the ability to accept new client equipment, software, and capabilities without compromising the efficiency of existing systems. It is not straightforward to add new processes and manage extra devices, especially when there are several hardware platforms and communication protocols to contend with. IoT applications must be built from the ground up to enable extendable services and operations.

Security and Privacy

On diverse networks, such as the Internet of Things, ensuring user security and privacy is strict. The fundamental functioning of the Internet of Things is built on data transmission between billions, if not trillions, of Internet-connected items. One great problem in IoT security left out of the standards is the key distribution between devices. The growing number of intelligent objects around us with sensitive data necessitates transparent and simple access control management, such as enabling one vendor to view the data. In contrast, another controls the device (Bahalul Haque et al. 2022).

Performance

The performance of IoT services is difficult to evaluate since it is based on the performance of many components and the underlying technology. The Internet of Things, like other programs, must constantly develop and expand its offerings in order to meet user expectations. To give the most value at the lowest cost to customers, IoT solutions must be monitored and validated. IoT performance may be measured using various criteria, including processing speed, connection speed, system form factor, and cost.

IoT also needs to manage the larger amount of information or data created in the ecosystem, ensuring the interoperability and quality of service.

Layered Architecture of IoT

Various designs have been suggested for IoT worlds. In general, such structures are divided into three categories. There are three types of architecture: three-layer architecture, four-layer architecture, and five-layer architecture. In this chapter, we will look at the three-layered architecture. It is organized keeping mid some specific tasks to accomplish by the system like executing service functions, transmitting data, and connection among service devices. It results in three layers, Application layer, Network/Transmission layer, and Perception/Edge layer.

Application Layer

In different implementations, this layer may include various services. Smart grids, healthcare, and autonomous automobiles are examples of IoT deployment in smart cities and homes. Because the application layer might serve as a service support middleware, a networking standard, or a cloud computing platform, security considerations vary depending on the application's environment and industry. The application layer provides customers with the services they require. The application layer, for example, should give temperature and relative humidity values to the client who has requested the information. This layer is critical for the IoT because it allows for creating high-quality smart services that fulfill user demands.

Network Layer

Acting as a bridge, the network layer controls data transfer to subsequent layers. This layer connects to the visual layer. Different smart devices are connected to the network layer following control function protocol (IEEE 802.x) and authentication standards (GPS, and Near-Field Connectivity (NFC)). A server backend architecture, smart devices, and the Internet protocol contribute to this tier. In addition, the network layer can be handled according to the peculiarities of the deployed environment. The transmission of data is highly prone to cyber-attacks. Intelligent intrusion detection key encryption with secured management-based IoT security framework is the most popular along with the latest adoption of blockchain technology.

Edge Layer

Edge layer manages the IoT devices or sensors like RFID, different actuators, cameras, intensity detectors, moisture and pressure sensors, etc., using gateways in a coordinating function to connect with the client or their working domains. Its main task is to collect data from the environment and transfer them forward for further processing. IPs like IPv6 or gateways can transmit this to follow protocol translation and traffic management. The sensors and actuators prohibit a common and standard security mechanism from protecting these devices. Hence, the inter-operability among devices and physical accessibility expose a handful of security threats. Researchers have proposed security solutions for this layer based on machine learning, multi-stepped authorization, secure channeling through anti-malware, etc.

Requirements for 5G Integrated IoT Architecture

5G-enabled IoT needs special attention for its heterogeneity, advancement, and application. However, there are some requirements that all the architecture should follow (Li et al. 2018b):

- 5G IoT must ensure a low latency of 1 ms considering the sensitive internet system and medical perspective.
- The architecture must ensure low energy consumption for low-battery life IoT devices but enough for 5G to transfer data.
- An advanced application like Virtual Reality or Augmented Reality needs a high speed of 25 Mbps, so the architecture must follow with the future needs.

- Security must be top-notch, considering massive data transmission at a very high speed.
- The devices with mobility factors will get priority for the 5G IoT infrastructure.

The fundamental 5G IoT architecture consists of five steps in general: sensors, IoT Gateway, 5G-based station, cloud storage, and application (Arsh et al. 2021). These steps can be comprised in IoT layers to bring up a general 5G IoT architecture.

Edge Layer of 5G IoT

The sensors and gateway of IoT can be comprised of 5G in this layer. For example, sensors for wearable ECG, temperature, smart manufacturing etc. will use this layer to transmit and process information using 5G technology (Shdefat et al. 2021).

Network Layer of 5G IoT

The network layer will hold the 5G base station and cloud storage to process data using IoT devices.

Application Layer of 5G IoT

The application layer will provide all the support for the end system like smart home, smart supply chain, etc. (Haque et al. 2021b).

Following the above-mentioned general architecture, 5G IoT can support millimeter-wave (Rahimi et al. 2018), D2D communication, nano-chip, wireless software (Huang et al. 2020), mobile edge computing, data analytics cloud computing (Mudigonda et al. 2020), and many more technologies and application. In Fig. 11.1, we have shown a generalized architecture for the 5G integrated IoT ecosystem.

Blockchain-Based 5G IoT

Blockchain (Haque and Bhushan 2021b) can bring trust and improved security to 5G IoT. It can accelerate data exchange at a lower cost by implementing a cryptographic encryption system to the architecture. The immutability and accountability that blockchain can ensure for the system are marvelous (Hewa et al. 2020). Blockchain integrated 5G IoT can bring revolution to industrial IoT, Unmanned Autonomous Vehicle (UAV), and so on (Haque et al. 2020). Blockchain and 5G IoT can also be

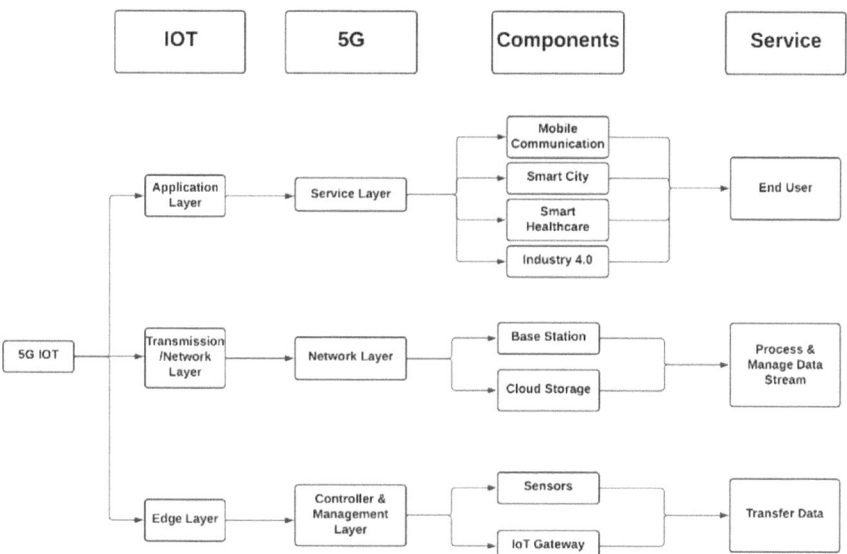

Fig. 11.1 A general 5G IoT-layered architecture

integrated with deep learning (Kaur and Shalu 2021). The architecture consists of the device layer, blockchain network, 5G mobile network and cloud network (Satpathy et al. 2021). It provides a data transmission using a smart contract with a 5G speed. Again, 5G IoT can be embedded with 5G mm-wave technology to build its processing center, object processor, sensing regions and application layer (Haque et al. 2021c). These layers work together using cloud storage and a 5G network to provide services like education, fire station, transportation, factories, etc.

5G IoT with Artificial Intelligence

Adversarial artificial intelligence can provide great security support towards 5G IoT (Bohara et al. 2021). It can enable technologies like massive MIMO, cloud RAN, multi-RAT to prevent security threats like fast gradient sign method, one-pixel attack, DeepFool etc. The architecture accepts machine learning methodology like logistic regression, naïve Bayes, Q learning, K-means, Markov decision model etc. (Haque et al. 2021d).

Opportunities of 5G Integrated IoT

5G, is a booming technology that has opened many windows of opportunities. High-speed and large-bandwidth capabilities will support more than 60,000 connections. Furthermore, 5G brings all networks together on a single platform. It also gives subscribers the ability to monitor their accounts and take swift action. 5G is backward compatible with earlier generations of networks. Moreover, 5G is designed to deliver the globally uninterrupted and constant connection. Enabling 5G with IoT will accelerate the development in many other sectors including technology, business, industry, etc.

Technological Advancement

IoT includes many aspects of technology that 5G can make the best use of. 5G can make these technologies overcome their shortcomings providing remarkable achievements. Here, we will discuss some technologies that 5G will change forever.

Network Function Virtualization (NFV)

NFV is used to develop network service resiliency and lessen the time it takes to adopt new systems and technologies. It separates the hardware and software requirements for complicated operations (Han et al. 2015). The NFV performs the role of a virtualization enabler and facilitates the dissemination of 5G-IoT. Virtualized load balancers, intrusion detection systems, and firewalls are all instances of NFV. Integrating 5G with IoT will allow NFV to detect threats more accurately and provide network services with flexibility and developed scalability.

Mobile Cloud Computing (MCC)

Cloud computing is a service that allows one to outsource his or her processing resources (Failed 2020). End-users can get authorized access to databases, datasets, and information over the Internet, including the ability to analyze and transfer with the power of 5G networks. Cloud computing is a new fundamental technology in IT architecture that allows users to compute or store data without building up an extensive infrastructure. The MCC combined cloud computing with mobile computing to give consumers elasticity and on-demand services. Many services now allow users to connect their mobile devices to the cloud. Data transmission using edge computing with 5G decreases the transfer time. Moreover, the inclusion of these technologies guarantees that context information has reduced latency and is more accessible.

Device-To-Device Communications (D2D)

5G network enables smooth D2D communication implementing direct communication without any intermediary (Goyal et al. 2021). There are four types of this D2D communication, including transmission among devices using the operator-controlled link, transmission among devices with the device-controlled link, direct D2D communication with an operator-controlled link and direct D2D communication with a device-controlled link (Mani Sekhar et al. 2021).

Software-Defined Network (SDN)

SDN opens up new network administration and design (Abdelwahab et al. 2016). It is emerging as the most promising answer for the Internet's future. It has two distinguishing features: data plane separation and advanced application development programming. This allows for more effective configuration, efficiency, and flexibility when creating network architectures (Xia 2014). The SDN prototype was used to create 5G networks in order to retain a flexible and quicker 5G-IoT topology (Xie et al. 2019).

Millimeter Wave Communication

Mobile networks continuously need increased retention to improve frequency. 5G mm wave can pave the way for newly developed radio wave frequency. It will enable IoT to work more efficiently providing high speed (up to 20Gbps) and high availability accompanied by a mobile network than 4G network. Moreover, the enhanced broadband in mobile networks due to this 5G evolution will initiate applications like VR and augmented video, live streaming, UHD video, etc.

Mobile Edge Computing (MEC)

A significant portion of cloud services has been possible because of the development of several advanced computer applications such as artificial intelligence and smart environments. Cloud computing has various needs, including low latency, location awareness, and mobility support. The MEC (Ahmed and Rehmani 2017) can bring the mentioned operations and resources nearer to the network edge. Because of MEC in 5G IoT, applications such as VR and AR will grow.

In addition to these technologies, many more are constantly being added to our day-to-day life all over the world. The network capability and coverage must be developed to mitigate this global traffic. 5G has the ability to evolve with these new innovations as well as enhance its ability (Nguyen et al. 2020).

Smart Cities

From supply chain management of all the necessary goods and needs of daily life to home automation system to improved communication, 5G will have a broader use in smart city programs. Smart Cities will benefit from 5G network with developed sensors advancing the urban infrastructure. 5G will be able to manage massive amounts of data and combine a variety of intelligent technologies that are continuously connecting with one another to bring a genuinely linked city even closer together (Minoli and Occhiogrosso 2019).

Smart Healthcare

Because 5G will have an impact on IoT, it will also have an impact on the areas touched by IoT. The Internet of Medical Things, or IoMT, is the most important of them. Rural and other comparable isolated places that lack proper health services can tremendously benefit from the Internet of Things connectivity. After a long crave of world-class health services to be remotely achievable like distant operations are becoming a possibility (Ahad et al. 2020).

Smart Vehicles

Automated cars collect various data on temperature, weather, traffic, GPS position, and other factors using modern sensors, resulting in a significant volume of data. A lot of energy is expended in the generation and processing of so much data. To deliver best services, these vehicles depend largely on real-time data transmission. The system that is built-in these vehicles can be initiated to collect every kind of data that are required including the crucial ones with high-speed connectivity and minimal latency. Eventually, it will enable the vehicles to autonomously monitor its operation and enhance future models including the system algorithm (Mistry et al. 2020b).

Smart Logistics

Advanced IoT tracking devices that can execute logistical activities will be able to use 5G connectivity. The real-time data transmission will be faster than ever before with the efficacy of high speed and low latency of 5G. Moreover, it will be energy efficient in case of long supply chain that takes time. For example, a consumer may learn where the fruit is grown, at what temperature it is kept during transit, and when it is delivered to a retailer (Wang et al. 2020).

Smart Grid

In day-to-day operations, the need for power is rapidly growing. Demand management can be aided by smart grids and virtual power plants. 5G technology is appropriate for real-time management in the energy and utility industry providing solutions that would ensure optimal operations and maintenance by quickly recognizing and responding to grid faults (Shahinzadeh et al. 2020). 5G can renovate the smart grid replacing wired technology establishing better deployment flexibility and cheaper expenditure.

Business

5G-enabled IoT is predicted to deliver more than just technical advancement; it is also estimated to support 22 million employments globally. The modernization of transportation, industry, agriculture, and other physical industries is likely to drive this rise. Areas like construction of oil miners, freight fleets, and railway that need faster transmission due to the nature of the product will be affected positively (Rong et al. 2020). 5G has the ability to advance smart manufacturing and intelligent equipment. In near future, 5G will enable IoT to do near-instantaneous network traffic management, resolution, increase security and public safety, and operate remotely.

Aerial and Satellite Research

The advancement of 5G network also opens the window of vast aerial and satellite communication and research. High Altitude Performance system (HAPs) is being investigated in accomplice with satellite. Modifying the 5G network with Narrowband-IoT (NB-IoT) makes it a seamless integration (Gineste et al. 2017). It can enable moderately structured satellites to communicate at low bitrate. It is also possible to enhance the 5G mobile network with combined satellite-terrestrial networks (Fang et al. 2020).

Mitigating Pandemic Situation

5G can renovate the technologies to mitigate issues in pandemic situation. The integration of 5G and IoT improves the telehealth service to check patients remotely through massive Machine Type Communication (mMTC). Moreover, 5G-supported Bluetooth Low Energy (BLE)-based IoT devices can manage COVID-19 patient detection and monitoring (Haque et al. 2021e). The massive connectivity of data does not need any gateway and provides long-time battery support for low-power IoT devices (Siriwardhana et al. 2020).

Industrial Usage with Other Technologies

5G-enabled IoT can revamp many industries by developing specific technologies as well as mitigate some of their core issues. Some of the technologies that will enable industrial IoT significantly are discussed here.

Blockchain Integrated 5G IoT

5G IoT can be integrated with decentralized blockchain technology to ensure better privacy and security of data in industrial IoT (IIoT) (Haque et al. 2022). Supply chain industries that store data on blockchain require real-time data tracking like timestamp, origin, shipment, amount, etc. (Haque and Bhushan 2021c). 5G-enabled IoT can ease the traceability ensuring public verifiability for stakeholders. In addition to this, combined with deep reinforcement learning (DRL), blockchain integrated with 5G IoT can provide a secure sharing scheme for IIoT (Liu et al. 2019).

Big Data Processing with Machine Learning (ML)

Various machine and deep learning approaches used for big data analysis like viz, Convolutional Neural Network, Recurrent Neural Network, Deep Neural Network, etc. need higher optimization and processing time. Along with them, complex mathematical models are used that take higher execution time. 5G-enabled IoT can reduce this time greatly and save energy ensuring efficient industrial resource management (Dai et al. 2019).

Artificial Intelligence (AI)

To build an AI-based industry, a lot of data are needed to train and test AI algorithm model. 5G-enabled IoT has the ability to provide the developed infrastructure to collect this huge amount of data (Kumari et al. 2020). Using this data, AI can generate valuable insights for the industry as well as give clear context of the network to further develop it.

Optimization System

5G-enabled IoT addresses network-related issues in IoT using many procedures of optimization. These optimization procedures include stochastic, heuristic, and computational approach along with genetic and evolutionary algorithms, etc. This aids in the effective monitoring and minimization of IoT device-generated network traffic.

Video Surveillance

Video surveillance is another application that is projected to operate well in the 5G ecosystem. Many industries as well as private sectors throughout the world are using surveillance systems. Wired connection is still used by many video surveillance systems today. Wireless communications like Wi-Fi and cellular are gaining popularity due to their ease, speed, and low cost of implementation. In near future, it is expected to increase at a higher rate to ensure better security. The utilization of 5G will support the necessary peed boost required for real-time video data analytics and the propagation of a large number of surveillance (Kumhar and Bhatia 2021).

Challenges of 5G Integrated IoT

5G integrated with IoT has provided us with extraordinary features ubiquitously. State-of-the-art research has shown that more applications and advancements are achievable through extensive connectivity ensuring reliable Quality of Service (QoS). With these advancements, come a lot more challenges that need to be taken care of. Here, we will discuss some of the crucial challenges of 5G integrated IoT.

General Challenges

5G IoT can be molded with other technologies but posed similar type of challenges irrespective of their applications.

Low Control on Data Usage and Storage

A massive quantity of data are created in a 5G IoT network. It is created through devices that are unique to numerous businesses. As a result, these data are frequently out of the control of all the stakeholders or users concerned (Sinha et al. 2017). The data can be subject to all suppliers whose equipment creates a network node, all service providers who use a common 5G network architecture, or all users who share a single cloud platform when used by all parties (Jaitly et al. 2017). As a result, it is a complex task to track which data come from whom, who is the creator and the processing system.

Scalability

Cloud-based architecture enables 5G IoT to control and manage the overall network. To put it another way, the nodes in the network create data for processing to a common cloud. Then, the network nodes send back the control signals to carry out tasks like storage reallocation, traffic management, fault management, routing, and so on. However, as the number of connected devices grows, so does the volume of data they generate, making scaling up the capacity and computational power of centralized cloud servers an approaching problem. Furthermore, devices connect to the cloud nodes via a gateway or an edge node in 5G and IoT ecosystem (Mehbodniya et al. 2022). Due to the large number of devices attempting to connect to the cloud, the fronthaul, midhaul, and backhaul networks directly close to gateway nodes frequently become narrowed reducing the scalability of the overall network.

Complicated Interaction Among Devices

The technological standards embedded in the devices that are used in 5G IoT ecosystem have varied signaling wave, different data bits, different PHY and MAC protocols, coding structure, user interfaces, etc. The operation of these devices is also backed by different operating systems. So, it is a very complicated task to initiate a mutual communication standard that will be followed throughout the 5G IoT infrastructure. Again, a common program or operating system that can keep up to different communication protocol for multiple devices is also very difficult to introduce. Hence, it can prohibit the extension of certain application and even restrict some devices to use in 5G IoT environment (Singla et al. 2021).

Data Auditability

Data created in 5G IoT networks have several owners and are extremely noncompliance, making it difficult to trace and audit. There are also situations when there are no common standards or protocols in place for data as discussed before to be shared across devices owned by various businesses (Bhushan and Sahoo 2017). Furthermore, data may be non-processable or not transmittable across various divisions of an organization due to differing communication protocols and also because of trust difficulties (Bhushan and Sahoo 2019). Similar sorts of data like meteorological and environmental data are sometimes not being impart or inter-operated by multiple entities in order to arrive at an agreed reasoning.

Heterogeneity of 5G and IoT Data

Both 5G and IoT ecosystem has a varied nature causing several compatibility concerns while implementing distinct applications. Because the data created in IoT networks and 5G cellular networks are diverse and multidimensional, it is almost impossible to forecast exact characteristics and outcome. As a result, early operations like as cleaning, ordering, and preprocessing are required to train this type of data since the integration of such a diverse set of data leads to incorrect calculation. Hence, test of datasets in diverse situations needs to give more attention to this aspect for dataset training and feature selection. Sensors and humans, for example, create data for IoT networks in smart homes. However, a common or central server must contain all the data utilized in this application (Palanisamy and Bhatia 2022). This server will be responsible to pool and train data from many sources so that it can cope with data variety and achieve higher prediction accuracy.

Blockchain Integrating 5G IoT Issues

Blockchain, integrating with 5G IoT, opens opportunities for many applications. But many of them come up with different challenges too. Some big challenges for blockchain integrated 5G IoT ecosystem are discussed here.

Processing Time

A few self-executing tasks like transaction and block verification are required to build the chain in the blockchain ecosystem. These computations follow some specific cryptographic procedures to maintain the authenticity of the blocks in blockchain that takes a lot of time. It can be a solution to lessen the amount of training data but there are specific computing restrictions in the IoT context that could lead to security problems. As a result, a big concern for the implementation of blockchain in the context of IoT is it needs fewer resource-intensive replacements to reduce processing time.

Privacy and Security

IoT alone brings many privacy issues to the blockchain integrated 5G network for its vast number of devices using numerous sensors connected worldwide. Moreover, blockchain prefers to public verification of transaction. Keeping up the encryption procedure of blockchain along with 5G data protection carries a significant challenge. There are some studies to overcome this issue proposing homomorphic computation to cover up the data at the time of access of any user (Zhou et al. 2018). But controlling over 51% of data from 5G IoT ecosystem can lead to reverse transaction or double-spending problem. So, combined solution is still a big challenge.

Storage Scalability

A major need of blockchain technology is the constant storing of transactions and blocks. Each node should theoretically have a copy of the ledger that grows in tandem with the transactions. From a scalability standpoint, the IoT ecosystem's storage effect will have an influence on the overall system's operation. The changing transactions that come with scaling up the system, in particular, need a lot of storage.

Cost

Blockchain has its own scalability issues, but on the other hand, throughput or cost is another big difficulty with this technology. For IoT, it is difficult to keep up with the growing number of transactions and their size. Two more issues that arise often are latency and transaction throughput. Due to the less data generation of private blockchain than public blockchain, it is more preferable to IoT environment. But 5G IoT generates big data and the analysis of this huge throughput by blockchain increases computational cost.

5G mm-Wave Issues

5G mm-wave application has promising significance in increasing robustness of the IoT ecosystem, low latency, and higher capacity allocation for Multiple Input Multiple Output (MIMO) technologies. But it has a range limitation in communication. Moreover, the core of 5G mm-wave is combined with devices that are different in characteristics like LTE and mm-wave band. Again, it is susceptible to blockage restricting its mobility for its rich scattering environment. It also consumes a lot of power due to extensive use of hardware. Hence, disturbance in connection for weather condition makes it challenging to use with IoT infrastructure (Dang 2022).

Threat Protection of 5G IoT

Even after mitigating issues of 5G and IoT distinctively, security threats will come again in a 5G integrated IoT infrastructure. There are lots of security threats like DoS attack, signaling storms, slice theft, penetration attacks, Man-in-the-middle attack (MITM), TCP level attack, security key exposure reset and IP spoofing attack, etc. are intended to technologies like SDN, NFV, Cloud server, etc. (Bhushan and Sahoo

2020). These attacks can not only expose the overall 5G IoT application but also make a big impact on privacy of data (Ahmad et al. 2020). There are some solutions to these attacks like usage of low-powered nodes and sensors, cloud and application security for SDN and NFV, etc. But it is still a concern for future prospective 5G IoT ("5G support for Industrial IoT Applications 2020).

Table 11.1 summarizes and provides a short overview of the prominent challenges and their possible solutions of 5G integrated IoT system.

Apart from the stated issues, there will come more shortcomings and challenges in 5G IoT systems and applications. Hence, it will provide future directions for a large number of domains of research and development.

Conclusion

This paper is focused on the opportunities and challenges of 5G IoT integration in several domains of application along with their possible solutions. 5G network has the ability to connect more than 100 billion of devices and exchange data at least 10 times quicker than LTE resulting a game changer in IoT environment. The aim should be the introduction of scalable, affordable, and efficient network architecture for 5G IoT connecting vast amounts of devices. The applications of 5G IoT in several domains to increase efficacy can bring a lot of opportunities along with challenges. The architectural and security aspects are needed much concern. Furthermore, manufacturers must conduct quality and maintenance testing to guarantee the adoption of future software and hardware to perform their functions under a variety of scenarios (Li et al. 2018c). 5G IoT can give huge economic benefits to enterprises eager to embrace this new world when combined with important technologies like cloud security, artificial intelligence, and remote computing. In this chapter, we try to analyze the opportunities of 5G integrated IoT ecosystem for eradicating the shortcomings of IoT describing the features of 5G. The 5G IoT architecture also shows that there are many challenges and forthcoming research direction. We hope these illustrations will provide thorough insights and inspire the development of 5G IoT applications.

Table 11.1 Summary of challenges and state-of-the-art solution of 5G IoT

Domain	Challenges	Description	Solution	References
General challenges	Low control on data usage and storage	Different types of data are used by different entities Difficult to track down source of malicious data	A BICM-ID-based NAND flash memory system optimize with EPEXIT algorithm to manage data storage	Fang et al. (2021)
	Scalability	Ability and computational power to scale up data nodes decreases with the growth of data routes Attempt of numerous data to access cloud results in narrowed gateway	A cost-efficient SDN with data optimization in several sectors of industrial IoT and 5G	Okwuibe et al. (2020)
	Complicated interaction among devices	Lack of common operating system among interacting devices	A protocol named oneM2M using ontology	Jin et al. (2021)
	Data auditability	Lack of compatibility among entities to process, transfer and share data	Kernel bypassing using RDMA and DPDK	Varga et al. (2020)
Heterogeneity	Data disparity	Variation in 5G and IoT technologies poses extra attention to preprocess data Dataset testing and feature selection get compromised	Proposed possible solutions for 5G HetNet mobility management based on paging, registration, and access procedure	Gures et al. (2020)
AI integrated system	Bandwidth demand	Requires high bandwidth to manage complex algorithm	Suggested a B5G framework capable of high-bandwidth functionality	Hossain et al. (2020)
	Cost	Optimization cost tends to be higher due to usage of vast amount of data for training and testing	A VNF-optimized low-cost adaptation for AI operations is proposed	Ibrahimpašić et al. (2021)
Blockchain integrated System	Processing time	Requires fewer resource-intensive 5G IoT structure to reduce overall processing time	Smart contract-based blockchain integration	Mehta and Gupta (2020)

(continued)

Table 11.1 (continued)

Domain	Challenges	Description	Solution	References
	Privacy and security	Encryption process for 5G data is difficult to establish	Combined with deep learning, a blockchain-based four-layered framework is proposed	Rathore et al. (2021)
	Storage scalability	Large chain of blocks reduces storage scalability	Utilizing fog computing and cloud manufacturing equipment, a Hyperledger Fabric framework is introduced	Hewa et al. (2020)
	Throughput	Less data generation is preferable for private blockchain but 5G IoT requires big data increasing computational cost	Blockchain with VNFs following a novel consensus algorithm	Hakiri and Dezfouli (2021)
Spectrum	5G mm-Wave issue	Lack of consistency among LTE and 5G-enabled devices Affected by weather condition	An antenna design is proposed using hybrid beam forming for mobile transmission	Chen (2020)
Device standard	Common protocols	Scarcity of common protocols and standards irrespective of devices and application	A ZigBee-based IoT system is proposed for multi-device operation platform	Wang et al. (2021)
Security threats	Cyber-attacks like TCP level attack, IP spoofing, etc	Threatens privacy of data and exposes overall system	A convolutional neural network-based malware detection system is introduced	Anand et al. (2021)
Infrastructure	Network scalability and interoperability	Limitations of hardware and advanced technology to combine 5G with IoT	Multiple frequency antennas with the power of front-end radio frequency, diplexer, and triplexer are designed	Sharma et al. (2021)

References

Abdelwahab S, Hamdaoui B, Guizani M, Znati T (2016) Network function virtualization in 5G. IEEE Commun Mag 54(4):84–91

Ahad A, Tahir M, Aman Sheikh M, Ahmed KI, Mughees A, Numani A (2020) Technologies trend towards 5G network for smart health-care using IoT: a review. Sensors 20(14):4047

Ahmad A, Bhushan B, Sharma N, Kaushik I, Arora S (2020) Importunity & evolution of IoT for 5G. In: 2020 IEEE 9th international conference on communication systems and network technologies (CSNT). https://doi.org/10.1109/csnt48778.2020.9115768

Ahmed E, Rehmani MH (2017) Mobile edge computing: opportunities, solutions, and challenges. Futur Gener Comput Syst 70:59–63

Anand A, Rani S, Anand D, Aljahdali HM, Kerr D (2021) An efficient CNN-based deep learning model to detect malware attacks (CNN-DMA) in 5G-IoT healthcare applications. Sensors 21:6346. https://doi.org/10.3390/s21196346

Arora S, Sharma N, Bhushan B, Kaushik I, Ahmad A (2020) Evolution of 5G wireless network in IoT. In: 2020 IEEE 9th international conference on communication systems and network technologies (CSNT). DOI: https://doi.org/10.1109/csnt48778.2020.9115773

Arsh M, Bhushan B, Uppal M (2021) Internet of Things (IoT) Toward 5G NETWORK: design requirements, integration trends, and future research directions. Adv Intell Syst Comput 887–899. https://doi.org/10.1007/978-981-15-9927-9_85

Bahalul Haque AKM (2019) Need for critical cyber defence, security strategy and privacy policy in Bangladesh—hype or reality? Int J Manag Inf Technol 11(01):37–50. https://doi.org/10.5121/ijmit.2019.11103

Bahalul Haque AKM, Bhushan B, Nawar A, Talha KR, Ayesha SJ (2022) Attacks and counter-measures in IoT based smart healthcare applications. In: Balas VE, Solanki VK, Kumar R (eds) Recent advances in internet of things and machine learning. Intelligent systems reference library, vol 215. Springer, Cham. https://doi.org/10.1007/978-3-030-90119-6_6

Bhushan B, Sahoo G (2017) A comprehensive survey of secure and energy efficient routing protocols and data collection approaches in wireless sensor networks. In: 2017 international conference on signal processing and communication (ICSPC). https://doi.org/10.1109/cspc.2017.8305856

Bhushan B, Sahoo G (2019) A hybrid secure and energy efficient cluster based intrusion detection system for wireless sensing environment. In: 2019 2nd international conference on signal processing and communication (ICSPC). https://doi.org/10.1109/icspc46172.2019.8976509

Bhushan B, Sahoo G (2020) Requirements, protocols, and security challenges in wireless sensor networks: an industrial perspective. In: Handbook of computer networks and cyber security, pp 683–713. https://doi.org/10.1007/978-3-030-22277-2_27

Bohara MH, Patel K, Saiyed A, Ganatra A (2021) Adversarial artificial intelligence assistance for secure 5G-enabled IoT. In: Tanwar S (eds) Blockchain for 5G-enabled IoT. Springer, Cham. https://doi.org/10.1007/978-3-030-67490-8_13

Bose SK, Brock S, Skeoch R, Rao S (2011) CloudSpider: combining replication with scheduling for optimizing live migration of virtual machines across wide area networks. In: 2011 11th IEEE/ACM international symposium on cluster cloud and grid computing, pp 13–22. https://doi.org/10.1109/CCGrid.2011.16

Chen WC (2020) 5G mm WAVE technology design challenges and development trends. In: 2020 international symposium on VLSI design, automation and test (VLSI-DAT), pp 1–4. https://doi.org/10.1109/VLSI-DAT49148.2020.9196316

Dai Y, Xu D, Maharjan S, Chen Z, He Q, Zhang Y (2019) Blockchain and deep reinforcement learning empowered intelligent 5G beyond. IEEE Netw 33(3):10–17. https://doi.org/10.1109/MNET.2019.1800376

Dang V (2022) Benefits of 5G millimeter-wave communication in IoT applications

Fang X, Wei T, Feng W, Wei H, Chen Y, Ge N, Wang CX (2020) 5G embraces satellites for 6G ubiquitous IoT: basic models for integrated satellite terrestrial networks. arXiv 2020. arXiv:2011.03182

Fang Y, Bu Y, Chen P, Mumtaz S, Lau F, Otaibi SA (2021) Irregular-mapped protograph LDPC-coded modulation: a bandwidth-efficient solution for 5G networks with massive data-storage requirement. arXiv preprint arXiv:2104.02856

Ficzere D, Soós G, Varga P, Szalay Z (2021) Real-life V2X measurement results for 5G NSA performance on a high-speed motorway. In: IFIP/IEEE international symposium on integrated network management (IM) pp 836–841

Gupta N, Sharma S, Juneja PK, Garg U (2020) SDNFV 5G-IoT: a framework for the next generation 5G enabled IoT. In: 2020 international conference on advances in computing, communication & materials (ICACCM), pp 289–294. https://doi.org/10.1109/ICACCM50413.2020.9213047

Gineste M, Deleu T, Cohen M, Chuberre N, Saravanan V, Frascolla V, Mueck M, Strinati EC, Dutkiewicz E (2017) Narrowband IoT service provision to 5G user equipment via a satellite component. In: Proceedings of the 2017 IEEE globecom workshops (GC Wkshps). Singapore, 4–8 December 2017, pp 1–4

Goyal S, Sharma N, Kaushik I, Bhushan B, Kumar N (2021) A green 6G network era: architecture and propitious technologies. Data Anal Manage 59–75. https://doi.org/10.1007/978-981-15-8335-3_7

Gures E, Shayea I, Alhammadi A, Ergen M, Mohamad H (2020) A comprehensive survey on mobility management in 5G heterogeneous networks: architectures, challenges and solutions. IEEE Access 8:195883–195913. https://doi.org/10.1109/ACCESS.2020.3030762

Hakiri A, Dezfouli B (2021) Towards a blockchain-SDN architecture for secure and trustworthy 5G massive IoT networks. In: Proceedings of the 2021 ACM international workshop on software defined networks & network function virtualization security (SDN-NFV Sec'21). Association for computing machinery, New York, NY, USA, pp 11–18. https://doi.org/10.1145/3445968.3452090

Han B, Gopalakrishnan V, Ji L, Lee S (2015) Network function virtualization: challenges and opportunities for innovations. IEEE Commun Mag 53(2):90–97

Haque AKMB, Bhushan B (2021a) Security attacks and countermeasures in wireless sensor networks. Integr WSNs Internet of Things 17–43. https://doi.org/10.1201/9781003107521-2

Haque AKMB, Bhushan B (2021b) Emergence of blockchain technology. Blockchain Technol Data Priv Manage 159–183. https://doi.org/10.1201/9781003133391-8

Haque AK, Bhushan B (2021c) Blockchain in a nutshell. Adv Data Min Database Manage 124–143. https://doi.org/10.4018/978-1-7998-6694-7.ch009

Haque AKMB, Muniat A, Ullah PR, Mushsharat S (2021a) An automated approach towards smart healthcare with blockchain and smart contracts. In: 2021a international conference on computing, communication, and intelligent systems (ICCCIS). https://doi.org/10.1109/icccis51004.2021.9397158

Haque B, Hasan R, Zihad OM (2021b) SmartOil: blockchain and smart contract-based oil supply chain management. IET Blockchain. https://doi.org/10.1049/blc2.12005

Haque AKM, Arifuzzaman BM, Siddik SAN, Kalam A, Shahjahan TS, Saleena TS, Alam M, Islam MR, Ahmmed F, Hossain MJ (2022) Semantic web in healthcare: a systematic literature review of application, research gap, and future research avenues. Int J Clin Practi 2022

Haque AKMB, Bhushan B, Dhiman G (2021d) Conceptualizing smart city applications: requirements, architecture, security issues, and emerging trends. Expert Syst 1–23. https://doi.org/10.1111/exsy.12753

Haque AKMB, Bhushan B, Nawar A, Talha KR, Ayesha SJ (2022) Attacks and countermeasures in IoT based smart healthcare applications. In: Recent Advances in Internet of Things and Machine Learning: Real-World Applications (pp. 67–90). Cham: Springer International Publishing

Haque AKM, Rahman M (2020) Blockchain technology: methodology, application and security issues. IJCSNS Int J Comput Sci Netw Secur 20(2)

Haque AKMB, Shurid S, Juha AT, Sadique MS, Asaduzzaman AS (2020) A novel design of gesture and voice controlled solar-powered smart wheel chair with obstacle detection. In: 2020 IEEE international conference on informatics, IoT, and enabling technologies (ICIoT). https://doi.org/10.1109/iciot48696.2020.9089652

Haque AKMB, Bhushan B, Hasan M, Zihad MM (2022) Revolutionizing the industrial internet of things using blockchain: an unified approach. In: Balas VE, Solanki VK, Kumar R (eds) Recent advances in internet of things and machine learning. Intelligent systems reference library, vol 215. Springer, Cham. https://doi.org/10.1007/978-3-030-90119-6_5

Hewa TM, Kalla A, Yliansttila ME, Liyanage M (2020) Blockchain for 5G and IoT: opportunities and challenges. In: IEEE eighth international conference on communications and networking (ComNet), pp 1–8. https://doi.org/10.1109/ComNet47917.2020.9306082

Hewa TM, Braeken A, Liyanage M, Ylianttila M (2022) Fog computing and blockchain based security service architecture for 5G industrial IoT enabled cloud manufacturing. IEEE Trans Ind Inf. https://doi.org/10.1109/TII.2022.3140792

Hossain MS, Muhammad G, Guizani N (2020) Explainable AI and mass surveillance system-based healthcare framework to combat COVID-I9 like pandemics. IEEE Netw 34(4):126–132. https://doi.org/10.1109/MNET.011.2000458

Huang M, Liu A, Xiong NN, Wang T, Vasilakos AV (2020) An effective service-oriented networking management architecture for 5G-enabled internet of things. Comput Netw 173:107208

Ibrahimpašić AL, Han B, Schotten HD (2021) AI-empowered VNF migration as a cost-loss-effective solution for network resilience. In: IEEE wireless communications and networking conference workshops (WCNCW) pp 1–6. https://doi.org/10.1109/WCNCW49093.2021.9420029

Improved Energy Based Multi-Sensor Object Detection in Wireless Sensor Networks T Palanisamy, D Alghazzawi, S Bhatia, AA Malibari (2022) Intelligent automation and soft computing. https://doi.org/10.32604/iasc.2022.023692

Jaitly S, Malhotra H, Bhushan B (2017) Security vulnerabilities and countermeasures against jamming attacks in wireless sensor networks: a survey. In: 2017 international conference on computer, communications and electronics (Comptelix). https://doi.org/10.1109/comptelix.2017.8004033

Jin W, Xu R, Lim S, Park D-H, Park C, Kim D (2021) Integrated service composition approach based on transparent access to heterogeneous IoT networks using multiple service providers. Mob Inf Syst. https://doi.org/10.1155/2021/5590605

Kaur UK, Shalu (2021) Deep learning approach for resource optimization in blockchain, cellular networks, and IoT: open challenges and current solutions. https://doi.org/10.1002/9781119785873.ch16

Kozma D, Varga P, Soós G (2019) Supporting digital production, product lifecycle and supply chain management in industry 4.0 by the arrowhead framework—a survey. In: 2019 IEEE 17th international conference on industrial informatics (INDIN), pp 126–131. https://doi.org/10.1109/INDIN41052.2019.8972216

Kumari A, Gupta R, Tanwar S, Kumar N (2020) Blockchain and AI amalgamation for energy cloud management: Challenges, solutions, and future directions. J Parallel Distrib Comput 143:148–166. ISSN 0743-7315. https://doi.org/10.1016/j.jpdc.2020.05.004. https://www.sciencedirect.com/science/article/pii/S074373152030277X

Kumhar M, Bhatia J (2021) Emerging communication technologies for 5G-enabled internet of things applications. In: Tanwar S (eds) Blockchain for 5G-enabled IoT. Springer, Cham. https://doi.org/10.1007/978-3-030-67490-8_6

Li S, Da Xu L, Zhao S (2018a) 5G internet of things: a survey. J Ind Inf Integr. https://doi.org/10.1016/j.jii.2018.01.005

Li S, Xu LD, Zhao S (2018b) 5g internet of things: a survey. J Ind Inf Integr 10:1–9. https://doi.org/10.1016/j.jii.2018.01.005

Li S, Xu LD, Zhao S (2018c) 5G internet of things: a survey. J Ind Inf Integr 10:1–9. ISSN 2452-414X. https://doi.org/10.1016/j.jii.2018.01.005

Liu CH, Lin Q, Wen S (2019) Blockchain-enabled data collection and sharing for industrial IoT with deep reinforcement learning. IEEE Trans Industr Inf 15(6):3516–3526. https://doi.org/10.1109/TII.2018.2890203

Llah et al (2019) 5G communication: an overview of vehicle-to everything, drones, and healthcare use-cases. IEEE Access 7:37251–37268. https://doi.org/10.1109/ACCESS.2019.2905347

Mani Sekhar SR, Nidhi Bhat G, Vaishnavi S, Siddesh GM (2021) Security and privacy in 5G-enabled internet of things: a data analysis perspective. In: Tanwar S (eds) Blockchain for 5G-enabled IoT. Springer, Cham. https://doi.org/10.1007/978-3-030-67490-8_12

Mehbodniya A, Bhatia S, Mashat A, Elangovan M, Sengan S (2022) Proportional fairness based energy efficient routing in wireless sensor network. Comput Syst Sci Eng 41(3):1071–1082. https://doi.org/10.32604/csse.2022.021529

Mehta P, Gupta R, Tanwar S (2020) Blockchain envisioned UAV networks: challenges, solutions, and comparisons, computer communications, 151:518–538. ISSN0140-3664. https://doi.org/10.1016/j.comcom.2020.01.023

Minoli D, Occhiogrosso B (2019) Practical aspects for the integration of 5G networks and IoT applications in smart cities environments. Wirel Commun Mobile Comput

Mistry ST, Tyagi S, Kumar N (2020a) Blockchain for 5G-enabled IoT for industrial automation: a systematic review, solutions, and challenges. Mech Syst Signal Process 135:106382. https://doi.org/10.1016/j.ymssp.2019.106382

Mistry I, Tanwar S, Tyagi S, Kumar N (2020b) Blockchain for 5G-enabled IoT for industrial automation: a systematic review, solutions, and challenges. Mech Syst Signal Process 135:106382. ISSN 0888-3270. https://doi.org/10.1016/j.ymssp.2019.106382. https://www.sciencedirect.com/science/article/pii/S088832701930603X

Mudigonda P, Abburi SK (2020) A survey: 5G in IoT is a boon for big data communication and its security. In: ICDSMLA 2019. Springer, Berlin/Heidelberg, Germany, pp 318–327

Ngo HQ (2021) Massive MIMO. In: Lin X, Lee N (eds) 5G and beyond. Springer, Cham. https://doi.org/10.1007/978-3-030-58197-8_4

Nguyen DC, Pathirana PN, Ding M, Seneviratne A (2020) Blockchain for 5G and beyond networks: a state of the art survey. J Netw Comput Appl 166:102693

Okwuibe J, Haavisto J, Harjula E, Ahmad I, Ylianttila M (2020) SDN enhanced resource orchestration of containerized edge applications for industrial IoT. IEEE Access 8:229117–229131. https://doi.org/10.1109/ACCESS.2020.3045563

Popovski P, Trillingsgaard KF, Simeone O, Durisi G (2018) 5G Wireless network slicing for eMBB, URLLC, and mMTC: a communication-theoretic view. IEEE Access 6:55765–55779. https://doi.org/10.1109/ACCESS.2018.2872781

Pranto TH, Noman AA, Mahmud A, Haque AKMB (2021) Blockchain and smart contract for IoT enabled smart agriculture. PeerJ Comput Sci 7. https://doi.org/10.7717/peerj-cs.407

Qualcomm (n.d.) What is 5G—everything you need to know about 5G. Retrieved Nov 9, 2020, from https://www.qualcomm.com/invention/5g/what-is-5g

Rahimi H, Zibaeenejad A, Safavi AA (2018) A novel IoT architecture based on 5G-IoT and next generation technologies. In: Proceedings of the 2018 IEEE 9th annual information technology, electronics and mobile communication conference (IEMCON), Vancouver, BC, Canada, 1–3 November 2018, pp 81–88

Rathore S, Park JH, Chang H (2021) Deep learning and blockchain-empowered security framework for intelligent 5G-enabled IoT. IEEE Access 9:90075–90083. https://doi.org/10.1109/ACCESS.2021.3077069

Rong B, Han S, Kadoch M, Chen X, Jara A (2020) Integration of 5G networks and internet of things for future smart city. Wirel Commun Mobile Comput. Article ID 2903525, 2 pp. https://doi.org/10.1155/2020/2903525

Satpathy S, Mahapatra S, Singh A (2021) Fusion of blockchain technology with 5G: a symmetric beginning. In: Tanwar S (eds) Blockchain for 5G-enabled IoT. Springer, Cham. https://doi.org/10.1007/978-3-030-67490-8_3

Shahinzadeh H, Mirhedayati AS, Shaneh M, Nafisi H, Gharehpetian GB, Moradi J (2020) Role of joint 5G-IoT framework for smart grid interoperability enhancement. In: 2020 15th international conference on protection and automation of power systems (IPAPS), pp 12–18. https://doi.org/10.1109/IPAPS52181.2020.9375539

Sharma S, Nigam P, Muduli A, Pal A (2021) Highly isolated self-multiplexing 5G antenna for IoT applications. In: Tanwar S (eds) Blockchain for 5G-enabled IoT. Springer, Cham. https://doi.org/10.1007/978-3-030-67490-8_23

Shdefat AY, Mostafa N, Saker L, Topcu A (2021) A survey study of the current challenges and opportunities of deploying the ECG biometric authentication method in IoT and 5G environments. Indones J Electr Eng Inf (IJEEI) 9(2):394–416. https://doi.org/10.52549/ijeei.v9i2.2890

Singla R, Kaur N, Koundal D, Lashari SA, Bhatia S, Imam Rahmani MK (2021) Optimized energy efficient secure routing protocol for wireless body area network. IEEE Access 9:116745–116759. https://doi.org/10.1109/ACCESS.2021.3105600

Sinha P, Jha VK, Rai AK, Bhushan B (2017) Security vulnerabilities, attacks and countermeasures in wireless sensor networks at various layers of OSI reference model: a survey. In: 2017 international conference on signal processing and communication (ICSPC). https://doi.org/10.1109/cspc.2017.8305855

Siriwardhana Y, De Alwis C, Gür G, Ylianttila M, Liyanage M (2020) The fight against the COVID-19 pandemic with 5G technologies. IEEE Eng Manage Rev 48(3):72–84. https://doi.org/10.1109/EMR.2020.3017451

Varga P, Peto J, Franko A, Balla D, Haja D, Janky F, Soos G, Ficzere D, Maliosz M, Toka L (2020) 5G support for industrial IoT applications— challenges, solutions, and research gaps. Sensors 20(3):828. https://doi.org/10.3390/s20030828

Varga P, Peto J, Franko A, Balla D, Haja D, Janky F, Soos G, Ficzere D, Maliosz M, Toka L (2020) 5G support for Industrial IoT applications—challenges, solutions, and research gaps. Sensors 20:828. https://doi.org/10.3390/s20030828

Wang J, Yang Z, Wang Z (2020) WITHDRAWN: intelligent logistics cost control based on 5G network and IOT hardware system. Microprocess Microsyst 103476. ISSN0141-9331. https://doi.org/10.1016/j.micpro.2020.103476. https://www.sciencedirect.com/science/article/pii/S0141933120306293

Wang X, Mao X, Khodaei H (2021) A multi-objective home energy management system based on internet of things and optimization algorithms. J Build Eng 33:101603. ISSN 2352-7102. https://doi.org/10.1016/j.jobe.2020.101603. https://www.sciencedirect.com/science/article/pii/S2352710219312719

Waring J (2018) China to take 40% of 5G connections in 2025. Mobile World Live. https://www.mobileworldlive.com/asia/asia-news/china-to-take-40-of-5g-connections-in-202

Xia W, Wen Y, Foh CH, Niyato D, Xie H (2014) A survey on software-defined networking. IEEE Commun Surv Tutor 17(1):27–51

Xie L, Ding Y, Yang H, Wang X (2019) Blockchain-based secure and trustworthy internet of things in SDN-enabled 5G-VANETs. IEEE Access 7:56656–56666. https://doi.org/10.1109/access.2019.2913682

Yang Z, Yue Y, Yang Y, Peng Y, Wang X, Liu W (2011) Study and application on the architecture and key technologies for IOT. In: International conference on multimedia technology, pp 747–751. https://doi.org/10.1109/ICMT.2011.6002149

Zhou L, Wang L, Sun Y, Lv P (2018) BeeKeeper: a blockchain-based IoT system with secure storage and homomorphic computation. IEEE Access 6:43472–43488. https://doi.org/10.1109/ACCESS.2018.2847632

Chapter 12
Post-Quantum Cryptographic Schemes for Security Enhancement in 5G and B5G (Beyond 5G) Cellular Networks

Saurabh Bhatt, Bharat Bhushan, Tanya Srivastava, and V. S. Anoop

Abstract 5G is the fifth generation of broadband cellular network and beyond 5G can be the 6G, which will be the sixth generation of broadband cellular network. Even though studies about 5G are still evolving, 6G has become a hot topic for cellular researchers these days. The expansion in the field of 5G and 6G is still in infancy stage as many problems still need to be solved. Out of these, security of data transmission is a premier concern. Therefore, cybersecurity is becoming increasingly important for these cellular networks. This paper is focused upon providing the in-depth overview of 5G and B5G networks. The paper aims to evaluate the insights of the security services of 6G networks and outlines various data security techniques used by 5G networks. The paper also provides introduction to quantum computing for cryptography and evaluates various post-quantum cryptography techniques. Finally, some novel research trends and directions in correlation of security of 5G and beyond 5G networks are listed to guide further research in the area.

Keywords Cellular networks · 5G · Quantum cryptography · Security

S. Bhatt · B. Bhushan (✉) · T. Srivastava
Department of Computer Science and Engineering, School of Engineering and Technology, Sharda University, Greater Noida, India
e-mail: bharat_bhushan1989@yahoo.com

S. Bhatt
e-mail: 2019002150.saurabh@ug.sharda.ac.in

T. Srivastava
e-mail: 2019622455.tanya@ug.sharda.ac.in

V. S. Anoop
Smith School of Business, Queen's University, Kingston, ON, Canada
e-mail: Anoop.vs@queensu.ca

© The Author(s) 2023   247
B. Bhushan et al. (eds.), *5G and Beyond*, Springer Tracts in Electrical and Electronics Engineering, https://doi.org/10.1007/978-981-99-3668-7_12

Introduction

Cellular network also known as mobile network is a type of communication network in which the to-and-from nodes are connected to each other wirelessly. Cellular network has played a very important role in the development of humankind and the evolution of the technologies used nowadays. Moreover, it plays a very significant role in day-to-day life of individuals nowadays, from day to day calling to using social media on cellular networks, it plays a very important role. As per time, cellular networks have evolved and are still evolving, from 1G to current 4G LTE, advancement towards the commercial rollout of 5G and then beyond 5G. Every successor of the cellular network generation evolves in such a way that it provides us with a lot more what its predecessor couldn't offer. From 4 to 5G, there's a huge change, from high data transfer speed to high level security to connectivity between objects and machines (Sutton 2015). 5G communication has been formally dispatched in July 2020, and the connected 5G industry is relied upon to drive the vivacious advancement of communication-related parts at home and abroad. At present, the 5G communication in work has a place with the FR1 frequency band, which is likewise a sub-6 GHz frequency band (Ahmad et al. 2020). The communication frequency band is around 1 GHz more than that of 4G LTE communication. In any case, the advancement of the connected communication design has caused incredible contrasts in communication hardware. 5G being still not commercially out yet, the research and study about beyond 5G that is 6G has already begun. Many countries have aimed for commercially rolling out 6G technology by year 2030. As expected, the 6G network is to coordinate the aerial, water and terrestrial communications into a powerful network. This network would be quicker, solid and could uphold an enormous number of gadgets with super low latency prerequisites (Akhtar et al. 2020). Throughout the world, researchers are proposing state of the art advancements as the key technologies in the acknowledgment of beyond 5G and 6G communications. Some of these technologies are tactile internet, mall cell communication, artificial intelligence (AI) and machine learning (ML), edge computing, quantum communication, etc.

In addition to all these features and benefits, cellular networks are susceptible to attackers causing security issues in the communications. With change in generation of cellular network, their security features also change. As the cellular networks evolve, they work on new technologies and these new technologies require new security features. With the advancement in generation security also increases. Cryptography, also referred as the foundation of modern security system, is widely used for the security of both 5G and beyond 5G systems. Many different cryptography algorithms are used for providing confidentiality, authentication and integrity for wireless networks (Qadir and Varol 2019; Arora et al. 2020).

So far only limited use of quantum computing is being studied for 5G networks and as it is expected that 6G will be working on quantum computers providing us with quantum communications, it is essential to keep an eye for quantum technology as it is expected to be the future of computing. Quantum computers, referred by many researchers as un-hackable computers, have proved to be the best replacement for the classical computers due to their high security features (Stubbs 2021). Post-quantum cryptographic algorithms will be used for maintaining the security in devices, communications and other technologies in the future (Malina et al. 2021).

The main aim of the paper is to present an updated conception of current scenario about the security features and review of 5G and B5G networks. The contribution of the paper is as follows:

- This work provides an in-depth overview as well as recent advancements in 5G and 6G networks.
- This work provides comparative study about the methodologies and security services related to B5G/6G networks.
- This work analyzes various data security techniques used by 5G cellular networks.
- This work highlights the role of quantum computing in securing 5G and B5G cellular systems.
- Finally, the work evaluates various post-quantum cryptography techniques that are suited for 5G and 6G networks.

The remainder of this work is organized as follows. Section 12.2 gives the depth overview of 5G and 6G networks highlighting the security features that are expected to be provided. Section 12.3 discusses the data security techniques used by the current 5G networks. The section highlights different cryptographic algorithms and key management techniques used by 5G networks. Section 12.4 is dedicated towards the introduction of quantum computing in cryptography for 5G and B5G networks. It provides an overview of quantum computing and discusses various associated attributes. Furthermore, Sect. 12.5 describes the various post-quantum cryptographic schemes highlighting their mathematical background. Finally, the conclusion is presented in Sect. 12.6.

Emerging Cellular Networks

Telecommunications and networking are few of the key technologies responsible for the evolution of human kind and technologies. Wireless communication and cellular communications have played a huge role in daily lives of people. In last few decades, the wireless cellular networks have evolved in various forms from 1G to current 4G LTE and 5G of near future. As we're advancing towards the commercial rollout of 5G all over the world, we'll be introduced with high data rates to low latency even these high speed and low latency will be dwarfed in front of the super high data rates and ultra-low latency of beyond 5G (B5G)/6G.

5G Overview

The world has observed a quick development of cellular communication technologies in last few years. From 2 to 3G to 4G, the cellular comminutions technologies have been advancing in a fast pace. The fifth-generation mobile technology standard is known as 5G. It is the next-in-line of the 4G network, which offers connectivity in majority of mobile phones these days. As compared to its predecessors, the 5G connection has broader bandwidth ultimately providing us with great download speed up to 10–20 Gg/s, because of this high speed 5G network will give a huge competition to internet service providers (ISP) of laptop and desktops such as cable internet providers (Malik and Bhushan 2022). Similar to its precursors, 5G also works on different cells. Cells are just the smaller divided geographical areas of the service area. Through a local antenna, these 5G devices are connected to the telephone network and internet. As per the International Telecommunications Union Radio-communication Sector (ITU-R), 5G has three major applications: Massive Machine Type Communications (mMTC), Enhanced Mobile Broadband (eMBB) and Ultra Reliable Low Latency (URLLC) (Zhang et al. 2019). As of now, only eMBB is deployed in real world. It provides faster download speed as it uses 5G as a development of 4G LTE communications having much higher capacity, quicker connections and higher throughput. The URLLC will be used for latency delicate applications or devices such as remote surgery and autonomous driving, as these applications require super low latency with errors lower than a single packet. As for the mMTC, it can support high connection mass of online devices (Popovski et al. 2018).

Evolution of 6G/B5G Overview

B5G (Beyond 5G) or 6G will be the sixth generation of mobile technology standards and will probably work over 6 GHz. 6G/B5G is currently under development and as being the successor of the 5G it'll be significantly faster than all of its predecessors. Similar to its predecessors, it'll be a broadband network working under cells. As compared to the 10–20 Gb/s download speed of 5G, 6G is supposed to have download speed around 1 Tb/s. The lowest latency a 5G network can get is in milliseconds (ms) level as compared to its predecessor, which is supposed to have latency below 1 ms. The traffic density of 10 Tb/s/km^2 of 5G will be diminished as compared to 1000 Tb/s/km^2 density of 6G. The energy efficiency of 6G will be 10 times relative to that of 5G. In almost every aspect 6G will be better than 5G network whether its spectrum efficiency, end-to-end reliability requirements, processing delays, mobility or radio only delay requirements (Khan et al. 2020). As expected, the 6G network will likely be able to support applications or devices beyond current situations even beyond 5G limits. The 5G network system will open our gates even further 4G with Augmented Reality (AR)/Virtual Reality (VR) and smart cities, but with 6G we'll be introduced to Internet of Everything (IoE), which is beyond IoT (Bhushan 2022), Edge AI (Chaccour and Saad 2021), AI-enabled smart cities and so on.

State-Of-The-Art

There have been various recent advancements in the fields of 5G and beyond 5G/6G networks. Zappone et al. (2020) suggested many different ML approaches that can help in supporting each target 5G network requirement by emphasizing its specific use cases, moreover, they also proposed future research directions on how ML can contribute for B5G networks. Ahmed et al. (2021) devised an incentive framework based on deep learning known as Deep-CRNet for detecting opportunistic spectrum access (OSA) problem in 5G and B5G cognitive radio. The accuracy of the proposed framework was calculated via simulated results and achieved 99.74%. Sekander et al. (2018) projected a brilliant study on multi-tier drone architecture for 5G and B5G networks by analyzing challenges related with multi-tier drone networks and their current advancements. The result of their study has shown the beneficial network load condition for drones. Huang et al. (2021) proposed the very first true data testbed for 5G and B5G intelligent network full for TTIN. It consists of 5G/B5G on location test networks, information obtaining and information distribution center. Mishra et al. (2021) envisaged a framework known as IoT High-end Autonomous Cooper-Ative framework (ITHACA) for 5G networks and communications beyond 5G. Furthermore, Letaief et al. (2019) have discussed the various probable technologies, which will allow mobile AI applications with AI-enabled technologies for 6G networks.

Taxonomy of 6G/B5G Wireless Systems

The taxonomy for 6G network includes key enablers, use cases, emerging ML schemes, communication technologies, network technologies and computing technologies.

Key Enablers

6G uses various types of technologies for operating and offering different applications. The key enablers of 6G are Homomorphic Encryption, Blockchain, AI and Photonics-based Cognitive Radio, Edge intelligence, Network Slicing, Ubiquitous Sensing and Space-Air-Ground Integrated Network (SAGIN) (Mahmood et al. 2020). Out of these Blockchain, Network Slicing, SAGIN and Ubiquitous Sensing are considered the major key enablers (Khan et al. 2020). Blockchain is simply a kind of a database, which makes it hard or even impossible to change alter or hack the data. It'll basically allow the 6G network to exchange huge amount of data securely.

With blockchains being one of the key enablers of 6G, it'll face few difficulties like high energy consumption and high latency (Hewa et al. 2020). Network Slicing is a process of creating logical and virtualized networks on a common physical infrastructure. As network slicing is already proposed via 5G technology, its actual working or realization will be shown off in 6G. SAGIN as the name suggests consists of satellite communication networks, aerial networks and ground networks. Few of the advantages of SAGIN are high throughput, much better resilience than its counterparts and large coverage areas. Finally, the ubiquitous sensing uses video-captured information for enabling smart decision-making and automated sensing.

Use Cases

5G networks provide us with many applications form AR/VR to smart cities. Generally, the use cases of 5G are divided into three main classes: eMBB, mMTC and URLLC. Several few new technologies require more than these so new use cases are defined for 6G connections. The use cases of 6G apart from that of 5G are: Human-centric services, Holographic communication-based services, Nano-Internet of things (N-IoT), Bio-Internet of things (B-IoT), Massive URLLC (mURLLC), Haptics communications and unmanned mobility (Khan et al. 2020). There is the need to implement more human centrical services than 5G such as the brain-computer interface, for which human physiology is used to measure its performance. The holographic communication-based services are totally based on super high accuracy remote connection. These cannot be obtained from a 5G network as these require high data rates than 5G can offer. The N-IoT and B-IoT as the name suggested are based on the communications of nanodevices and biodevices over a network. Much like 5G, the 6G network will also use URLLC but on a massive scale thus having mURLLC (Zhang et al. 2021a). Based on URLLC, mURLLC denotes IoE applications. It'll basically merge the 5G URLLC with the machine massive machine-type communications (Mahmood et al. 2021). Last but not the least, Haptics communications are a type of non-verbal communications which works from a remote place with enabling sense of touch.

Emerging Machine Learning Schemes

Machine Learning (ML) is contemplated to play an important role in development and working of 6G network. Recently, in past few years, ML has evoked great attention in various smart applications from self-driving cars to voice assistants. As for 6G, ML is not only expected to provide smart applications but also smart transceivers and smart access control techniques and schemes. This makes ML a fundamental pillar of 6G network. For 6G purposes, ML is basically divided into three categories: Federated learning (Yang et al. 2021), Meta-learning (Jung and Saad 2021) and Quantum Machine Learning (Kashyap et al. 2022). To overcome challenges of the original ML processes, recently federated learning is being adopted. Federated leaning applies ML via a distributed means by allowing on-device ML without drifting data through end devices to the cloud. Meta-learning helps the models to learn with complex designs. Quantum ML is the combination of ML and quantum physics, which ultimately results in fast raining speed of models.

Communication Technologies

6G communications will use many different communication technologies for providing a wide variety of smart applications. The communication technologies that the 6G network will use are quantum communications (Kashyap et al. 2022; Wang and Rahman 2021), visible light communications (Ariyanti and Suryanegara 2020), terahertz communications, 3D wireless communications, holographic communications and nanoscale communications (Khan et al. 2020). Quantum communications is the field of communication that uses the fundamentals of quantum physics and quantum computing for providing protection to data. Basically, the photons of light are used for transmission of data through optic cables. This security feature of quantum communication makes it suitable for 6G networks. Visible light communication full for VLC is a type of data transmission method, which uses visible lights for the transmission of data. It uses the visible light spectrum from 430 to 790 THz. The transmission of data through illuminous sources is the main advantage of VLC. With addition of mmWave bands as used in 5G networks, 6G will use Terahertz communications too. Terahertz communication is also a different type of wireless data transmission technology which provides ultra-high-speed wireless extensions of fiber optics for 6G networks. Another technology that 6G will use is the nanoscale communications, as the name suggests this technology is used for communication as nano level distances of 1 m or cm. It uses very short wavelengths for the transmission of data (Yuan et al. 2020).

Network Technologies

Networking technologies that will be used in 6G networks are bio-networking, 3D networking, nano-networking and optical networking. The N-IoT relies on molecular communication to function. Nanometer-scale devices can be made with a variety of material, such graphene and metamaterials. B-IoT is a type of IoT technology that utilizes biological cells. B-IoT and N-IoT appear to be critical components of future 6G smart services but they face a number of implementation hurdles. Physical layer technology design for molecular communication is a difficult task. Because B-IoT and N-IoT are basically different from traditional IoT, unique routing algorithms must be planned in addition to physical layer techniques. Similarly, new 3D communication models must be developed as they are different in nature than 2D communication network (Calvanese Strinati et al. 2020).

Computing Technologies

As the 6G system will include a huge variety different smart applications and devices, it'll require different types of computing technologies as well for generating the humongous amounts of data. Quantum computing, high-performance computing and intelligent edge computing will be used for the analysis of such data. The quantum computing is said to change the whole field of computing by providing with much higher speeds we haven't experienced yet. The key factor of quantum computing is the security it provides. As for huge amount of analyzing and computing of huge loads of data high-performance computing is required (Blog: Samsung Research 2021). Apart from these, intelligent edge computing is required for providing intelligent on-demand computing and storage abilities (Hui et al. 2021) (Table 12.1).

Security Services in 6G/B5G Wireless Networks

The 6G/B5G is a new technology with new use cases, features and architecture with these it also brings necessities for new security services. Basically, there are four security services required for 6G network, they are: Confidentiality, Availability, Authentication and Integrity (Bhushan and Sahoo 2017).

Confidentiality

Confidentiality consists of two things: privacy and data confidentiality. Privacy helps in the protection of the traffic flow from an attacker as an attacker can study the traffic flow and can identify sensitive information. As 5G and B5G, both will be used throughout various applications loads of user's data will be associated with their privacy (Saxena et al. 2021). Data confidentiality on the other hand limits the data access only to the authorized users and prevents the data leakage or disclosure to unauthorized users. Data encryption is widely used for securing the data confidentiality by stopping unlicensed users from gaining sensitive information.

Availability

Availability as the name suggests defines, to which extent some data or service is available or accessible. It basically estimates the strength of the system or network. Availability attacks are most common types of attacks happening over systems or networks. Denial of Service (DoS) is the most common type of availability attack. In this, a particular service or series or service made inaccessible to the user by flooding the network resulting in crashing of the service. As a huge number of IoT and IoE devices will be connected to the 6G network, it'll be a challenge for the network to

Table 12.1 Various B5G/6G taxonomies

Key Enablers	Use Cases	ML schemes	Communications Technologies	Network Technologies	Computing Technologies
• Homomorphic encryption • Blockchain • AI and photonics-based cognitive radio • Edge intelligence • Network slicing • Ubiquitous sensing • SAGIN	• eMBB • mMTC • URLLC	• Federated learning • Meta-learning • Quantum Machine Learning	• Quantum communications • Visible light communications • Terahertz communications • 3D wireless communications • Holographic communications • Nanoscale communications	• Bio-networking • 3D networking • Nano-networking • Optical networking	• Quantum computing • High-performance computing • Intelligent edge computing

prevent availability attacks such as DoS and Distributed Denial of Service (DDoS) (Bhushan and Sahoo 2017).

Authentication

Authenticity is of two types: Message authenticity and Entity authenticity. Message authenticity makes sure that the message has not been changed or modified, while in between the transmit and that the receiver is getting the same message as which the sender has sent. On the other hand, entity authenticity makes sure that the sending party or the receiving party is the same that they claim to be and not someone else. 6G having data speeds up to 1 Tb/s with ultra-low latency will have much faster authenticity as compared to its predecessors. Various public key-based Authenticated Key Agreements (AKA) are proposed for better security (Goyal et al. 2021).

Integrity

Integrity means protection against inappropriate modification or deletion of information or data. Authentication makes sure about the source of the message is not altered, integrity protects from the alteration or replication of the message by unauthorized entities. As cellular networks are aimed towards more and more connectivity easing human lives by supporting and connecting to applications used by humans in daily lives, integrity of user's data is a key security necessity as more and more of user's data is being used. Integrity security can be delivered by the mutual authentication of mobility management entity (MME) and user equipment (UE).

Data Security Techniques for 5G Heterogeneous Networks

As we are moving towards a more digital era, come digital attackers. No network is safe from cyber-attackers. Security is of the data that is very substantial. Hence, numerous data security techniques are used in 5G networks for providing the safest and securest communication of data.

Visual Secret Sharing

Cryptography is an art of sharing data secretly. Visual secret sharing also known as visual cryptography is a method of sharing data by encysting visual media such as text, image, etc. in a way such that the final decrypted data are in the form of a visual image. In this, the secret data are divided into many different shares or parts, for decryption and getting the secret message the user must have all the shares of the original image. This technique provides high security as all the shares are required and even if one share is missing no information or the data can be decrypted. Another advantage of visual secret sharing is that it requires low computational complexity (Liu and Chang 2018).

Steganography

Steganography is the method of obscuring data or a message inside another message or file like an image, audio or video file to avoid exposure of the data. The data are extracted by the one for whom the data are intended to receive. For enhanced protection, steganography is used with encryption techniques for concealing and protecting of the data. It basically works by replacing some useless or vacant parts of the file with the bits of data which is to be hidden. There are five major types of steganography: Steganography in images, videos, audio, text and network. Three of the most used approaches for steganography are: Least Significant bit (LSB) (Singh et al. 2016), secure cover selection (Qin et al. 2021) and palette-based technique (Hao et al. 2021). In LSB, the useless bits of the transporter file are identified, which are then replaced with the secret data. In secure cover selection, the blocks of carrier files are compared to find the perfect match to carry the secret data. Lastly, in palette-based technique, digital images are used. First, the secret data are encrypted then it is hidden among the wide palette of cover image. In network steganography, the data are hidden in the network control protocols like UDP, TCP, etc.

Cryptographic Algorithms

Cryptography is a method of securely transmitting the data from sender to receiver with the use of secret writing. The data are sent to the receiver in form of cipher text, for which cryptographic algorithms are used for conversion pf the pain text to cipher text. The main applications of these algorithms are digital signatures, data encryption and authentication.

Elliptic-Curve Cryptography

Elliptic-curve cryptography (ECC) is a type of public key cryptography method or cryptosystem based on mathematical elliptic curves. ECC is popular for creating smaller, quicker and more efficient cryptographic keys (Wikimedia Foundation 2021a). ECC has all the properties that an asymmetric cryptosystem has from encryption decryption to key exchange and signatures. Mostly the ECC is used for encryption of internet traffic. Private Keys in ECC are integer values mostly a 256-bit integer. While the public keys are the coordinates of the curve, these points are known as EC points. The key generation in ECC is very simple and easy as just random integers within a range are generated, any integer within that range can be used as a valid private key. There are significant overheads in ECC. The size of the blocks is also key dependent here. ECC also does provide resistance towards mutual authentication and replay attacks. It also provides differential fault analysis. Apart from these, ECC also provides various different features like key provisioning, key monitoring, key maintenance and management. ECC is also resilient and scalable method of cryptography. Different algorithms are used by ECC such as EdDSA and ECDSA for digital signatures, FHMQV, X25519 and ECDH for key agreement and EEECC and ECIES for encryption (ECC keys 2021).

RSA

RSA stands for Rivest-Shamir-Adleman. It is also a type of public key cryptography system used for secure transmission of data. It is one of the oldest cryptosystems. As being a traditional public key cryptosystem, it can perform various tasks such as encryption, decryption, key exchange and signatures. Similar to ECC, it's also a type of asymmetric cryptography (Wikimedia Foundation 2021b). The block size is of 86 bytes. There are less overheads in RSA cryptosystem. It also provides partial resistance towards mutual authentication and replay attacks. Similar to ECC, the RSA is also provided key provisioning, key monitoring, key maintenance and management and is also scalable and resilient. RSA provides partial analysis of differential fault. In RSA, the keys are generated usually using two large prime numbers. RSA is used in many services like VPN, web browsers, email services and many more communication services. As compared to ECC, RSA is slow thus, it is not used directly for encryption of data.

Diffie-Helman

Diffie-Helman is a technique of exchanging cryptographic keys securely. It is the first broadly used technique of exchanging keys over an insecure channel. Hence, it is called Diffie-Helman key exchange. As two different organizations or parties need to exchange keys for a successful encrypted communication (Li 2010). The channel between these parties needs to be secure for usual security reasons, here the Diffie-Helman exchange is used for exchange of keys between two unknown or known parties by forming a shared secret key in an uncertain channel. Here the cipher overheads are significant. Diffie-Helman also shows resistance towards mutual authentication and replay attacks up to some degree. The cipher block size here is variable, as it is dependent upon the selected prime number. As compared to ECC and RSA, the Diffie-Helman is partially resilient and scalable, and it also provides partial key provisioning, key monitoring, key maintenance and management.

ElGamal

ElGamal is a type of asymmetric encryption system used for public key cryptosystems. It is based on Diffie-Helman key exchange system (Wikimedia Foundation 2021c). It comprises of three parts the key generation, encryption and decryption algorithm (Tsiounis and Yung 1998). This cryptosystem depends on the trouble of discovering discrete logarithm inside a cyclic group. ElGamal is a type of probabilistic encryption, it means that many different ciphertexts of a plaintext can be generated. As compared to RSA and ECC, they both worked on integer factorization while ElGamal works with discrete logarithm. There are moderate number of overheads in ElGamal. The block size of ciphertext is variable and depends upon the key length. This doesn't show resiliency and scalability. Key provisioning, key monitoring, key maintenance and management are also absent here. The resistance towards replay attacks and mutual authentication is not effective as much. Furthermore, the differential fault analysis provided by ElGamal is not as effect as much too.

DES

DES, short for Data Encryption Standard, is a type of symmetric cryptosystem method meaning it requires a single key for encryption and decryption of data. It is based on LUCIFER (Wikimedia Foundation 2021d), a Feistel block cipher. It has a 64-bit block size. Here, 64-bit blocks of ciphertexts are transformed into 48-bit keys. There are total of 16 rounds of encryption in DES and each round executes the essential attributes of cryptography which are transposition and substitution. This offers important and generic tools for watermarking digital videos and streams for tamper detection and protection against watermark copy attacks. There are very few overheads of ciphertexts here. DES is not scalable or resilient and doesn't provide differential fault analysis, key provisioning, key monitoring, key maintenance and management. Furthermore, it doesn't show resistance towards mutual authentication and replay attacks. Because of these problems, there are concerns about security and speed of DES is making users to shift towards newer block cipher designs or reusing DES like Triple DES (TDES).

AES

AES stands for Advanced Encryption Standard. Similar to DES, it is also a type of symmetric block cipher system. It is well known for its use by the U.S. government for the protection of classified data (Bernstein and Cobb 2021). It was developed as an alternative for DES. It has the cipher blocks of 128-bit and key lengths of 128, 192 and 256 bits. The cipher overhead is also noteworthy in AES. As being an improvement of DES, it does have resistance against mutual authentication and replay attacks and also does provide users with differential fault analysis and was

Table 12.2 Different types of cryptographic algorithms used in 5G networks

Functionality	ECC	RSA	Diffie-Helman	ElGamal	DES	AES
Block size	Depends on key	86 bytes	Variable depends upon the selected prime no	Variable, key length dependent	64 bits	128 bits
Number of keys	1/1	1/1	1/1	1/1	1/1	3/3
Overheads	Noteworthy	Less	Noteworthy	Moderate	Very less	Noteworthy
Resilience, scalability, key management and provisioning	Yes	Yes	Partially	No	No	Partially, key provisioning depends upon computing speed
Resistance towards mutual authentication and replay attacks	Yes	Yes	Up to an extent	Not effective	No	Yes

also faster and more reliable. It does also provide partial resiliency and scalability with key management and provisioning but the key provisioning is dependent upon the computational speed of the system. Due to so many advantages and high security, it is one of the most popular encryption algorithms used today, used in many ways from wireless security to web browsers (Table 12.2).

Key Management

CKSM stands for Cryptographic Key Management System. It is a system that is used to protect key data. Key management as the name suggests is the process of managing or handling the cryptographic keys inside a cryptosystem. It basically includes storing, generating and exchanging keys as per user's requirement.

Key Escrow

Key escrow is a method in which keys that are required to decrypt an encrypted data are stored in an escrow. Escrow is basically a bond kept in safekeeping of third party and engaged only when certain conditions are met. In simple words, key escrow is nothing more than a process of storing cryptographic keys. As a third party is involved key escrow system is not much on the safer side and does include some risks (Sugumar and Ramakrishnan 2018; Foundation 2020). Apart from this, there are other problems related to this too like the mutual authentication of both parties is not always satisfactory and key escrow also have shown some problem with identity-based encryption.

Identity-Based Encryption (IBE)

IBE is public key-based encryption (PKE) that uses identifiers as the source for encryption. In this, the public key of the user is the distinctive detail about the identity of the user. Here, the public key of the user is created with the public key of a third party. Similarly, the private key of the user is also computed in conjunction of the private key of the third party, making it secure as no one else than the third party can access the user's private key (Khan and Niemi 2017). These third parties are known as private key generators (PKG).

Attribute-Based Encryption (ABE)

ABE is also a type of PKE in which the secret key and the ciphertext both are reliant upon attributes such as the country of the user or the particular type of services they've enrolled for. Here, decryption of cyphertext is only possible if the attributes of the user key match that of the ciphertext (Zhang et al. 2021b). ABE is divided into two different types those are: Key-policy attribute-based encryption (KP-ABE) and Ciphertext-policy attribute-based encryption (CP-ABE) (Wikimedia Foundation 2021e). User's secret keys in KP-ABE are produced by an access tree that describes the user's privilege and encrypts data over set of attributes. CP-ABE, on the other hand, encrypts data using an access tree and secret key is encrypted based on set of attributes.

Overview and Fundamentals of Quantum Computing

In causal and simple words, the use of computers is known as computing. There are two major types of computing: classical and quantum. Classical being the one which we all use in our day to day lives. Other is quantum computing, it is a type of computing that combined concepts of computer science and quantum physics. Quantum computing is still a progressively growing research area (Elsevier. (n.d.). 2021).

Architecture of Classical Computing Versus Quantum Computing

There are huge differences between classical computing and quantum computing. With the increment of quantum properties in quantum computers, it makes the difference between the two even greater.

Classical Computing

Classical computing also known as binary computing is the traditional approach in computing. In classical computing, the information is kept in bits. Everything words, integers, audio, video, image, all are represented in the form of bits. These bits are represented either in 0 or 1. The information is fragmented into simple Boolean logical gates. Every logical gate in classical computing takes input as 1 or 2 bits and as an output, it gives new bits as a result (Navaneeth and Dileep 2020). There are a total of seven logical gates in classical computing, each takes and gives output in unique way. The computation is done by specially arranging the logical gates accordingly.

Quantum Computing

Quantum computing is the type of computing as the name suggests which uses the properties of quantum mechanics to provide a huge leap over classical computation for solving problems and calculations. Quantum computers follow the probabilistic approach for calculations, meaning they solve problem upon the most probable outcome, simultaneously using several other dimensions. As compared to classical computing which uses 0's and 1's for representing the data, quantum computing offers many new ways of data representation. In quantum computing, quantum bits are used. These quantum bits are known as qubits. The operations which the qubits contain are sensitive and are unstable (Haller 2021), due to which the qubits require very specific requirements for working correctly. For functioning efficiently, vacuum and temperature very close to absolute zero are required by the qubits. Furthermore, they endure no interference, which turns out to be exceptionally muddled while working on a nanoscale individual electrons and photons. For differentiating bits and qubits, lets take an example—as classic computer uses 8 bits to denote a number between 0 and 255, instead of bits 8 qubits can represent all the number between 0 and 255, that too simultaneously.

Quantum computing offers superposition (Khrennikov 2021), due to which between these 0's and 1's, infinite other states can also be present there. In these states, there are infinite number of qubits. Apart from superposition quantum, computers also offer other quantum mechanics-based phenomenon such as quantum entanglement (Duarte 2019). It is a phenomenon that occurs when particles are created. It is a property between two or more qubits, which allows the qubits to have higher amount of correlation with each other. These properties like superposition, quantum entanglement, quantum interference, No-cloning theorem and destructive measurement are not present in the classical computing. Hence, making a huge difference between quantum computing and classical computing.

Mathematical Representation for Quantum Computing

In quantum physics, the qubits work upon the spin of the qubits and their direction of spin. Mathematically, the use of calculus is not required for qubits. Hence, the concept of vectors is highly useful for describing and analyzing the spin of qubits. Apart from this, fundamentals of matrices are also used for the measurement of qubit spin in some cases (Lam 2019). The likelihood of getting a particular estimation for spin, as far as one might be concerned, can be depicted utilizing probability. In this way, a comprehension of probability hypothesis is somewhat valuable as it relates with quantum mechanics. Hence, matrices, complex number, vectors and probability play an extremely huge part in mathematics and calculations of quantum computing (Chamola et al. 2021). The states zero and one in qubit are presented by $|0\rangle$ and $|1\rangle$, respectively. Entanglement of two qubits can be represented as:

$$|0\rangle \otimes |1\rangle = |01\rangle \tag{12.1}$$

In quantum computing, vectors are represented as a list of numbers. A single qubit is considered a two-dimensional vector. The dimension of vector is represented as the rows, i.e., the dimension of the vector is labeled as number of rows of the vector. Representation of qubit in vector is as follows:

$$|0\rangle : \begin{bmatrix} 1 \\ 0 \end{bmatrix} \tag{12.2}$$

$$|1\rangle : \begin{bmatrix} 0 \\ 1 \end{bmatrix} \tag{12.3}$$

While in the state of a superposition, the qubit has 50–50% probability of becoming a zero and one respectively. For mathematical representation, let's take an example of superposition:

$$\sqrt{2/3} \cdot |0\rangle + \sqrt{1/3} \cdot |1\rangle \rightarrow \begin{bmatrix} \sqrt{2/3} \\ \sqrt{1/3} \end{bmatrix} \tag{12.4}$$

Above, superposition is shown in which $|0\rangle$ has the probability of 2/3 and the probability of $|1\rangle$ is 1/3. The square roots are there because the unit circle where the sum of probabilities is always one can be represented by vectors.

The Impact of Quantum Computing on Modern Cryptography

The rapid increment in the field of quantum computing can also disturb many big and small organizations. As quantum computing is a lot different and superior than that of classical computing in many different aspects, it has several impacts on cryptographic algorithms made for classical computers.

Cryptanalysis

As quantum computers are expected to provide a huge number amount of leap over the common classical computers, it will be very easy for the quantum computers to break the security provided to us by the classical methods of cryptography. With the large-scale rollout of quantum computers for people, havoc will be caused regarding the security, as majority of the technology uses the classical cryptography algorithms. Quantum computing algorithms have proved to be efficient against both symmetric and asymmetric cryptography algorithms (Mitchell 2020; Schanck 2020). As asymmetric algorithms rely on the huge amount of time taken by classical computers to factorize huge integers for their security, this can easily be tackled with use of Shor's algorithm (Devitt et al. 2005). On the other hand, for finding the key of symmetric cryptographic algorithms, computers take around $k/2$ operations, with quantum algorithms like Grove's algorithm (Mandviwalla et al. 2018), this time could be further reduced to $\sqrt{k}$ (k being the size of key) operations.

Security Impacts

Cryptography is one of the most significant techniques used in modern technology for securing a communication either between two users or a user and its machine. As current cryptographic algorithms are considered insecure in comparison to quantum computers, it can cause catastrophic impacts on the digital world. For algorithms that are only used for affirming the integrity of sent data, for instance whose use has no extended-out influence, there will no big issues, to the extent that until new and secure algorithms are introduced. Comparing to key establishment algorithms and encryption algorithms, here the impact will be tremendous (Fernandez-Carames 2020).

Replacement Algorithms

As these quantum computers will have such a destructive impact on the security given by current cryptographic algorithms, many organizations have already started development, research and studies on new cryptographic algorithms and standards. As for symmetric cryptographic algorithms, no promising new technologies and advancements are made as current symmetric cryptographic algorithms work on using 265-bit keys apart from moving towards using more longer keys for encryption. As for asymmetric cryptographic algorithms, it relies on the factorization of huge integers. It has been proved to be a piece of cake for the quantum computers to factorize huge integers. So many other factors are also needed to be taken care for the development of new asymmetric cryptographic algorithms, which can withstand the quantum computers. Many agencies such as ISO, NIST, IEC, etc. have already started their development of quantum-resistant asymmetric cryptographic algorithms (Mitchell 2020).

Quantum Algorithms Affecting Cryptosystems

As stated above that the quantum algorithms will have a significant amount of impact upon the current cryptographic systems. Whether its symmetric or asymmetric cryptographic algorithms, some quantum algorithms with the help of quantum computers can easily surpass them.

Shor's Algorithm

Peter Williston Shor, in 1994, created an algorithm for integer factorization known as Shor's algorithm. It's a polynomial-time quantum computer algorithm. In his research, he proposed that large factorization of large integers can be done via quantum computers (Wikimedia Foundation 2021f). Modern cryptographic algorithms such as RSA provide security on the basis that classical computers have slower computational speed for factorizing huge integers and could take huge amount of time basically millions or billions of years for the factorization (Devitt et al. 2005; Bhatia and Ramkumar 2020). Shor guaranteed that it is feasible to change the factorization issue over to another issue of finding the time of an integer $0 < x < N$. A periodic function where $a \geq 0$, x is an integer coprime to N:

$$F(a) = x^a \, mod \, N \tag{12.5}$$

Shor's Algorithm works as:

- Firstly, the factorization problem is converted into the problem of period finding:

  Since period is r and it is periodic with F(a)

$$x^0 mod N = 1 \tag{12.6}$$

So,

$$x^r mod N = 1 \tag{12.7}$$

Then,

$$X^r = 1 mod N \tag{12.8}$$

Mathematically,

$$(x^{r/2} + 1)(x^{r/2} - 1) = 0 mod N \tag{12.9}$$

Following equation is a multiple of N

$$(x^{\frac{r}{2}} + 1)(x^{\frac{r}{2}} - 1) = 0 \tag{12.10}$$

Between $(x^{r/2} + 1)$ and $(x^{r/2} - 1)$ at least one of these should have a non-trivial factor common to N, for not being a multiple of N.

Now obtaining a factor of N using:

$$GCD((x^{\frac{r}{2}} + 1), N) \tag{12.11}$$

$$GCD((x^{r/2} - 1), N) \tag{12.12}$$

Above GCD stands for Greatest Common Divisor.

- Now finding the period using Quantum Fourier Transformation.

Initialize the qubits in superposition and compute the modular exponentiation. Now perform the Quantum Fourier Transformation multiple times for getting the good probabilistic result. The Quantum Fourier Transformation uses amplitude amplification.

- Finally, finding with the use of periods finding the factors, after r is recognized.

Out of:

$$GCD((x^{r/2} + 1), N) \tag{12.13}$$

$$GCD((x^{r/2} - 1), N) \tag{12.14}$$

At least, one factor of these equations will be a non-trivial factor of N. Hence, the factor is found.

Quantum Annealing

Quantum annealing is a process of finding the most efficient solution. It is a process used for solving optimization problems. It normally gives low-energy solutions for applications that necessitate the genuine least energy and others require great low-energy tests. Several companies use quantum annealing such as IBM, Microsoft, D-Wave, etc. (Mitchell 2020; Hauke et al. 2020).

As for cryptography, quantum annealing is used for factorizing integers into prime numbers, same as Shor's algorithm. Similarly, like Shor's algorithm, it can be used for factorizing large integer values using quantum computers. The quantum annealing approach is as follows:

Take $N = pq$, here both p and q are the prime numbers. These two prime numbers can be written into binary form as,

$$p = 1 + \sum_{i=1...s_p} 2^i p_i \tag{12.15}$$

and,

$$q = 1 + \sum_{i=1...s_q} 2^i q_i \tag{12.16}$$

So, the function can be,

$$f\left(p1, p2, \ldots, p_{sp}, q1, q2, \ldots q_{sp}\right) = (N - pq)^2 \tag{12.17}$$

If the p_i, q_i are found for which the value of f is minimum or 0, the problem for factorization is solved. For such situations, quantum annealing such as D-Wave is used for finding the minimum value.

For example, suppose,

$$N = 15 = pq \tag{12.18}$$

And let the binary depiction of p has 2 bits: $(x_1, 1)$
So,

$$p(x_1, 1) = 2x_1 + 1 \tag{12.19}$$

Similarly,

$$q(x_2, x_3, 1) = 2^2 x_2 + 2x_3 + 1 \tag{12.20}$$

Likewise, the binary depiction of a prime number will always comprehend a 1 as its least important number. The function to minimize is f,

$$f(x_1, x_2, x_3) = (N - pq)^2 \tag{12.21}$$

$$= (15 - (2x_1 + 1)(2^2 x_2 + 2x_3 + 1))^2 \tag{12.22}$$

Further solving for (x_1, x_2, x_3) and we will get the factors.

As quantum computing is based upon the probabilistic approach, it is always a must to run these algorithms multiple times for getting the best probabilistic result.

Grover's Algorithm

In 1996, Lov Grover came up with a quantum search algorithm called as Grover's algorithm. It is used for improved efficiency of searching data over an unstructured database. For a database of N number of items in it and we want to search for a specific item, it would take a classical computer $N/2$ opertaions to find the specified item at average case and at worst case it would take N operations (Devitt et al. 2005; Brickman et al. 2005). For quantum computers, the time required for searching is very less as it would only take $\sqrt{N}$ opertains to find the specified item.

For example, we have to find an item s from a set S having 2^n items in it.

Denoting every time in the set with a number so,

$$x \in \{0, 1, \ldots 2^n - 1\} \tag{12.23}$$

Let $f(x)$ be a function in which $f(x)$ checks for the required item and checks if it is or not,

$$f(x) = \begin{cases} 1 \ if \ x = s \\ 0 \ otherwise \end{cases} \tag{12.24}$$

For computing the problem in a quantum computer, few changes are needed to be made, such as:

x converted to a qubit, $x \rightarrow |x\rangle$

f converted into an operator, $f \rightarrow \hat{\theta}$

Now, $\hat{\theta}|x\rangle$ can be represented as,

$$\hat{\theta}|x\rangle = \begin{cases} |x\rangle \ if x = s \\ -|x\rangle \ otherwise \end{cases} \tag{12.25}$$

In a simplified manner, it can be written as,

$$\hat{\theta}|x\rangle = (-1)^{f(x)}|x\rangle \tag{12.26}$$

Now, we can modify the problem and say that we want to find the qubit $|s\rangle$ from the set of qubits $S = \{|x\rangle : x \in \{0, 1, \ldots, \}\}$.

As for solving the above problem, we will have to start with from superposition of all the possible solutions:

$$|E\rangle = \frac{1}{\sqrt{2^n}} \sum_{x=0}^{2^n-1} |x\rangle \tag{12.27}$$

The $|E\rangle$ contains all the solutions including $|s\rangle$. There are $\frac{1}{2^n}$ probability that $|E\rangle$ will give the required solution. As all the solutions have same amplitude, we have to grow the amount of amplitude for $|s\rangle$ in $|E\rangle$. For that we have to take an intermediate state. So,

Let us take $|\psi_1\rangle$ as intermediate state.

$$|\psi_1\rangle = |E\rangle \tag{12.28}$$

$$= \hat{\theta}|E\rangle \tag{12.29}$$

$$= (2|E\rangle|E\rangle - 1)\hat{\theta}|E\rangle \tag{12.30}$$

Similarly,

$$|\psi_2\rangle = |\psi_1\rangle \tag{12.31}$$

$$= \hat{\theta}|\psi_1\rangle \tag{12.32}$$

$$= (2|E\rangle|E\rangle - 1)\hat{\theta}|\psi_1\rangle \tag{12.33}$$

Generally,

$$|\psi_1\rangle = ((2|E\rangle|E\rangle - 1)\hat{\theta})^t |E\rangle \tag{12.34}$$

Hence, with each iteration of $|\psi\rangle$, amplitude of the solution also increases.

Post-Quantum Cryptography

Post-quantum cryptography also known as quantum resistant or quantum proof cryptography is a type of cryptography, which is thought to protect us from attacks conducted via quantum computers (Wikimedia Foundation 2021g; Bernstein n.d). Cryptography is one of the most important techniques used in modern technology for secure communication either between two users or a user and its machine. According to several surveys, more than 90% of the currency is digital, all of it uses the concept of cryptography for security purposes. As quantum computers are the possible future, they offer huge advantages and new technologies as compared to classical computers. They will break the current infrastructure of computations and current cryptographic techniques will not be powerful enough to counter them. Hence, we'll be requiring new cryptographic techniques or algorithms to protect us from such scenarios, where we've to protect our data from hacker or attacks coming from quantum computers. The National Security Agency (NSA) of USA has already transitioned to post-quantum cryptographic algorithms as it is not known when a powerful enough quantum computer will be there which can break current cryptographic techniques. Dissimilar to quantum-based cryptography, post-quantum cryptosystems depend on some numerical issues that are not difficult to figure out for the receiving end, however, harder for the attacker (Chen et al. 2016).

Mathematical View

There are various quantum resistant cryptographic techniques that have already been proven to work effectively to provide security that current cryptography schemes couldn't provide while facing quantum computers. The working of these techniques is heavily relied upon their mathematical backgrounds.

Lattice-Based Cryptography

Lattice-based cryptography (Pradhan et al. 2019; Yao et al. 2021) as the name suggests is a type of cryptographic primitive, which uses the lattices for creation of cryptographic algorithms. Unlike common cryptographic schemes, some lattice-based constructions seem, by all accounts, to be impervious to assault by both traditional and quantum computers (Nejatollahi et al. 2019). Moreover, numerous lattice-based developments are viewed as secure under the supposition that specific all around concentrated on computational lattice issues can't be tackled proficiently.

- SVP (Shortest Vector Problem) is the widely used Lattice-based cryptosystem. The key generation in SVP is done:

$$(p, q, n, m), B \in Z^{n \times m} \tag{12.35}$$

$$\text{Theencryptioniscarriedoutby} : H_B : 1, \ldots, d^m \to Z_p^n \tag{12.36}$$

$$\text{Thedecryptionofkeyisdoneby} : H_B(x) = B_x mod p \tag{12.37}$$

- DGS (Discrete Gaussian sampling) is another type of lattice-based scheme. Key generation in DGS is done using:

$$M \geq u^2 \times \left(Q D^2 Q^{-1}\right) = u^2 \left(V.V^c\right), a \epsilon \Lambda + \text{Pc} = (P - a2) \tag{12.38}$$

$$\text{The encryption equations are}: a_2 \leftarrow D_{\Lambda 2 + P2, \sqrt{M_2}} \tag{12.39}$$

$$a_1 \leftarrow D_{\Lambda 1 + P1 - a_2, \sqrt{M_1} + a_2} \tag{12.40}$$

$$M = (M_1 + M_2) > 0 \tag{12.41}$$

$$V_2 = \sqrt{M_2} \tag{12.42}$$

$$a_2 \rightarrow D_{\sqrt{M_2}} = V_2 \times S_1 \tag{12.43}$$

Decryption is done by:

$$a = a_2 + (P + a_2) - V_1 \left[V_1^{-1} \times (P - a_2)\right] u \tag{12.44}$$

$$a = a_2 + c - V_1 [V_a^{-1} \times c] u \tag{12.45}$$

- NTRU is another cryptosystem based on Lattice-based cryptography. It is a public key-based cryptosystem. The key generation in NRTU is done:

$$K \equiv p I_q \times J (mod q) \tag{12.46}$$

$$\text{Encryptionofthedataisdoneby}: C \equiv R \times K + M (mod q) \tag{12.47}$$

$$\text{Finally, thedecryptionofthedataisdoneby}: u = I \times C (mod q) \tag{12.48}$$

$$u = I (R p I_q J + M)(mod q) \tag{12.49}$$

$$u = [R p I_q J + M](mod q) \tag{12.50}$$

$$v = I_p I M (mod q) = M \tag{12.51}$$

Multivariate Cryptography

Multivariate cryptography (Ding and Petzoldt 2017; Carenzo and Polak 2019), as the name suggests, is a cryptographic primitive that used multivariate polynomial equations. Multivariate means multiple variables. Multivariate public key cryptosystems have an arrangement of nonlinear multivariate polynomials.

$$p^{(1)}(x_1, x_2, \ldots x_n) = \sum_{i=1}^{n} \sum_{j=1}^{n} p_{ij}^{(1)} \cdot x_i x_j + \sum_{i=j}^{n} \cdot x_i + P_0^{(1)} \tag{12.52}$$

$$p^{(2)}(x_1, x_2, \ldots x_n) == \sum_{i=1}^{n} \sum_{j=1}^{n} p_{ij}^{(2)} \cdot x_i x_j + \sum_{i=j}^{n} \cdot x_i + P_0^{(2)} \tag{12.53}$$

$$p^{(n)}(x_1, x_2, \ldots x_n) == \sum_{i=1}^{n} \sum_{j=1}^{n} p_{ij}^{(n)} \cdot x_i x_j + \sum_{i=j}^{n} \cdot x_i + P_0^{(n)} \tag{12.54}$$

The size of the public key is roughly around: $m \cdot (\frac{n+d}{d})$.

Here d is the degree of the polynomials in the equation. Most of the time d is taken as 2 for better efficiency. The security of the system is based on the problem:

The m multivariate quadratic polynomials $p^{(1)}(x), \ldots, p^{(m)}(x)$. Now we have to find the vector $\bar{x} = \bar{x}_1, \ldots, \bar{x}_n$ such that $p^1(\bar{x}) = p^{(m)}(\bar{x})$.

This problem is ended up being a NP-hard and resolving system of multivariate polynomials is demonstrated as NP-complete (Garey and Johnson 2009). This being the reason it is viewed as a good contender for post-quantum cryptography. The above-stated problem is supposed to be hard for classical as well as quantum computers. For the construction of multivariate cryptosystem, let $\mathcal{F}$ be an effectively invertible quadratic such that,

$$\mathcal{F} : \mathbb{F}^n \rightarrow \mathbb{F}^m \tag{12.55}$$

Two invertible linear maps,

$$\mathcal{S} : \mathbb{F}^m \rightarrow \mathbb{F}^m \tag{12.56}$$

$$\mathcal{T} : \mathbb{F}^n \rightarrow \mathbb{F}^n \tag{12.57}$$

Public key will be,

$$\mathcal{P} : \mathcal{S} \circ \mathcal{F} \circ \mathcal{T} \tag{12.58}$$

Private key will be,

$$\mathcal{S}, \mathcal{F}, \mathcal{T} \tag{12.59}$$

The $\mathscr{F}$ is combined with $\mathscr{T}$ and $\mathscr{S}$ and is well hid in the public key. If the public key is $\mathscr{P}$, find linear maps $\overline{\mathscr{S}}$ and $\overline{\mathscr{T}}$ as well as a simply invertible quadratic map $\overline{\mathscr{F}}$ such that $\mathscr{P} : \overline{\mathscr{S}} \circ \overline{\mathscr{F}} \circ \overline{\mathscr{T}}$.

The encryption will be:

$$w = \mathscr{P}(z) \tag{12.60}$$

$$z \in \mathbb{F}^n \tag{12.61}$$

$$w \in \mathbb{F}^m \tag{12.62}$$

Here, z is the message and the decryption is done by:

$$x = \mathscr{S}^{-1}(w) \tag{12.63}$$

$$y = \mathscr{F}^{-1}(x) \tag{12.64}$$

$$z = \mathscr{T}^{-1}(y) \tag{12.65}$$

And $m \geq n$, as it ensures that the ciphertext has only one possible plain text.

Hash-Based Cryptography

Hash-based cryptography (Potii et al. 2017) is the cryptographic primitive that uses hash functions for security of the message. As of now hash-based cryptography is used for creation of digital signatures. It is used in almost all the digital signature algorithms (Wikimedia Foundation 2021h). The hash-based cryptography algorithm is as follows.

Let $\mathscr{H}$ be the hash function so that,

$$\mathscr{H} : \{0, 1\}* \rightarrow \{0, 1\}n \tag{12.66}$$

Make 2 random strings X_0 and X_1

$$S = (X_0, X_1) \tag{12.67}$$

Here S is the secret key. Let P be the public key,

$$P = (\mathscr{H}(X_0), \mathscr{H}(X_1)) \tag{12.68}$$

Now the public key is published. Now to sign a bit let's say 0 the signer has to make the string X_0 public. Then the verifier will calculate the $\mathscr{H}(X_0)$ and match it with the value of public key. Similarly, to sign the bit 1 the signer has to make the string X_1 public then the verifier need to calculate the $\mathcal{H}(X_1)$ and match the value with the public key. Bit string of length b, for signing it, the secret key will be generated by the signer of length 2b.

$$S = (X_{10}, X_{11}, X_{20}, X_{21}, \ldots, X_{m0}, X_{m1})$$

The public key will be:

$$P = (\mathcal{H}(X_{10}), \mathcal{H}(X_{11}), \ldots, \mathcal{H}(X_{m0}), \mathcal{H}(X_{m1}))$$

Some of the hash-based cryptography schemes are newer versions of XMSS, Lamport signatures, SPHINCS and Merkle signature scheme.

Code-Based Cryptography

Code-based cryptography (Cohen et al. 2021; Samokhina and Trushina 2017) uses error correctional and detection algorithms for security. In these, the errors are used for encrypting the message and for decryption the errors are removed from the message. While moving data, at least one or more bits may get flicked. To recuperate the original message, the error detection and corrections are utilized. Linear error correction codes are widely used code-based cryptography schemes as they can also be used for creating one-way functions (Alagic et al. 2019). A bounded distance decoding problem is as follows:

Linear code:

$$C \subseteq F_2^n$$

$$y \in F_2^n$$

$$t \in \mathbb{N}$$

In this linear code, we've to find,

$$x \in C \text{ such that } dist(x, y) \leq t$$

This problem is proved to be a NP-Complete problem. McEliece encryption scheme can be used as a probable solution for the above-stated problem. In this scheme, the sender adds up error in the message with the help of receiver's public key. The errors are added in such a way that only the receiver can find and correct them as they have the private key.

Mathematically, let's take S, G and P be the matrices over F. G here is the generator matrix for the Goppa code. These codes are used for error correction efficiently.

So, public key is

$$G' = S \circ G \circ P, t \tag{12.69}$$

And secret key,

$$P, S, G \tag{12.70}$$

Encryption can be done by,

$$c = mG' + z \in F^n \tag{12.71}$$

Here, the message is multiplied using the receiver's public key and z is the error, which has been added. Now for decryption,

$$x = cP^{-1} = mSG + zP^{-1} \tag{12.72}$$

That's how the receiver detects and corrects the error.

Post-Quantum Analysis

It is a must to analyze how security of the 5G and B5G systems would affect in the post-quantum era. The 5G systems will have huge impact on their security schemes as they follow modern/classical cryptography. The USIM and unique identifier of 5G which is Subscription Permanent Identifier (SUPI) will majorly affect as these will be the focus of the attackers. As for B5G or 6G networks, the quantum computers won't have much effect as of now because these networks will use quantum computing approaches for development. Hence, providing security in post-quantum era.

Authentication and Key Establishment

In post-quantum era, the 5G systems will be heavily under vulnerability. If an attacker interrupts the matching pair of authentication response message and authentication request of the authentication and key agreement, both the messages will be carrying 128-bit values. For cracking them, Groover's algorithm can be used as it will reduce the time to 2^{64} operations only. This can provide enough data to the attacker about the key of the USIM. Furthermore, if the SUPI of the network is also discovered, it can lead to cloning of the whole USIM (Wang et al. 2021). As for B5G or 6G networks, the studies are still undergoing and so far, these networks are supposed to be working fine in the post-quantum era because of much superiority as compared to their predecessors and higher security measures.

Symmetric Encryption and Integrity Protection

In case of symmetric encryption, when an attacker interrupts the traffic which is sent over the network, he gets 32-bit MAC that addresses along with encrypted protocol data units. Using this information, an attack could be initiated, though it will be harder as compared to when the attacker has both the authentication and key agreement messages (Munilla et al. 2021). Here, the attacker would also need to know other information such as the plaintext, ciphertext and counter values. Thus, due to these many complications, conducting attacks using the information of authentication and key agreement message is much easier as compared to conducting attacks by intercepting the traffic of the network.

Asymmetric Encryption

As for security of asymmetric encryption, it is used for the protection of the user equipment's permanent identity, which, in the case of 5G, is SUPI. Usually, the USIM stores the home network public key that can be obtained using the user equipment (What next in the world of post-quantum cryptography 2021). As stated above that the Shor's algorithm has already proved to be decimating the modern asymmetric cryptographic schemes easily, the private key can be easily discovered just by knowing the public key in post-quantum era. This can cause menace by canceling the service of mobile identity confidentiality in all the USIMs having public key.

Conclusion

As an evolving study of 5G and beyond 5G networks, it has proved to be the technology of future which can revolutionize our daily lives. The 5G systems will open gates to new technologies such as AR/VR and smart cities. Furthermore, Internet of Everything (IoE) that is beyond IoT, Edge AI, AI-enabled smart cities and so on will be brought to us by 6G or beyond 5G networks, resulting in the total change of the way of living for humans. With these functionalities, the network systems will also need proper data security standards. By applying proper security techniques into the cellular networks, the users can be rest assured about security, and these future cellular networking technologies will come out to be an outstanding communication technology. In this paper, the evolution of 5G and 6G networks is discussed. Further, this paper explores the taxonomy used in 6G networks, security services provided by 6G networks and data security methods used in 5G networks such as steganography, RSA, AES, etc. Furthermore, the paper describes the architecture of quantum computers and presents the mathematical background of various quantum cryptographic schemes. Finally, the work describes the post-quantum cryptographic algorithms such as lattice-based cryptosystems, code-based cryptography, etc.

References

Ahmad A, Bhushan B, Sharma N, Kaushik I, Arora S (2020) Importunity & evolution of IoT for 5G. In: 2020 IEEE 9th international conference on communication systems and network technologies (CSNT). https://doi.org/10.1109/csnt48778.2020.9115768

Ahmed R, Chen Y, Hassan B (2021) Deep learning-driven opportunistic spectrum access (OSA) framework for cognitive 5G and beyond 5G (B5G) networks. Ad Hoc Netw 123:102632. https://doi.org/10.1016/j.adhoc.2021.102632

Akhtar MW, Hassan SA, Ghaffar R, Jung H, Garg S, Hossain MS (2020) The shift to 6G communications: vision and requirements. Human-Centric Comput Inform Sci 10(1). https://doi.org/10.1186/s13673-020-00258-2

Alagic G, Alperin-Sheriff J, Apon D, Cooper D, Dang Q, Liu Y-K, Miller C, Moody D, Peralta R, Perlner R, Robinson A, Smith-Tone D (2019) Status report on the first round of the NIST post-quantum cryptography standardization process. https://doi.org/10.6028/nist.ir.8240

Ariyanti S, Suryanegara M (2020) Visible light communication (VLC) for 6G technology: the potency and research challenges. In: 2020 fourth world conference on smart trends in systems, security and sustainability (WorldS4). https://doi.org/10.1109/worlds450073.2020.9210383

Arora S, Sharma N, Bhushan B, Kaushik I, Ahmad A (2020) Evolution of 5G wireless network in IoT. In: 2020 IEEE 9th international conference on communication systems and network technologies (CSNT). https://doi.org/10.1109/csnt48778.2020.9115773

Bernstein C, Cobb M (2021) What is the advanced encryption standard (AES)? definition from search security. Retrieved 8 Nov 2021, from https://searchsecurity.techtarget.com/definition/Advanced-Encryption-Standard#:~:text=The%20Advanced%20Encryption%20Standard%20(AES)%20is%20a%20symmetric%20block%20cipher,cybersecurity%20and%20electronic%20data%20protection

Bernstein DJ (n.d.) Introduction to post-quantum cryptography. Post-Quantum Crypt 1–14. https://doi.org/10.1007/978-3-540-88702-7_1

Bhatia V, Ramkumar KR (2020) An efficient quantum computing technique for cracking RSA using Shor's algorithm. In: 2020 IEEE 5th international conference on computing communication and automation (ICCCA). https://doi.org/10.1109/iccca49541.2020.9250806

Bhushan B (2022) Middleware and security requirements for internet of things. In: Sharma DK, Peng SL, Sharma R, Zaitsev DA (eds) Micro-electronics and telecommunication engineering. Lecture notes in networks and systems, vol 373. Springer, Singapore. https://doi.org/10.1007/978-981-16-8721-1_30

Bhushan B, Sahoo G (2017) Recent advances in attacks, technical challenges, vulnerabilities and their countermeasures in wireless sensor networks. Wireless Pers Commun 98(2):2037–2077. https://doi.org/10.1007/s11277-017-4962-0

Blog: Samsung Research. BLOG | Samsung Research. (n.d.). Retrieved 6 Nov 2021, from https://research.samsung.com/blog/Towards-6G-Security-Technology-Trends-Threats-and-Solutions

Brickman K-A, Haljan PC, Lee PJ, Acton M, Deslauriers L, Monroe C (2005) Implementation of Grover's quantum search algorithm in a scalable system. Phys Rev A 72(5). https://doi.org/10.1103/physreva.72.050306

Calvanese Strinati E, Barbarossa S, Choi T, Pietrabissa A, Giuseppi A, De Santis E, Vidal J, Becvar Z, Haustein T, Cassiau N, Costanzo F, Kim J, Kim I (2020) 6G in the sky: On-demand intelligence at the edge of 3D Networks (invited paper). ETRI J 42(5):643–657. https://doi.org/10.4218/etrij.2020-0205

Carenzo M, Polak M (2019) Accelerating multivariate cryptography with constructive affine stream transformations. In: Proceedings of the 2019 federated conference on computer science and information systems. https://doi.org/10.15439/2019f277

Chaccour C, Saad W (2021) Edge intelligence in 6G systems. Comput Commun Netw. https://doi.org/10.1007/978-3-030-72777-2_12

Chamola V, Jolfaei A, Chanana V, Parashari P, Hassija V (2021) Information security in the post quantum era for 5G and beyond networks: threats to existing cryptography, and post-quantum cryptography. Comput Commun 176:99–118. https://doi.org/10.1016/j.comcom.2021.05.019

Chen L, Jordan S, Liu Y-K, Moody D, Peralta R, Perlner R, Smith-Tone D (2016). Report on post-quantum cryptography. https://doi.org/10.6028/nist.ir.8105

Cohen A, D'Oliveira RG, Salamatian S, Medard M (2021) Network coding-based post-quantum cryptography. IEEE J Select Areas Inform Theory 2(1):49–64. https://doi.org/10.1109/jsait.2021.3054598

Devitt SJ, Fowler AG, Hollenberg LC (2005) Investigating the practical implementation of Shor's algorithm. In: Micro- and nanotechnology: materials, processes, packaging, and systems II. https://doi.org/10.1117/12.583191

Ding J, Petzoldt A (2017) Current state of multivariate cryptography. IEEE Secur Priv 15(4):28–36. https://doi.org/10.1109/msp.2017.3151328

Duarte FJ (2019) Cryptography via quantum entanglement. Fundam Quantum Entanglement. https://doi.org/10.1088/2053-2563/ab2b33ch22

ECC keys. (n.d.) Elliptic Curve Cryptography (ECC)—practical cryptography for developers. Retrieved 8 Nov 2021, from https://cryptobook.nakov.com/asymmetric-key-ciphers/elliptic-curve-cryptography-ecc

Elsevier. (n.d.) Quantum computing research trends report. Elsevier.com. Retrieved 8 Nov 2021, from https://www.elsevier.com/solutions/scopus/who-uses/research-and-development/quantum-computing-report

Fernandez-Carames TM (2020) From pre-quantum to post-quantum IOT Security: a survey on quantum-resistant cryptosystems for the internet of things. IEEE Internet of Things J 7(7):6457–6480. https://doi.org/10.1109/jiot.2019.2958788

Garey MR, Johnson DS (2009) Computers and intractability: a guide to the theory of Np-completeness. W.H. Freeman and Company

Goyal S, Sharma N, Kaushik I, Bhushan B, Kumar N (2021) A green 6g network era: architecture and propitious technologies. Data Analytics Manage. https://doi.org/10.1007/978-981-15-8335-3_7

Haller J (2021) An introduction to quantum computing architecture. Enable Architect. Retrieved 8 Nov 2021, from https://www.redhat.com/architect/quantum-computing

Hao Y, Yan X, Wu J, Wang H, Yuan L (2021) Multimedia communication security in 5G/6G cover-less steganography based on image text semantic association. Secur Commun Netw 2021:1–12. https://doi.org/10.1155/2021/6628034

Hauke P, Katzgraber HG, Lechner W, Nishimori H, Oliver WD (2020) Perspectives of quantum annealing: methods and implementations. Rep Prog Phys 83(5):054401. https://doi.org/10.1088/1361-6633/ab85b8

Hewa T, Gur G, Kalla A, Ylianttila M, Bracken A, Liyanage M (2020) The role of blockchain in 6G: challenges, opportunities and research directions. 2020 2nd 6G Wireless Summit (6G SUMMIT). https://doi.org/10.1109/6gsummit49458.2020.9083784

Huang Y, Liu S, Zhang C, You X, Wu H (2021) True-data testbed for 5G/B5G intelligent network. Intell Converged Netw 2(2):133–149. https://doi.org/10.23919/icn.2021.0002

Hui Y, Cheng N, Su Z, Huang Y, Zhao P, Luan TH, Li C (2021) Secure and personalized edge computing services in 6G heterogeneous vehicular networks. IEEE Internet Things J. https://doi.org/10.1109/jiot.2021.3065970

Jung M, Saad W (2021) Meta-learning for 6G communication networks with reconfigurable intelligent surfaces. In: ICASSP 2021—2021 IEEE international conference on acoustics, speech and signal processing (ICASSP). https://doi.org/10.1109/icassp39728.2021.9413598

Kashyap S, Bhushan B, Kumar A, Nand P (2022) Quantum blockchain approach for security enhancement in cyberworld. In: Kumar R, Sharma R, Pattnaik PK (eds) Multimedia technologies in the internet of things environment, vol 3. Studies in Big Data, vol 108. Springer, Singapore. https://doi.org/10.1007/978-981-19-0924-5_1

Khan M, Niemi V (2017) Concealing IMSI in 5G network using identity based encryption. Netw Syst Secur. https://doi.org/10.1007/978-3-319-64701-2_41

Khan LU, Yaqoob I, Imran M, Han Z, Hong CS (2020) 6G wireless systems: a vision, architectural elements, and Future Directions. IEEE Access 8:147029–147044. https://doi.org/10.1109/acc ess.2020.3015289

Khrennikov A (2021) Roots of quantum computing supremacy: superposition, entanglement, or complementarity? Eur Phys J Special Topics 230(4):1053–1057. https://doi.org/10.1140/epjs/ s11734-021-00061-9

Lam R (2019) The math behind quantum computing - qubits and superposition. Medium. Retrieved 8 Nov 2021, from https://medium.datadriveninvestor.com/the-math-behind-quantum-computing-qubits-and-superposition-f7a871668125

Letaief KB, Chen W, Shi Y, Zhang J, Zhang Y-JA (2019) The roadmap to 6G: Ai empowered wireless networks. IEEE Commun Mag 57(8):84–90. https://doi.org/10.1109/mcom.2019.190 0271

Li N (2010) Research on Diffie-Hellman key exchange protocol. In: 2010 2nd international conference on computer engineering and technology. https://doi.org/10.1109/iccet.2010.548 5276

Liu Y, Chang C-C (2018) A turtle shell-based visual secret sharing scheme with reversibility and authentication. Multimedia Tools Appl 77(19):25295–25310. https://doi.org/10.1007/s11042-018-5785-z

Mahmood NH, Alves H, Lopez OA, Shehab M, Osorio DP, Latva-Aho M (2020) Six key features of machine type communication in 6G. 2020 2nd 6G Wireless Summit (6G SUMMIT). https:// doi.org/10.1109/6gsummit49458.2020.9083794

Mahmood NH, Böcker S, Moerman I, López OA, Munari A, Mikhaylov K, Clazzer F, Bartz H, Park O-S, Mercier E, Saidi S, Osorio DM, Jäntti R, Pragada R, Annanperä E, Ma Y, Wietfeld C, Andraud M, Liva G, Seppänen P (2021) Machine type communications: Key drivers and enablers towards the 6G era. EURASIP J Wireless Commun Netw 2021(1). https://doi.org/10. 1186/s13638-021-02010-5

Malik A, Bhushan B (2022)Challenges, standards, and solutions for secure and intelligent 5G internet of things (IoT) scenarios. In: Smart and sustainable approaches for optimizing performance of wireless networks: real-time applications. Wiley, pp 139–165. https://doi.org/10.1002/ 9781119682554.ch7

Malina L, Dzurenda P, Ricci S, Hajny J, Srivastava G, Matulevicius R, Affia A-AO, Laurent M, Sultan NH, Tang Q (2021) Post-quantum ERA privacy protection for intelligent infrastructures. IEEE Access 9:36038–36077. https://doi.org/10.1109/access.2021.3062201

Mandviwalla A, Ohshiro K, Ji B (2018) Implementing Grover's algorithm on the IBM quantum computers. In: 2018 IEEE international conference on big data (big data). https://doi.org/10. 1109/bigdata.2018.8622457

Mishra D, Zema NR, Natalizio E (2021) A high-end IOT devices framework to foster beyond-connectivity capabilities in 5G/B5G architecture. IEEE Commun Mag 59(1):55–61. https://doi. org/10.1109/mcom.001.2000504

Mitchell CJ (2020) The impact of quantum computing on real-world security: a 5G case study. Comput Secur 93:101825. https://doi.org/10.1016/j.cose.2020.101825

Munilla J, Burmester M, Barco R (2021) An enhanced symmetric-key based 5G-aka protocol. Comput Netw 198:108373. https://doi.org/10.1016/j.comnet.2021.108373

Navaneeth AV, Dileep MR (2020) A study and analysis of applications of classical computing and quantum computing: a survey. ICT Anal Appl. https://doi.org/10.1007/978-981-15-8354-4_25

Nejatollahi H, Dutt N, Ray S, Regazzoni F, Banerjee I, Cammarota R (2019) Post-quantum lattice-based cryptography implementations. ACM Comput Surv 51(6):1–41. https://doi.org/10.1145/3292548

Popovski P, Trillingsgaard KF, Simeone O, Durisi G (2018) 5G wireless network slicing for embb, URLLC, and MMTC: a communication-theoretic view. IEEE Access 6:55765–55779. https://doi.org/10.1109/access.2018.2872781

Potii O, Gorbenko Y, Isirova K (2017) Post quantum Hash based digital signatures comparative analysis. features of their implementation and using in public key infrastructure. In: 2017 4th international scientific-practical conference problems of info communications. Science and Technology (PIC S&T). https://doi.org/10.1109/infocommst.2017.8246360

Pradhan PK, Rakshit S, Datta S (2019) Lattice based cryptography: its applications, areas of Interest & future scope. In: 2019 3rd international conference on computing methodologies and communication (ICCMC). https://doi.org/10.1109/iccmc.2019.8819706

Qadir AM, Varol N (2019) A review paper on Cryptography. In: 2019 7th international symposium on digital forensics and security (ISDFS). https://doi.org/10.1109/isdfs.2019.8757514

Qin S, Tan Z, Zhou F, Xu J, Zhang Z (2021) A verifiable steganography-based secret image sharing scheme in 5G Networks. Secur Commun Netw 2021:1–14. https://doi.org/10.1155/2021/6629726

Samokhina M, Trushina O (2017) Code-based cryptosystems evolution. In: 2017 IVth international conference on engineering and telecommunication (EnT). https://doi.org/10.1109/icent.2017.10

Saxena S, Bhushan B, Ahad MA (2021) Blockchain based solutions to secure Iot: Background, integration trends and a way forward. J Netw Comput Appl. https://doi.org/10.1016/j.jnca.2021.103050

Schanck J (2020) Improving post-quantum cryptography through cryptanalysis. UWSpace. Retrieved 8 Nov 2021, from https://uwspace.uwaterloo.ca/handle/10012/16060

Sekander S, Tabassum H, Hossain E (2018) Multi-tier drone architecture for 5G/B5G cellular networks: challenges, trends, and prospects. IEEE Commun Mag 56(3):96–103. https://doi.org/10.1109/mcom.2018.1700666

Singh T, Verma S, Parashar V (2016) Securing internet of things in 5G using audio steganography. Commun Comput Inform Sci. https://doi.org/10.1007/978-981-10-3433-6_44

Stubbs R (2021) Quantum computing and its impact on cryptography. Cryptomathic. Retrieved 6 Nov 2021, from https://www.cryptomathic.com/news-events/blog/quantum-computing-and-its-impact-on-cryptography

Sugumar B, Ramakrishnan M (2018) Key escrow with elliptic curve cryptography—conceptual framework for distributed mobile networks. Indonesian J Electr Eng Comput Sci 11(3):1060. https://doi.org/10.11591/ijeecs.v11.i3.pp1060-1067

Sutton A (2015) 4G to 5G: evolution or revolution. 5G radio technology seminar. Exploring technical challenges in the emerging 5G ecosystem. https://doi.org/10.1049/ic.2015.0032

Tsiounis Y, Yung M (1998) On the security of elgamal based encryption. Public Key Cryptography. https://doi.org/10.1007/bfb0054019

Wang C, Rahman A (2021) Quantum-enabled 6G wireless networks: opportunities and challenges. https://doi.org/10.36227/techrxiv.14785737

Wang L-J, Zhang K-Y, Wang J-Y, Cheng J, Yang Y-H, Tang S-B, Yan D, Tang Y-L, Liu Z, Yu Y, Zhang Q, Pan J-W (2021) Experimental authentication of quantum key distribution with post-quantum cryptography. NPJ Quantum Inform 7(1). https://doi.org/10.1038/s41534-021-00400-7

What next in the world of post-quantum cryptography ... (n.d.). Retrieved 8 Nov 2021, from https://www.ericsson.com/en/blog/2020/3/post-quantum-cryptography-symmetric-asymmetric-algorithms

Wikimedia Foundation (2020) Key escrow. Wikipedia. Retrieved 8 Nov 2021, from https://en.wikipedia.org/wiki/Key_escrow

Wikimedia Foundation (2021a) Elliptic-curve cryptography. Wikipedia. Retrieved 8 Nov 2021a, from https://en.wikipedia.org/wiki/Elliptic-curve_cryptography

Wikimedia Foundation (2021b) RSA (cryptosystem). Wikipedia. Retrieved 8 Nov 2021b, from https://en.wikipedia.org/wiki/RSA_(cryptosystem)

Wikimedia Foundation (2021c) ElGamal encryption. Wikipedia. Retrieved 8 Nov 2021c, from https://en.wikipedia.org/wiki/ElGamal_encryption

Wikimedia Foundation (2021d) Lucifer (cipher). Wikipedia. Retrieved 8 Nov 2021d, from https://en.wikipedia.org/wiki/Lucifer_(cipher)

Wikimedia Foundation (2021e) Attribute-based encryption. Wikipedia. Retrieved 8 Nov 2021e, from https://en.wikipedia.org/wiki/Attribute-based_encryption

Wikimedia Foundation (2021f) Shor's algorithm. Wikipedia. Retrieved 8 Nov 2021f, from https://en.wikipedia.org/wiki/Shor%27s_algorithm

Wikimedia Foundation (2021g) Post-quantum cryptography. Wikipedia. Retrieved 8 Nov 2021g, from https://en.wikipedia.org/wiki/Post-quantum_cryptography

Wikimedia Foundation (2021h) Hash-based cryptography. Wikipedia. Retrieved 8 Nov 2021h, from https://en.wikipedia.org/wiki/Hash-based_cryptography#:~:text=Hash%2Dbased%20cryptography%20is%20the,type%20of%20post%2Dquantum%20cryptography.&text=Hash%2Dbased%20signature%20schemes%20combine,with%20a%20Merkle%20tree%20structure

Yang Z, Chen M, Saad W, Shikh-Bahaei M, Poor HV, Cui S (2021) Federated learning in 6G mobile wireless networks. Comput Commun Netw. https://doi.org/10.1007/978-3-030-72777-2_16

Yao B, Wang H, Su J, Zhang W (2021) Graph-based lattices cryptosystem as new technique of post-quantum cryptography. In: 2021 IEEE 5th advanced information technology, electronic and automation control conference (IAEAC). https://doi.org/10.1109/iaeac50856.2021.9390858

Yuan Y, Zhao Y, Zong B, Parolari S (2020) Potential key technologies for 6G Mobile communications. Sci China Inform Sci 63(8). https://doi.org/10.1007/s11432-019-2789-y

Zappone A, Di Renzo M, Debbah M (2020) Deep learning for energy-efficient beyond 5G networks. Green Commun Energy-Efficient Wireless Syst Netw. https://doi.org/10.1049/pbte091e_ch3

Zhang S, Wang Y, Zhou W (2019) Towards secure 5G networks: a survey. Comput Netw 162:106871. https://doi.org/10.1016/j.comnet.2019.106871

Zhang Z, Cao S, Yang X, Liu X, Han L (2021b) An efficient outsourcing attribute-based encryption scheme in 5G mobile network environments. Peer-to-Peer Networking Appl 14(6):3488–3501. https://doi.org/10.1007/s12083-021-01195-2

Zhang X, Zhu Q, Poor HV (2021a) Age-of-information for MURLLC over 6G multimedia wireless networks. In: 2021a 55th annual conference on information sciences and systems (CISS). https://doi.org/10.1109/ciss50987.2021.9400300

Chapter 13
Enhanced Energy Efficiency and Scalability in Cellular Networks for Massive IoT

Husam Rajab and Tibor Cinkler

Abstract The significant expansion of cellular networks has increased their potential to support a wide range of use cases beyond their original purpose of providing broadband access. One such development is using cellular networks to support the Internet of Things (IoT), called Cellular IoT (CIoT). The growth of CIoT is an important trend in the evolution of cellular networks, it leads to broader and more comprehensive ecosystem circumstances. The extensive IoT business evolution is transforming a diverse sector, including health, smart cities, security, and agriculture. Nevertheless, a large scale with very different characteristics and use cases struggle with connectivity challenges due to the unique traffic features of massive IoT and the tremendous density of IoT devices. This study aims to identify the critical obstacles that hinder the widespread deployment of IoT over cellular networks and suggest an innovative algorithm to mitigate them effectively. We discovered that the primary challenges revolve around three specific areas: connection setup, network resource management, and energy consumption. In this regard, we investigate the integration of massive Machine-Type Communication (mMTC) into cellular networks, focusing on the performance of Narrowband IoT (NB-IoT) in supporting mMTC.

Keywords IoT · CIoT · mMTC · NB-IoT

Introduction

Wireless communication networks have significantly impacted various aspects of our lives, including healthcare, professional networking, and accessing information. Over the past few decades, these networks have evolved from being expensive to becoming pervasive and accessible to many of the population. The development of

H. Rajab (✉) · T. Cinkler
Budapest University of Technology and Economics, Budapest, Hungary
e-mail: husamrajab@tmit.bme.hu

T. Cinkler
e-mail: cinkler@tmit.bme.hu

© The Author(s) 2023
B. Bhushan et al. (eds.), *5G and Beyond*, Springer Tracts in Electrical and Electronics Engineering, https://doi.org/10.1007/978-981-99-3668-7_13

cellular networks has been particularly revolutionary, leading to the emergence of new use cases and challenges. There are four generations of cellular networks (1G–4G) shown in Fig. 13.1, and the fifth generation has recently been introduced. The first and second generations were primarily focused on voice-based communication. The first and second generations of cellular networks were aimed to perform voice-centric operations. Whereas the third and fourth generations of cellular networks involved data packets with new data rate and frequency. Each new generation of cellular networks has built upon the services of the previous generation and has seen significant technological advancements. However, one drawback of the first four generations was that their communication design was primarily focused on human-centric services.

In other words, the first four generations of cellular networks were developed primarily to support voice and data communication between individuals. While these networks have evolved to offer faster speeds and more advanced features, their basic design remained centered around human communication needs. With the advent of the fifth generation of cellular networks, there is a greater emphasis on designing networks to support a broader range of use cases, including machine-to-machine communication and the Internet of Things (IoT). Before the 5G, cellular networks' concentration was to perform more services to human users (Liberg et al. 2017). The Internet of Things (IoT) has emerged as a significant technological revolution in recent years. IoT refers to the interconnection of many devices embedded in everyday objects, enabling them to access the internet and share data without requiring human intervention. This idea embodies the concept of everything being connected. Any smart device can connect to the internet through physical items such as machines, devices, and vehicles.

IoT devices can serve various applications, including healthcare, agriculture, security, smart cities, industrial automation, and autonomous driving. New applications are emerging daily, showcasing this technology's vast potential. The growth of IoT has led to the development of new business models and opportunities for innovation.

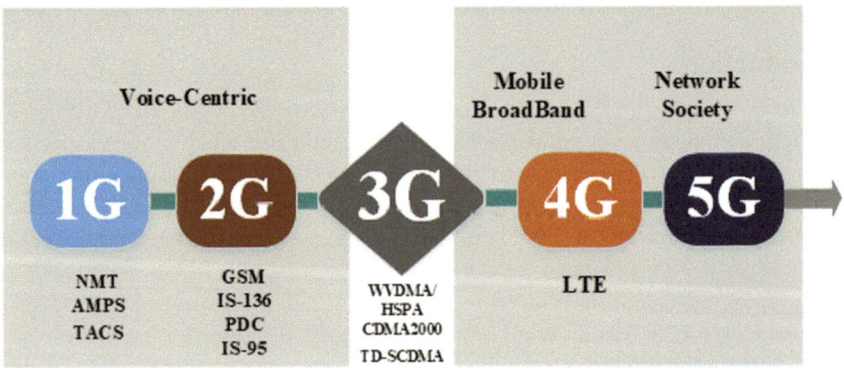

Fig. 13.1 Cellular generation

With the increasing use of connected devices, IoT is set to become an integral part of our daily lives in the years to come. This diversity of IoT applications has caused transcendent freedom to users, and recently we have seen an enormous accession in their numbers. Massive devices are already implemented, and these numbers are expected to grow shortly, with predictions reporting that most equal 29 billion IoT devices will be functioning by 2023 (Ericsson Mobility Report). To adapt to the new requirements for device connectivity to further assist IoT, previous cellular networks must be restructured.

This chapter will focus on identifying the core issues that limit IoT implementation over cellular networks at a large scale and a novel solution to mitigate them. The majority of the problems arise in three distinct aspects, i.e., the establishment of connection, utilization of network resources, and efficiency. In this context, we examined the containment of massive Machine-type Communications (mMTC) into cellular networks. Apart from that, the performance of Narrowband-IoT (NB-IoT) within cellular networks will be improved.

This chapter is divided into several sections. Firstly, we provide an overview of related work in "Literature Review" section. Next, in "Narrowband-Internet of Things (NB-IoT)" section, we discuss the Narrowband-Internet of Things (NB-IoT) and explain the Power Saving Mode (PSM) and extended Discontinuous Reception (eDRX). We then present our methodology and performance analysis in "Methodology" section. Validation results for the proposed algorithm and analytical NB-IoT model are provided in "Result" section and "Discussion" section, respectively. Finally, we offer concluding remarks and outline future work in "Conclusion" section.

Literature Review

In this section, we examine the literature related to the issues discussed in this chapter. We review the significant works related to each key challenge and explore how research in this area has evolved in recent years.

The diverse range of IoT applications has varying requirements, such as stringent latency, unbiased transmissions, static or large mobility devices, and small or high volumes of data. Therefore, only some approaches can cater to all IoT applications. In such cases, Low-Power Wide-Area Networks (LPWANs) are becoming the preferred option for many IoT use cases (Masoudi et al. 2021). A Low-Power Wide-Area (LPWA) wireless IoT radio access network faces four Performance Indicators (KPIs) inconsistencies: Coverage Area A, Battery Lifetime, Device Capacity, and Estimation Cost. Traditional cellular networks cannot meet these KPIs. As IoT devices started to be deployed in large numbers, various surveys were conducted to address these issues (Langat and Musyoki 2022; Singh et al. 2021a; Amodu and Othman 2018). Numerous surveys have identified significant inefficiencies in LPWA networks and emphasized critical research directions. One of the primary concerns highlighted in most research is the high number of collisions in the random-access channel (RACH) during the RA process. As a result, several surveys have been conducted to address this issue (Andrade et al. 2018; Althumali and Othman 2018; Kafi 2021) concentrated mainly on the RA process, and the consequence of IoT traffic on HTC. At that time, IoT devices were mostly recognized of lower priority and importance corresponded to HTC devices. Recently, deploying IoT devices is massively increasing over cellular networks, recent surveys (Mahmood et al. 2020; Wu et al. 2020; El-Tanab and Hamouda 2021) exposed additional issues and recognized new research directions.

In recent years, the research community has recognized the equal importance of IoT devices compared to HTC devices due to their significant development. As a result, several recent surveys have been conducted to address this issue (Al-Dulaimi et al. 2018; Li et al. 2021; Suma 2021). Recent research has focused on addressing issues affecting IoT and HTC devices in 4G and upcoming 5G networks. One proposed solution is the Enhanced Access Barring (EAB) scheme, which dynamically adjusts the preventing parameters to balance the number of collisions and network access delay for both devices (El-Tanab and Hamouda 2021; Vidal et al. 2019; Bui et al. 2019; Tello-Oquendo et al. 2018; Zhan et al. 2021; Leyva-Mayorga et al. 2019; Haile et al. 2021; Singh et al. 2021b). Nonetheless, few research has focused on the broadcast transmission functionality and network utilization, and energy consumption for IoT devices. Previous suggestions aimed to improve the efficiency of multicasting transmissions by modifying the Modulation and Coding Scheme (MCS) used (Zuhra et al. 2019; Fuad et al. 2021; Rinaldi et al. 2020; Chen et al. 2020). As the number of devices using multi-cast services increased, previous schemes were found to need to be more sufficient in providing satisfactory services without affecting unicast traffic or adding significant processing at the base station. As a result, parallel research was conducted to improve the quality of service experienced by users.

In Ghandri et al. (2018), The authors distinguish between two types of services: bandwidth-intensive streaming services and file delivery, while in Park et al. (2018) and Guangzhi (2021), the devices are categorized into different groups based on the services they receive. The authors in Guangzhi (2021) users were categorized into groups based on the feedback received about the channel quality. Similar approaches have been proposed in related works, where devices are categorized into groups based on their experienced Quality of Service (QoS) (Zuhra et al. 2019; Chen et al. 2021; Li et al. 2018; Gong et al. 2022; Saily et al. 2019).

Since the advent of the IoT era, energy conservation has been recognized as a crucial objective for both device and network aspects in current and future cellular networks. Several research studies (Liu et al., 2019; Tang et al., 2020; Ferranti et al., 2019; Jahid et al., 2019) have highlighted the challenging issues and addressed various research directions, while other research (Uppal and Gangadharappa 2021; Pham et al. 2020) Initial research on energy consumption focused on the network side, with efforts to investigate proposed energy-saving practices and approaches and identify suitable parameters based on the development prospects of both the network and devices. Reducing the energy consumption of both the network and the devices has been a crucial goal since the beginning of the IoT era. Initially, most research on energy consumption focused on the network side, aiming to reduce the BS's energy consumption due to the large number of devices they had to serve simultaneously. Several parallel research areas followed, which can be broadly divided into the following categories:

1. Resource allocation schemes

Researchers have investigated various transmission strategies and resource allocation schemes that depend on the transmission models of the devices (either HTC or IoT) and the total traffic load in the cell (Wu et al. 2015; Hou et al. 2020).

2. Dynamic adaptation of transmission parameters

These proposals (Bonnefoi et al. 2019; Jahid et al. 2021; Ozyurt et al. 2021; Lassoued and Boujnah 2020; Lv et al. 2020) aim to optimize the transmission and operation parameters of the BSs dynamically to minimize their energy consumption by utilizing on/off switching and irregular transmission schemes.

3. Collision mitigation methods

These approaches aim to reduce the energy consumption of BSs by mitigating interference from different transmissions and preventing the retransmission of previous messages. Several approaches have been proposed in this area, as highlighted in Khazali et al. (2018), Nikjoo et al. (2018) and Ghosh (2020).

With the massive increase in device numbers, the research community began to focus on the energy consumption of the devices. This led to the development of many parallel directions in research. Several works (Chang and Tsai 2018; Sehati and Ghaderi 2018; Verma et al. 2019, 2018; Bithas et al. 2019; Li and Chen 2019; Mughees et al. 2021; Sanislav et al. 2018) aimed to optimize the Discontinuous Reception (DRX) configuration requirements of IoT devices to raise the sleeping period to minimize the energy consumption.

Several studies (e.g., Sanislav et al. 2018; Al Homssi et al. 2018; Chen et al. 2018; Malik et al. 2018; Wang et al. 2020; Tsoukaneri et al. 2020; Liang et al. 2018) have proposed optimizing transmission parameters to reduce the energy consumption of IoT devices during operation, by adjusting settings such as duty cycles or transmission numbers (Himeur et al. 2020). Another direction in reducing energy consumption of IoT devices is optimizing resource allocation and data transmission parameters, such as adjusting the data rate (Chen et al. 2018; Malik et al. 2018). Most energy-related research hasn't focused on cellular network technology in recent years. As a result, the unique characteristics of individual devices were not considered, and some studies attempted to establish general models for energy consumption in IoT devices (Tsoukaneri et al. 2020; Finnegan and Brown 2018; Azoidou et al. 2018; Sadowski and Spachos 2020; Duhovnikov et al. 2019; Lan et al. 2019).

As a result, various works such as Finnegan and Brown (2018), Andres-Maldonado et al. (2017), Yeoh et al. (2018), Lauridsen et al. (2018), Bello et al. (2019), Soussi et al. (2018) and Sinha et al. (2017) have focused on evaluating the impact of NB-IoT technology on energy consumption, examining its distinct operating modes and associated energy costs. In addition, Yeoh et al. (2018) conducts experimental research on the energy consumption of a commercial prefab NB-IoT module for necessary communication services.

Narrowband-Internet of Things (NB-IoT)

NB-IoT is a Low Power Wide Area Network (LPWAN) radio technology licensed and designed for enhanced indoor coverage for many low-cost, low-capability, and low-power IoT devices. It eliminates dual connectivity and mobility features, further reducing device costs.

Currently, two significant cellular IoT technologies are NB-IoT and LTE-M, which target IoT use cases.

NB-IoT is designed to cater to low-cost Machine-Type Communication (MTC) UEs with lower power consumption and higher coverage area than conventional enhanced Mobile Broadband (eMBB) UEs. This is achieved by utilizing a small portion of the spectrum, a distinct radio interface design, and simplified LTE network functions. NB-IoT is a new 3GPP radio-access technology that is partially backwards compatible with previous generations of cellular networks, meaning existing devices cannot immediately use it. NB-IoT has been designed to be backwards compatible with previous generations of cellular networks, leveraging the existing physical layer design to a great extent for coexistence with legacy designs (Wang et al. 2017).

NB-IoT physical channels utilize the exciting cellular network to extensive coverage that allows seamless coexistence and interoperability. NB-IoT is a half-duplex technology and supports OFDMA transmissions in the downlink and SCFDMA transmissions in the uplink, similar to 4G. The technology requires a minimum channel bandwidth of 180 kHz, equivalent to one Physical Resource Block (PRB). This means the UE does not need to listen to the DL while transmitting in the UL and vice versa, regardless of the deployment mode. Figure 13.2 illustrates the design of NB-IoT subframes, which support half-duplex operation and use OFDMA transmissions in the downlink and SCFDMA transmissions in the uplink. The technology requires a minimum channel bandwidth of 180 kHz, equivalent to one Physical Resource Block (PRB). The physical channels specified in the NB-IoT standard include the Narrowband Physical Broadcast Channel (NPBCH), which is used for broadcasting master information for regularity access (i.e., Master Information Block or MIB), the Narrowband Physical Downlink Control Channel (NPDCCH) for uplink and downlink scheduling information, the Narrowband Physical Downlink Shared Channel (NPDSCH) for downlink dedicated and standard data, the Narrowband Physical Random-Access Channel (NPRACH) for uplink dedicated and standard data, and the Narrowband Physical Uplink Shared Channel (NPUSCH) for uplink data. The NPUSCH channel has two formats: NPUSCH format 1 for UL data transmissions and NPUSCH format 2 for Hybrid Automatic Repeat Request (HARQ) feedback for NPDSCH.

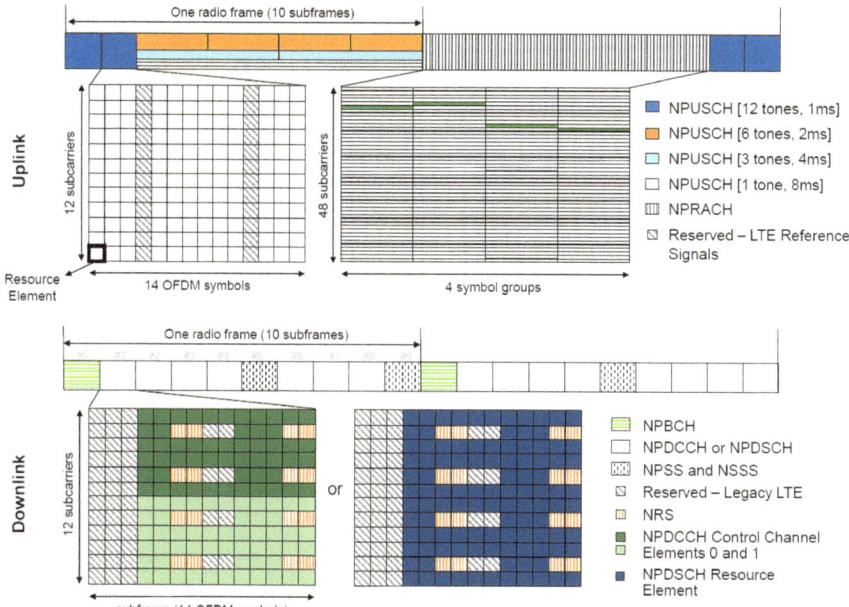

Fig. 13.2 NB-IoT in-band physical channels time multiplexing (Rastogi et al. 2020)

Power Saving Features

To ensure an extended battery life of more than 10 years on a single battery charge, NB-IoT employs two power-saving techniques:

1. The Power Saving Mode (PSM)
2. The extended Discontinuous Reception (eDRX)

Both approaches enable the UE to enter a power-saving mode in which monitoring for paging/scheduling information is not required.

1. **PSM**: The Power Saving Mode (PSM) in NB-IoT allows devices to enter a deep sleep mode by disconnecting from most of their connections while remaining connected to the network, which can be seen in Fig. 13.3. This mode allows the device to save power while not connected to the network, but still wake up whenever necessary to send data. The PSM technique is specifically designed to help IoT devices conserve battery power and potentially achieve a battery life of over 10 years.

   PSM is a power-off mode that keeps the device connected to the network, according to the 3GPP TS 23.682 specification. Curiously, the PSM mode seemed in 3GPP specifications earlier than the NB-IoT in 3GPP Release 12. In PSM, the device turns to a sort of power-off mode for a suitable amount of time. If the device needs to transmit data, it can wake up without required to register in the network and the necessary signaling.

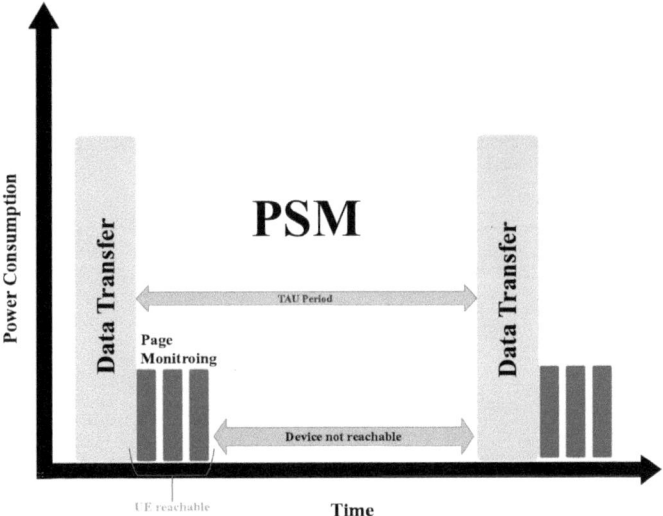

Fig. 13.3 Power saving mode (PSM)

2. **eDRX**: NB-IoT employs an extended discontinuous reception (eDRX) technique that puts devices in an idle mode where they do not receive radio signals
   for a specific period. This allows the devices to conserve power and prolong
   battery life. Periodically, the devices wake up to receive paging messages from
   the network and check for incoming information before returning to deep sleep
   mode. The period of intermittent eDRX reception in the NB-IoT mode ranges
   from 20.48 to 10485.76 s, with the eDRX mode allowing the receiving path of
   the device to remain shut off for a more extended period. Including eDRX in
   the 3GPP Release, 13 specifications enable an additional power-saving mode for
   IoT devices.

   In summary, after the Active Timer expires, the UE switches to PSM mode,
   completely disconnecting the radio and only maintaining a primary oscillator for
   time reference. In PSM, energy consumption is similar to the power-off state.
   eDRX is a technique that prolongs the sleep time of the I-DRX. When using
   eDRX, an active phase is controlled by a Paging Time Window (PTW) timer
   during each eDRX cycle, where the UE can be reached using I-DRX cycles,
   followed by a sleep phase for the rest of the eDRX process (Fig. 13.4). This
   cycle continues until the Active Timer expires

Methodology

In a computer system, all components require energy to function. While desktop
computers have Power Supply Units (PSUs), laptops typically rely on batteries.
This chapter explicitly focuses on how LPWA technologies optimize power usage,

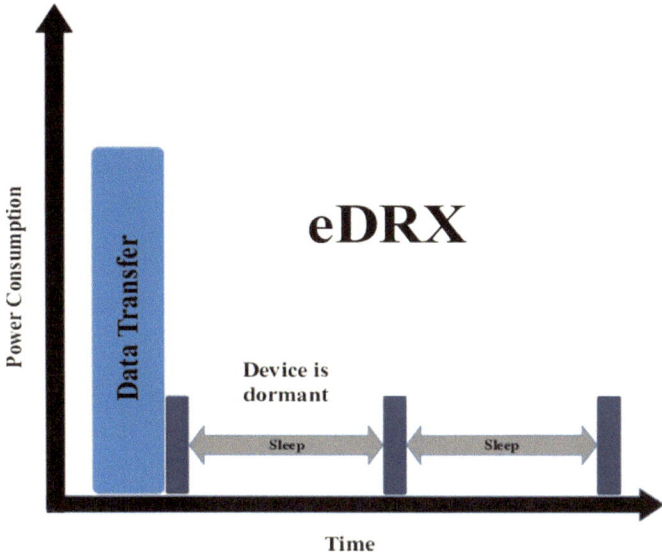

Fig. 13.4 Extended discontinuous reception (eDRX)

particularly NB-IoT. Micro-controllers and sensors in LPWA networks are small computer systems prioritizing energy consumption optimization to achieve longer battery life rather than performance analyses. While components utilize power to perform computations, some energy dissipates from the system. This chapter will provide a basic understanding of energy (E) and power (P) concepts and how they relate to consumption.

To discuss the concept of energy in computer systems, it is essential to understand the two primary forms of energy—potential and dynamic. The law of conservation of energy states that the total energy in a closed system remains constant and cannot be created or destroyed; instead, it transforms into other forms of energy. Computer systems change the energy required to perform tasks into different shapes, mainly heat. Energy is also measured as the amount of work done on an object per unit of time, referred to as the energy consumption rate. The primary parameters used to measure energy in computer systems include voltage (V), current (A), power (W), energy (Wh), and time (s).

A fundamental formula for calculating power will be used and modified to compute these factors. Equation 13.1 is used to calculate the average power by dividing the energy by the time elapsed:

$$P = \frac{E}{t} \qquad (13.1)$$

To calculate power, we need to measure the voltage provided by a constant source like a battery and the current flowing through it. This can be done using the following equation, which is a modified form of Ohm's law:

$$P(t) = V(t) \times I(t) \tag{13.2}$$

It is important to note that power consumption can be optimized by controlling either the voltage, current, or both. In the case of battery-powered devices, reducing the voltage or current can help to extend the battery life. However, this may come at the cost of reduced performance or functionality. Therefore, finding the optimal balance between power consumption and performance is crucial for designing efficient and effective computer systems.

When using a computer system, the energy consumed can be calculated by determining the power usage over a specific period. Equations 13.1 and 13.2 can be used to derive a function that calculates the energy consumed by a computer system based on the voltage, current, and time elapsed. This can be expressed as:

$$E = \int_{t_1}^{t_2} V(t) \times I(t) \times dt = \int_{t_1}^{t_2} P(t) \times dt \tag{13.3}$$

Power Management in 3GPP Standard

LPWA technologies are designed to optimize power usage and ensure a longer device battery life. In the 3GPP standard, power management techniques sustain low energy consumption while maintaining a reliable connection. Some of these techniques include:

- Low power mode allows devices to enter a deep sleep state while still connected to the network, thereby conserving power.
- Lightweight MAC protocols: These protocols are designed to be simple and efficient, reducing the energy required for communication.
- Topology: The topology of LPWA networks is optimized to reduce the energy required for communication by using fewer hops and minimizing interference.
- Utilization of more complex base stations: By using more complex base stations, LPWA networks can achieve better coverage and reduce the energy required for communication.

To conserve power in LPWA technology, User Equipment (UE) does not require continuous data transmission. Instead, it wakes up from sleep mode to send requested data and utilizes power-hungry components for a short time. Lightweight MAC protocols are also needed to reduce complex overhead for LPWA UEs. Network topology options include Mesh topology, commonly used in standard cellular networks and WLAN. UEs should aim to connect directly to the base station to avoid unnecessary jumps, which can improve battery life. In 3GPP standardized technologies, only the user can initiate the low power mode. Discharging unnecessary operations on base stations can further extend the battery life of UEs.

Low Power Mode

LPWA technologies, including cellular networks, use a low-power mode to optimize power consumption and extend battery life. This mode involves powering down heavy elements like the processor. The low-power mode can be implemented differently, depending on the application. For example, a device that only transmits information using the uplink can be scheduled to send information twice a day or manually trigger transfer messages. If the device can receive notifications through the downlink, it must listen to the network for these messages. There are several ways to achieve this, and the most suitable approach depends on the use case and how often the device wakes up from low-power mode. If the device periodically transmits messages, it can simultaneously listen for messages on the downlink.

In eMTC and NB-IoT, the low power mode is implemented differently, but both use power-efficient techniques like PSM and eDRX. The difference between these techniques is that eDRX allows the modem to listen to incoming signals, whereas PSM requires the modem to wake up to send data before receiving any data. Although energy-efficient communication is crucial for the successful deployment of MTC over existing cellular networks, there needs to be more research focused on energy-efficient uplink MTC scheduling.

Algorithm 13.1 illustrates the procedure to apply the Prediction Energy Saving Technique (PEST) on the User Equipment (UE) side.

To improve energy efficiency, the UE in cellular networks can store a prescheduling command transmitted via a narrowband physical uplink shared channel (NPUSCH) and check it when an uplink packet occurs. The UE follows the legacy scheduling request procedure if there is no prescheduling request. However, if there is a prescheduling request, the UE delays the uplink packet transmission until the scheduled time without triggering the scheduling request procedure.

In some cases, there may be a radical change in traffic arrival, such as when some micro-BSs are turned off to save energy during low-traffic periods. This situation requires neighboring BSs to cover the coverage areas of the turned-off BSs, which is called cell zooming. During low-traffic periods, the density of active BSs decreases, and communication distances increase. Legacy UEs, such as smartphones, are designed for daily recharging and are more efficient in such patterns. This proposed algorithm will be evaluated using analytical and simulation models.

Algorithm 13.1 Proposed UE procedure in an NB-IoT network

Scheduling request procedure:
  if scheduling request is triggered then **then**
    store up link packet in the buffer
  else if scheduling command is already stored then **then**
    delay up link transmission by prescheduled time
  else
    process scheduling request procedure with RA
  end if
  Uplink Scheduling Procedure:
  if TX time equals scheduling command time **then**
    if buffer is not empty **then**
      process NPUSCH transmission
    else
      ignore NPUSCH
    end if
  end if

Battery Lifetime Estimation

We adopt a methodology similar to that used to assess the UE battery life by measuring energy consumption. Our study considers a smart utility sensor, which sends periodic uplink reports with a predetermined inter-arrival time (IAT) as per the traffic profile. Before the start of periodic reporting, the UE needs to re-establish the RRC connection and thus perform the CP procedure.

We divided the battery lifetime approximation into four phases for modeling the periodic traffic pattern:

- P1: The UE wakes up from Power Saving Mode (PSM), establishes the RRC connection, and transmits data using the CP procedure.
- P2: The UE continuously monitors the Narrowband Physical Downlink Control Channel (NPDCCH) until the RRC connection is released.
- P3: The UE utilizes extended/enhanced Discontinuous Reception (eDRX) until the Active Timer expires.
- P4: The UE enters sleep mode using PSM until the next transmission period begins.

To estimate the energy consumption for transferring one UL report, denoted as E_{report}, we used a similar methodology to the one described in. This method assumes a smart utility sensor and periodic UL reporting with a predefined Inter-Arrival Time (IAT). Before the periodic reporting, the UE must reestablish the RRC connection, which involves performing the CP procedure.

$$E_{report} = E_{conn} + E_{rel} + E_{idle} + P_{standby} + T_{sleep} \qquad (13.4)$$

$$T_{sleep} = \text{IAT} - T_{conn} - T_{rel} - T_{idle} \qquad (13.5)$$

The energy consumption for transferring one UL report, E_{report}, was estimated using a similar methodology. As described above, the four phases for modeling the periodic traffic pattern were P1, P2, P3, and P4. The energy consumed in joules within the phases P1, P2, and P3 is denoted as E_{conn}, E_{rel}, and E_{idle}, respectively. $P_{standby}$ represents the average power consumption in PSM, and T_{conn}, T_{rel}, T_{idle}, and T_{sleep} represent the duration in seconds of the phases P1, P2, P3, and P4, respectively. Finally, the energy consumed per day, denoted as E_{day}, and the battery lifetime in years indicated as B_{life} can be determined as follows:

$$E_{day} = \frac{D_{day}}{\text{IAT}} \times E_{report} \qquad (13.6)$$

$$B_{life} = \frac{\text{Bat}_C}{\frac{E_{day}}{3600} \times 365 \times 25} \qquad (13.7)$$

In our simulation, we consider periodic UL reports as UDP packets with 50 B of payload and a battery capacity of $\text{Bat}_C = 5$ Wh (Wang et al. 2017). We use the value of D_{day} to represent the duration of 1 day in seconds.

Results

This section presents the validation results of our proposed analytical NB-IoT model. We used Eqs. 13.1, 13.2, and 13.3 to calculate the modules' energy consumption and average power. The validation was performed based on two metrics: battery lifetime and latency for performing Control Plane (CP) optimization. Our proposed algorithm aims to reduce the total energy consumption of NB-IoT devices by exploiting their lack of mobility and minimizing the number of costly and unnecessary procedures. The module was tested with a voltage of 3.7 V and a signal strength of −75 dBm. The energy consumption and average power were calculated based on the voltage and current for a particular period. It should be noted that the module is still under development.

PSM

Figure 13.5 displays the results of a 1-h Power Saving Mode (PSM) test where the system only woke up from sleep once. The initial peak in the current was measured

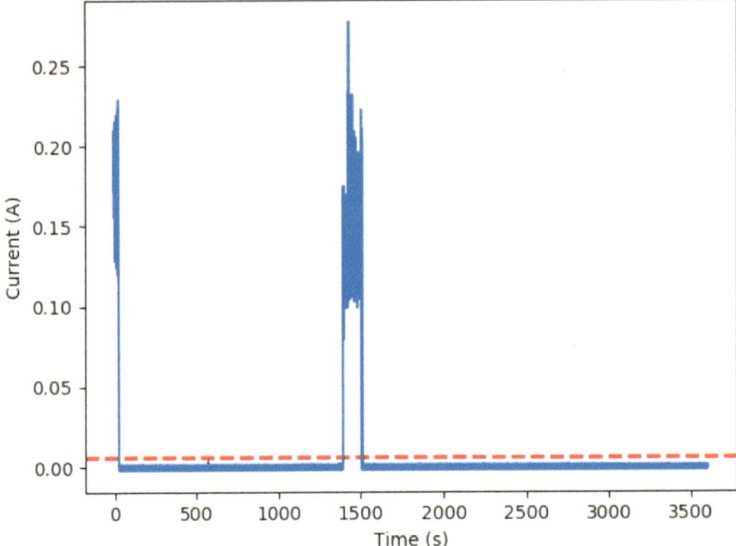

Fig. 13.5 1-h PSM test

Table 13.1 PSM active and sleep

Average working current	101.498 mA
Average working power	375.544 mW
Average standby current	143.428 μA
Average standby power	530.684 μW

before the system entered sleep mode for the first time. It is worth noting that the module used in the test is still under development.

Furthermore, we performed Timer and Button analyses, looking at the cycles individually to achieve better results for both modes. Table 13.1 shows the results of these analyses and is illustrated in Fig. 13.6, indicating a significant decrease in both current and power when beginning the sleep mode.

eDRX

Figure 13.7 displays the results of a 20-min eDRX test, where a request was sent to the system at the beginning of the analysis, resulting in the first spike in current. This spike reached a peak of over 0.25 A. Following this, the system woke up periodically to listen to the downlink, with an average current of approximately 0.2 A. During idle cycles, the system was in sleep mode, with the current varying between 0.1 and 0.15 A. Figure 13.8 shows multiple spikes in the current readings during each

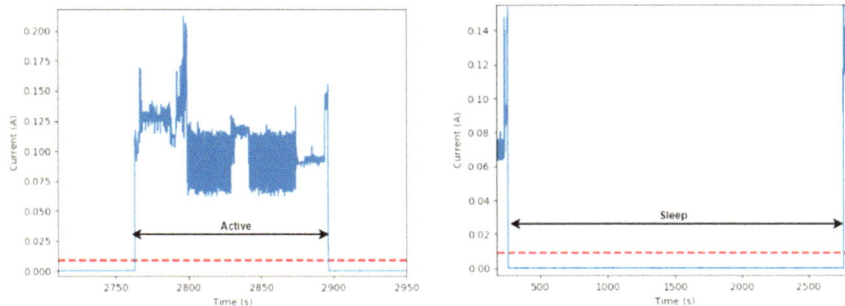

Fig. 13.6 PSM active and sleep cycles

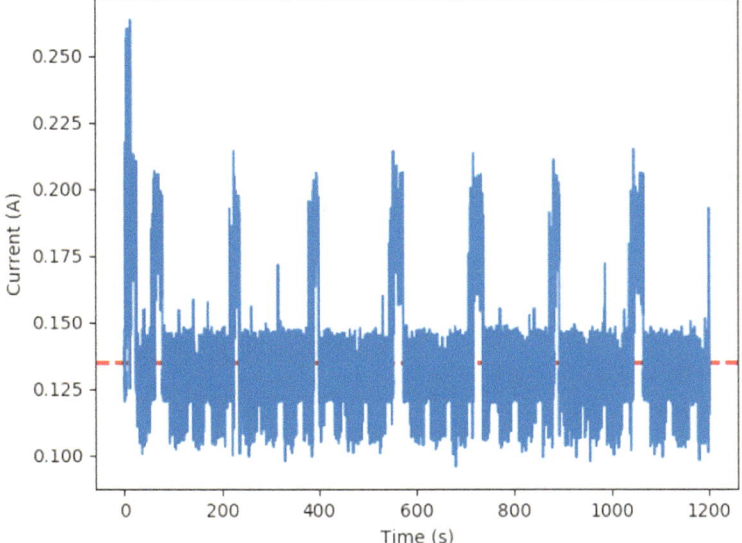

Fig. 13.7 20 min eDRX test

cycle, which may be because the current is not measured from the system alone, as applications are also running on the application processor. The results for each cycle are provided in Table 13.2.

Discussion

This section describes a comprehensive evaluation of energy consumption in NB-IoT devices based on measurements, focusing on identifying optimization targets for network operations. Furthermore, we proposed an energy consumption model that

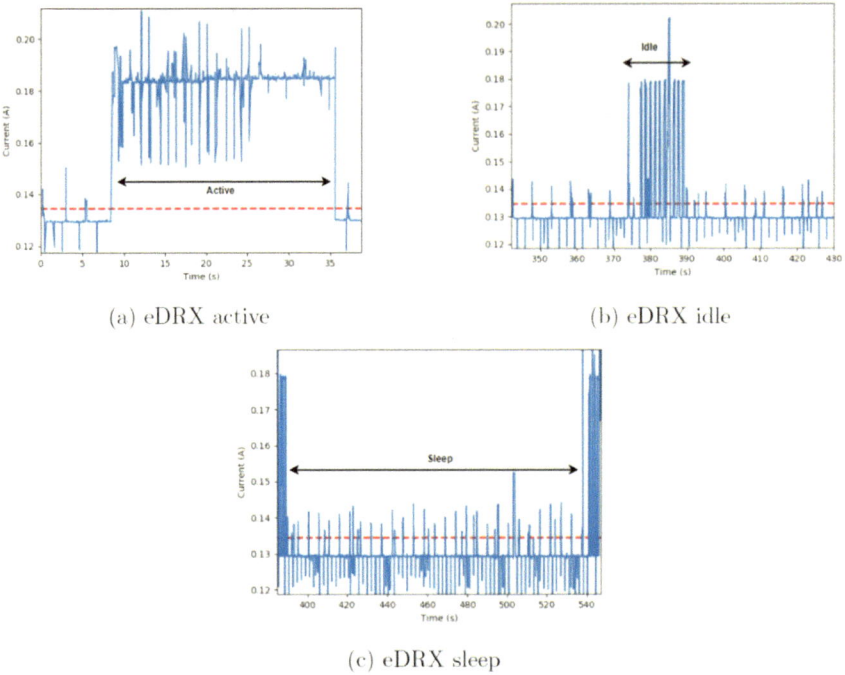

(a) eDRX active   (b) eDRX idle

(c) eDRX sleep

Fig. 13.8 eDRX active, idle, and sleep cycles

Table 13.2 The result of each separate cycle

Average working current	183.313 mA
Average working power	678.257 mW
Average standby current	144.356 mA
Average standby power	534.118 mW
Average sleep current	129.786 mA
Average sleep power	480.206 mW

we utilized in simulation experiments and our empirical observations to estimate the battery requirements for NB-IoT devices to achieve the desired battery life.

Estimating an NB-IoT device's lifetime can be accomplished by using the battery's capacity and network configurations. In the case of Power Saving Mode (PSM), we can calculate the battery life using four different timers: 1 h, 1 day, 1 week, and 1 month (30 days). Table 13.3 provides each cycle's average power and respective times. To determine the battery's lifetime, we divide the battery's energy by the average power according to the following formula:

$$\text{Lifetime} = \frac{5\,\text{Wh}}{P_{tot}} \tag{13.8}$$

Table 13.3 Lifetime of 5 Wh battery (PSM)

	Once/hour	Once/day	Once/week	Once/month
Operator A NB-IoT	2528 h (105 days)	40,463 h (1686 days)	93,449 h (3894 days)	112,239 h (4677 days)
Operator B NB-IoT	3443 h (143 days)	49,351 h (2056 days)	99,301 h (4138 days)	114,060 h (4753 days)
Operator A eMTC	398 h (17 days)	8614 h (356 days)	42,130 h (1755 days)	83,905 h (3496 days)

Table 13.4 Average power summary (eDRX)

	Average power (mW)	Time (s)
Operator A NB-IoT (active)	210.729	11.41
Operator A NB-IoT (idle)	96.088	15.36
Operator A NB-IoT (sleep)	59.039	148.56

Table 13.3 illustrates a 5 Wh battery life when utilizing Power Saving Mode (PSM) with NB-IoT and eMTC. As anticipated, the battery life is considerably shorter for eMTC compared to NB-IoT. The battery life varies from 17 to 3496 days, depending on the duration of the sleep mode.

Table 13.4 depicts the power usage for each eDRX cycle and their respective times for the same battery with 5 Wh. Similar to PSM, we can utilize the same equations to estimate the battery's lifetime.

Conclusions

The proliferation of the Internet of Things (IoT) has revolutionized various domains of our lives by extending network connectivity to everyday objects, enabling them to communicate with each other without human intervention. IoT devices have a wide range of applications, including digital health, smart homes, autonomous driving, and industrial automation, with new applications being developed daily.

Cellular networks have emerged as a strong candidate to support IoT devices, mainly due to their extensive deployment, large coverage area, and varying data rates. However, traditional cellular networks were historically designed to help high-throughput communication (HTC) devices, exhibiting considerably different traffic patterns than IoT devices, leading to inefficiencies at both the network and device levels. These challenges are not limited to a single area but span various operational areas of cellular networks, such as the connection establishment process, network resource utilization, and device energy consumption.

LPWAN technology selection for IoT applications should be determined case-by-case, considering the device's data transmission requirements, desired lifetime, and access to a charging source. PSM may be more appropriate for devices that only need to send data infrequently, while eDRX may be more suitable for devices that listen to incoming information frequently. Considering the power-saving feature utilized is crucial since neither PSM nor eDRX is a one-size-fits-all technology.

The current chapter focuses on the challenges IoT devices face in cellular networks, focusing on their unique communication patterns and requirements. To address these challenges, the chapter presents an analytical model that enables estimating the energy consumption of an NB-IoT device. The results obtained from the analysis indicate that the use cases for eDRX and PSM differ and that devices that need to listen to the downlink frequently may require more frequent battery recharging.

Future Studies

The rising traffic generated by emerging IoT applications presents a significant challenge for cellular networks. Machine-to-Machine (M2M) communications, known as MTC, are crucial for current and future cellular networks. Therefore, cellular networks must continually evolve and adapt to new requirements. NB-IoT is an example of this evolution as it leverages LTE technology to provide IoT support. However, new technologies like NB-IoT and eMTC still present unresolved research issues and uncertainties. Further research is needed to address these challenges and support the growth of IoT in cellular networks.

Various unknown factors, such as range and configurations, make it difficult to provide accurate details on the energy consumption of NB-IoT devices. Additionally, no experiments were conducted to analyze how the amount of data sent affects energy consumption, an area for future development.

Based on the findings of this chapter, there are several open issues and potential improvements that need to be addressed, including:

1. Expanding the NB-IoT model proposed in this study to analyze the enhanced Discontinuous Reception (eDRX) and PSM performance.
2. Extending the analysis to cover extended coverage areas.
3. Conducting experimental measurements of battery lifetime, considering the non-ideal characteristics of actual batteries, such as self-discharge and temperature variations.
4. Investigating alternative antenna schemes that can improve the Signal-to-Noise Ratio (SNR) without increasing the User Equipment (UE) complexity.

Acknowledgements This research is supported by Tempus Public Foundation, Stipendium Hungaricum Scholarship Programme and High Speed Networks Lab, Department of Telecommunications and Media Informatics, Budapest University of Technology and Economics. "This work was supported by the CHIST-ERA grant SAMBAS (CHIST-ERA-20-SICT-003) funded by FWO, ANR, NKFIH, and UKRI."

Funding This work was supported by the Ericsson—BME 5G joint research and cooperation project, partly funded by the National Research, Development and Innovation Office, Hungary with project number 2018-1.3.1-VKE-2018-00005.

Declaration Availability of Data and Materials The data used to support the findings of this study are available from the corresponding author upon request.

Competing Interests The authors declare that they have no competing interests.

References

Al Homssi B, Ai-Hourani A, Chavez KG, Chandrasekharan S, Kandeepan S (2019) Energy-efficient IoT for 5G: a framework for adaptive power and rate control. In: 2018, 12th international conference on signal processing and communication systems, ICSPCS 2018—proceedings. https://doi.org/10.1109/ICSPCS.2018.8631733

Al-Dulaimi A et al (2018) 5G networks: fundamental requirements, enabling technologies, and operations—Google Books. https://books.google.hu/books?hl=en&lr=&id=RTtpDwAAQ BAJ&oi=fnd&pg=PR21&dq=M2M+Communications+in+5G:+State-of-the-Art+Architect ure,+Re-cent+Advances,+and+Research+Challenges.+&ots=1zwrVzJf3s&sig=ho_i7OdUE-VgrdgVx4m3yyzCSjU&redir_esc=y#v=onepage&q&f=false

Althumali H, Othman M (2018) A survey of random access control techniques for machine-to-machine communications in LTE/LTE-A networks. IEEE Access 6:74961–74983

Amodu OA, Othman M (2018) Machine-to-machine communication: an overview of opportunities. Comput Netw 145:255–276

de Andrade TPC, Sekijima LR, da Fonseca NLS (2018) A cluster-based random-access scheme for LTE/LTE-A networks supporting massive machine-type communications. In: 2018 IEEE international conference on communications

Andres-Maldonado P, Ameigeiras P, Prados-Garzon J, Navarro-Ortiz J, Lopez-Soler JM (2017) Narrowband IoT data transmission procedures for massive machine-type communications. IEEE Netw 31:8–15

Azoidou E et al (2018) Battery lifetime modeling and validation of wireless building automation devices in thread. IEEE Trans Ind Inf 14:2869–2880

Bello H, Jian X, Wei Y, Chen M (2019) Energy-delay evaluation and optimization for NB-IoT PSM with periodic uplink reporting. IEEE Access 7:3074–3081

Bithas PS, Maliatsos K, Foukalas F (2019) An SINR-aware joint mode selection, scheduling, and resource allocation scheme for D2D communications. IEEE Trans Veh Technol 68:4949–4963

Bonnefoi R, Farès H, Bélis P, Louët Y (201) Optimal power allocation for minimizing the energy consumption of a NOMA base station with cell DTX. In: 2019 URSI Asia-Pacific radio science conference, AP-RASC 2019. https://doi.org/10.23919/URSIAP-RASC.2019.8738765

Bui ATH, Nguyen CT, Thang TC, Pham AT (2019) A comprehensive distributed queue-based random framework for mMTC in LTE/LTE-A networks with mixed-type traffic. IEEE Trans Veh Technol 68:12107–12120

Chang HL, Tsai MH (2018) Optimistic DRX for machine-type communications in LTE-A network. IEEE Access 6:9887–9897

Chen S, Yang B, Yang J, Hanzo L (2020) Dynamic resource allocation for scalable video multirate multicast over wireless networks. IEEE Trans Veh Technol 69:10227–10241

Chen CY, Huang ACS, Huang SY, Chen JY (2018) Energy-saving scheduling in the 3GPP narrowband internet of things (NB-IoT) using energy-aware machine-to-machine relays. In: 2018 27th wireless and optical communication conference, WOCC 2018, pp 1–3. https://doi.org/10.1109/ WOCC.2018.8373791

Chen J et al (2021) Joint resource allocation and cache placement for location-aware multi-user mobile edge computing. https://doi.org/10.48550/arxiv.2103.11220

Dahlman E, Parkvall S, Sköld J (2016) 4G, LTE-advanced pro and the road to 5G. Academic Press

Duhovnikov S, Baltaci A, Gera D, Schupke DA (2019) Power consumption analysis of NB-IoT technology for low-power aircraft applications. In: IEEE 5th world forum on internet of things, WF-IoT 2019—conference proceedings, pp 719–723. https://doi.org/10.1109/WF-IOT.2019. 8767234

El-Tanab M, Hamouda W (2021) An overview of uplink access techniques in machine-type communications. IEEE Netw 35:246–251

Ericsson Mobility Report. https://www.ericsson.com/en/reports-and-papers/mobility-report? gclid=Cj0KCQjw3IqSBhCoARIsAMBkTb1WwCdGUR6d95bQmwGFv9JZbC40-c9ZwnL HtIH7xXXPEuqLT8ygNkAaAsScEALw_wcB&gclsrc=aw.ds

Ferranti L et al (2019) HIRO-NET: self-organized robotic mesh networking for internet sharing in disaster scenarios. In: 20th IEEE international symposium on a world of wireless, mobile and multimedia networks, WoWMoM 2019. https://doi.org/10.1109/WOWMOM.2019.8793029

Finnegan J, Brown S (2018) An analysis of the energy consumption of LPWA-based IoT devices. In: 2018 International symposium on networks, computers and communications, ISNCC 2018. https://doi.org/10.1109/ISNCC.2018.8531068

Fuad M, bin Jahangir H, Razu Ahmed M, Zafar Md Imran A, Humayun Kabir M (2021) Resource allocation & energy consumption reduction for 5G (NR-new radio) wireless communication. 1–15. https://doi.org/10.3390/xxxxx

Ghandri A, Boujelben Y, Jemaa MB (2018) A low-complexity scheduling for joint unicast and multicast transmissions in LTE—a network. In: 2018 14th international wireless communications and mobile computing conference, IWCMC 2018, pp 136–141. https://doi.org/10.1109/ IWCMC.2018.8450493

Ghosh J (2020) A trade-off between energy efficiency and spectral efficiency in macro-femtocell networks. IEEE Trans Veh Technol 69:10914–10924

Gong P-Y, Wang C-H, Sheu J-P, Yang D-N (2022) Distributed DRL-based resource allocation for multicast D2D communications, pp 01–06. https://doi.org/10.1109/GLOBECOM46510.2021. 9685485

Guangzhi W (2021) Application of adaptive resource allocation algorithm and communication network security in improving educational video transmission quality. Alex Eng J 60:4231–4241

Haile H, Grinnemo KJ, Ferlin S, Hurtig P, Brunstrom A (2021) End-to-end congestion control approaches for high throughput and low delay in 4G/5G cellular networks. Comput Netw 186:107692

Himeur Y, Alsalemi A, Bensaali F, Amira A (2020) Building power consumption datasets: survey, taxonomy and future directions. Energy Build 227:110404

Hou R, Huang K, Xie H, Lui KS, Li H (2020) Caching and resource allocation in small cell networks. Comput Netw 172:107100

Jahid A et al (2019) Toward energy efficiency aware renewable energy management in green cellular networks with joint coordination. IEEE Access 7:75782–75797

Jahid A, Alsharif MH, Uthansakul P, Nebhen J, Aly AA (2021) Energy efficient throughput aware traffic load balancing in green cellular networks. IEEE Access 9:90587–90602

Kafi MH (2021) Prioritised random access channel protocols for delay critical M2M communication over cellular networks—White Rose eTheses Online. https://etheses.whiterose.ac.uk/30168/

Khazali A, Sobhi-Givi S, Kalbkhani H, Shayesteh MG (2018) Energy-spectral efficient resource allocation and power control in heterogeneous networks with D2D communication. Wirel Netw 26(1):253–267

Lan D et al (2019) Latency analysis of wireless networks for proximity services in smart home and building automation: the case of thread. IEEE Access 7:4856–4867

Langat K, Musyoki S (2022) Towards device driven 5G: radio resource allocation perspective. In: Proceedings of the sustainable research and innovation conference. https://sri.jkuat.ac.ke/jku atsri/index.php/sri/article/view/169

Lassoued N, Boujnah N (2020) Power saving approach in LTE using switching ON/OFF eNodeB and power UP/DOWN of neighbors. Smart Innov Syst Technol 147:337–349

Lauridsen M, Krigslund R, Rohr M, Madueno G (2018) An empirical NB-IoT power consumption model for battery lifetime estimation. In: IEEE vehicular technology conference 2018, pp 1–5

Leyva-Mayorga I, Stefanovic C, Popovski P, Pla V, Martinez-Bauset J (2019) Random access for machine-type. Communications. https://doi.org/10.1002/9781119471509.w5GRef031

Li M, Chen HL (2019) Energy-efficient traffic regulation and scheduling for video streaming services over LTE-A networks. IEEE Trans Mob Comput 18:334–347

Li S, da Xu L, Zhao S (2018) 5G internet of things: a survey. J Ind Inf Integr 10:1–9

Li X et al (2021) Physical layer security of cognitive ambient backscatter communications for green internet-of-things. IEEE Trans Green Commun Netw 5:1066–1076

Liang JM, Wu KR, Chen JJ, Liu PY, Tseng YC (2018) Energy-Efficient uplink resource units scheduling for ultra-reliable communications in NB-IoT networks. In: Wireless communications and mobile computing

Liberg O, Sundberg M, Wang Y-PE, Bergman J, Sachs J (2017) Cellular internet of things: technologies, standards, and performance, p 382

Liu Q, Han T, Ansari N (2019) Energy-efficient on-demand resource provisioning in cloud radio access networks. IEEE Trans Green Commun Netw 3:1142–1151

Lv Z, Hu B, Lv H (2020) Infrastructure monitoring and operation for smart cities based on IoT system. IEEE Trans Ind Inf 16:1957–1962

Mahmood NH et al (2020) White paper on critical and massive machine type communication towards 6G. https://doi.org/10.48550/arxiv.2004.14146

Malik H et al (2018) Radio resource management scheme in NB-IoT systems. IEEE Access 6:15051–15064

Masoudi M, Azari A, Cavdar C (2021) Low power wide area IoT networks: reliability analysis in coexisting scenarios. IEEE Wirel Commun Lett 10:1405–1409

Mughees A, Tahir M, Sheikh MA, Ahad A (2021) Energy-efficient load-aware user association in ultra-dense wireless network. In: Proceeding—2021 26th IEEE Asia-Pacific conference on communications, APCC 2021, pp 254–259. https://doi.org/10.1109/APCC49754.2021.9609810

Nikjoo F, Mirzaei A, Mohajer A (2018) A novel approach to efficient resource allocation in NOMA heterogeneous networks: multi-criteria green resource management. https://doi.org/10.1080/08839514.2018.148613232,583-612

Ozyurt AB, Basaran M, Ardanuc M, Durak-Ata L, Yanikomeroglu H (2021) Intracell frequency band exiling for green wireless networks: implementation, performance metrics, and use cases. IEEE Veh Technol Mag 16:31–39

Park J, Hwang JN, Li Q, Xu Y, Huang W (2018) Optimal DASH-multicasting over LTE. IEEE Trans Veh Technol 67:4487–4500

Pham QV, Mirjalili S, Kumar N, Alazab M, Hwang WJ (2020) Whale optimization algorithm with applications to resource allocation in wireless networks. IEEE Trans Veh Technol 69:4285–4297

Rastogi E, Saxena N, Roy A, Shin DR (2020) Narrowband internet of things: a comprehensive study. Comput Netw 173:107209

Rinaldi F et al (2020) A Novel approach for MBSFN area formation aided by D2D communications for eMBB service delivery in 5G NR systems. IEEE Trans Veh Technol 69:2058–2070

Sadowski S, Spachos P (2020) Wireless technologies for smart agricultural monitoring using internet of things devices with energy harvesting capabilities. Comput Electron Agric 172:105338

Saily M et al (2019) 5G radio access networks: enabling efficient point-to-multipoint transmissions. IEEE Veh Technol Mag. https://doi.org/10.1109/MVT.2019.2936657

Sanislav T, Zeadally S, Mois GD, Folea SC (2018) Wireless energy harvesting: empirical results and practical considerations for internet of things. J Netw Comput Appl 121:149–158

Sehati A, Ghaderi M (2018) Online energy management in IoT applications. In: 2018 proceedings—IEEE INFOCOM, pp 1286–1294

Singh U, Dua A, Tanwar S, Kumar N, Alazab M (2021a) A survey on LTE/LTE-A radio resource allocation techniques for machine-to-machine communication for B5G networks. IEEE Access 9:107976–107997

Singh U, Dua A, Kumar N, Guizani M (2021b) QoS aware uplink scheduling for M2M communication in LTE/LTE-A network: a game theoretic approach. IEEE Trans Veh Technol. https://doi.org/10.1109/TVT.2021.3132535

Sinha RS, Wei Y, Hwang SH (2017) A survey on LPWA technology: LoRa and NB-IoT. ICT Express 3:14–21

El Soussi M, Zand P, Pasveer F et al (2018) Evaluating the performance of eMTC and NB-IoT for smart city applications. In: 2018 IEEE international conference. ieeexplore.ieee.org

Suma V (2021) Internet of things (IoT) based smart agriculture in India: an overview. J ISMAC. https://doi.org/10.36548/jismac.2021.1.001

Tang W et al (2020) Wireless communications with programmable metasurface: new paradigms, opportunities, and challenges on transceiver design. IEEE Wirel Commun 27:180–187

Tello-Oquendo L, Vidal JR, Pla V, Guijarro L (2018) Dynamic access class barring parameter tuning in LTE-A networks with massive M2M traffic. In: 2018 17th annual mediterranean ad hoc networking workshop, Med-Hoc-Net 2018, pp 1–8. https://doi.org/10.23919/MEDHOC NET.2018.8407086

Tsoukaneri G, Garcia F, Marina MK (2020) Narrowband IoT device energy consumption characterization and optimizations

Uppal N, Gangadharappa M (2021) A survey on spectral and energy efficient next-generation wireless networks. In: Proceedings of the 2021 8th international conference on computing for sustainable global development, INDIACom 2021, pp 155–160. https://doi.org/10.1109/IND IACOM51348.2021.00028

Verma S, Kawamoto Y, Kato N (2019) Energy-efficient group paging mechanism for QoS constrained mobile IoT devices over LTE-A pro networks under 5G. IEEE Internet Things J 6:9187–9199

Verma S, Kawamoto Y, Nishiyama H, Kato N, Huang CW (2018) Novel group paging scheme for improving energy efficiency of IoT devices over LTE-A pro networks with QoS considerations. In: 2018 IEEE international conference on communications

Vidal JR, Tello-Oquendo L, Pla V, Guijarro L (2019) Performance study and enhancement of access barring for massive machine-type communications. IEEE Access 7:63745–63759

Wang YPE et al (2017) A primer on 3GPP narrowband internet of things. IEEE Commun Mag 55:117–123

Wang L, Wang W, Hu X, Xie T (2020) Optimization of large-scaled random access congestion control oriented to narrow band internet of things. J Phys Conf Ser 1570:012089

Wu Y, Zhang N, Rong K (2020) Non-orthogonal random access and data transmission scheme for machine-to-machine communications in cellular networks. IEEE Access 8:27687–27704

Wu Q et al (2015) Energy efficient resource allocation for wireless powered communication networks—White Rose eTheses Online. https://etheses.whiterose.ac.uk/29656/

Yeoh CY, bin Man A, Ashraf QM, Samingan AK (2018) Experimental assessment of battery lifetime for commercial off-the-shelf NB-IoT module. In: International conference on advanced communication technology, ICACT, 2018-February, pp 223–228

Zhan W, Sun X, Wang X, Fu Y, Li Y (2021) Performance optimization for massive random access of mMTC in cellular networks with preamble retransmission limit. IEEE Trans Veh Technol 70:8854–8867

Zuhra SU, Chaporkar P, Karandikar A (2019) Toward optimal grouping and resource allocation for multicast streaming in LTE. IEEE Trans Veh Technol 68:12239–12255

Correction to: Wireless Backhaul Optimization Algorithm in 5G Communication

Astha Sharma, Mukesh Soni, Abhaya Nand, Suryabhan Pratap Singh, and Sumit Kumar

Correction to:
Chapter 4 in: B. Bhushan et al. (eds.), *5G and Beyond*,
Springer Tracts in Electrical and Electronics Engineering,
https://doi.org/10.1007/978-981-99-3668-7_4

The original version of the book was inadvertently published with incorrect author name and affiliation in chapter 4. The erratum chapter and the book have been updated with the changes.

The updated version of this chapter can be found at
https://doi.org/10.1007/978-981-99-3668-7_4

© The Author(s) 2023
B. Bhushan et al. (eds.), *5G and Beyond*, Springer Tracts in Electrical
and Electronics Engineering, https://doi.org/10.1007/978-981-99-3668-7_14